D1171947

SUPPORTED METAL COMPLEXES

CATALYSIS BY METAL COMPLEXES

F. R. HARTLEY

Royal Military College of Science, Shrivenham,
Wiltshire, SN6 8LA, England

SUPPORTED
METAL COMPLEXES

A New Generation of Catalysts

D. REIDEL PUBLISHING COMPANY

A MEMBER OF THE KLUWER ACADEMIC PUBLISHERS GROUP

DORDRECHT / BOSTON / LANCASTER / TOKYO

Library of Congress Cataloging in Publication Data

CIP

Hartley, F. R.
    Supported metal complexes.

    (Catalysis by metal complexes)
    1. Metal catalysts.   2.   Organometallic catalysts.   3.   Complex
compounds.   I.   Title.   II.   Series.
QD505.H37   1985   541.3'95   85-19362
ISBN 90-277-1855-5

Published by D. Reidel Publishing Company,
P.O. Box 17, 3300 AA Dordrecht, Holland.

Sold and distributed in the U.S.A. and Canada
by Kluwer Academic Publishers
190 Old Derby Street, Hingham, MA 02043, U.S.A.

In all other countries, sold and distributed
by Kluwer Academic Publishers Group,
P.O. Box 322, 3300 AH Dordrecht, Holland.

Printed in the Netherlands

To
V.H., S.M.H., J.A.H. and E.J.H.

CONTENTS

PREFACE   xiii

ABBREVIATIONS   xv

CHAPTER 1. *Introduction*   1
    1.1. Catalysis   1
    1.2. Reasons for Supporting Metal Complexes   3
    1.3. Catalyst Requirements   9
    1.4. Types of Support   9
       1.4.1. Organic Polymers   11
       1.4.2. Inorganic Supports   14
    1.5. Chemically Modified Electrodes   16
    1.6. Immobilised Enzymes and Reagents for Organic Syntheses   16
    1.7. Triphase Catalysis   16
    1.8. Heterogenisation of Metal Complex Catalysts   18
       1.8.1. Phase Transfer   19
       1.8.2. Supported Liquid and Gas Phase Catalysts   19
       1.8.3. Use of Melts   23
       1.8.4. Lattice Metal Complexes   23
       1.8.5. Water Soluble Complexes   24
    1.9. Polymer Supported Metal Catalysts   25
References   26

CHAPTER 2. *Preparation of the Supports*   34
    2.1. General Considerations for Organic Polymers   34
    2.2. Styrene Based Systems   38
       2.2.1. Functionalisation of Preformed Polystyrene   38
       2.2.2. Copolymerisation of Functionalised Styrenes   48
    2.3. Non-Styrene Polymers   51
    2.4. Radiation Grafting   55
       2.4.1. Techniques of Radiation Grafting   57
       2.4.2. Reactions Occurring under the Influence of Radiation   60
       2.4.3. Factors Affecting Radiation Grafting   62

2.5.   Silica-Based Systems   67
2.6.   Other Inorganic Supports   71
References   72

CHAPTER 3. *Introduction of Metals onto Supports*   80
3.1.   Ion-Exchange-Based Catalysts   80
3.2.   Functionalised Supports   81
3.3.   Metal Complexes Bound to Polymeric Supports Through Metal-
       Carbon Bonds   87
3.4.   Polymerisation of Functionalised Monomers   89
3.5.   Direct Reaction Between Organometallic Compounds and
       Inorganic Oxide Surfaces   92
3.6.   Surface Bonding of Metal Carbonyls on Inorganic Oxides   95
3.7.   Supported Ziegler-Natta Catalysts   103
3.8.   Surface Supported Metal Salts   105
3.9.   Surface Complexes of Transition Metal Oxides on Oxide
       Supports   106
References   107

CHAPTER 4. *Characterisation of Supported Catalysts*   118
4.1.   Microanalysis   118
4.2.   Chromatographic Methods   119
       4.2.1. Gel Chromatography   119
       4.2.2. Temperature Programmed Decomposition Chromato-
               graphy   120
4.3.   Spectroscopic Methods   121
       4.3.1. Infrared   121
       4.3.2. Raman   124
       4.3.3. Inelastic Electron Tunnelling   125
       4.3.4. Ultraviolet and Visible   125
       4.3.5. Nuclear Magnetic Resonance   125
       4.3.6. Electron Spin Resonance   128
       4.3.7. Mössbauer   131
       4.3.8. Mass Spectrometry   132
       4.3.9. ESCA   133
       4.3.10. Extended X-ray Absorption Fine Structure   134
4.4.   Electron Microscopy   135
References   136

CHAPTER 5. *The Use of Supported Metal Complex Catalysts*   141
5.1.   Introduction   141
5.2.   Optimisation of Conditions   142

5.3. Laboratory Application   145
5.4. Industrial Application   145
References   147

CHAPTER 6. *Hydrogenation*   149
6.1. Introduction   149
6.2. Nature of the Support   150
6.3. Effect of Cross-Linking   151
6.4. Nature of the Solvent   151
6.5. Nature of the Metal Complex   152
6.6. Activity of Supported as Compared to Homogeneous Catalysts 152
6.7. Selectivity   153
6.8. Stability   154
6.9. Survey of Supported Hydrogenation Catalysts   152
    6.9.1. Titanium, Zirconium and Hafnium   154
    6.9.2. Chromium, Molybdenum and Tungsten   156
    6.9.3. Iron, Ruthenium and Osmium   157
    6.9.4. Cobalt   162
    6.9.5. Rhodium   163
        6.9.5.1. Analogues of $[Rh(PPh_3)_3Cl]$ 163
        6.9.5.2. Other Rhodium(I)-Phosphine Complexes   170
        6.9.5.3. Rhodium(I)-Phosphinite Complexes   173
        6.9.5.4. Rhodium Carbonyl Complexes   173
        6.9.5.5. Organometallic Rhodium Complexes   174
        6.9.5.6. Rhodium Carboxylate Complexes   176
        6.9.5.7. Rhodium Amide and Imidazole Complexes 177
        6.9.5.8. Rhodium Thioether Complexes   177
        6.9.5.9. Organorhodium(III) Complexes   177
    6.9.6. Asymmetric Hydrogenation   178
    6.9.7. Iridium   185
    6.9.8. Nickel   187
    6.9.9. Palladium and Platinum   189
    6.9.10. Actinides   194
6.10. Reduction of Inorganic Molecules   194
6.11. Michael Addition   195
References   196

CHAPTER 7. *Hydrosilylation*   204

CHAPTER 8. *Reactions Involving Carbon Monoxide*   216
8.1. Introduction   216

8.2. Hydroformylation 218
8.2.1. Cobalt Hydroformylation Catalysts 220
8.2.2. Rhodium(I) Hydroformylation Catalysts 223
8.2.3. Asymmetric Hydroformylation 228
8.2.4. Other Transition Metal Hydroformylation Catalysts 228
8.3. Carbonylation of Methanol 229
8.4. Fischer-Tropsch Reaction 235
8.4.1. Fischer-Tropsch Formation of Paraffins 236
8.4.2. Fischer-Tropsch Formation of Olefins 237
8.4.3. Fischer-Tropsch Formation of Alcohols 240
8.5. Water Gas Shift Reaction 240
8.6. Alkoxycarbonylation of Olefins 243
8.7. Isocyanates Formed by Carbonylation of Nitro Compounds and Azides 243
8.8. Syntheses of Aldehydes and Ketones 244
8.9. Substitution of Carbonyl Ligands in Metal Carbonyls 245
References 246

CHAPTER 9. *Dimerisation, Oligomerisation, Polymerisation, Disproportionation and Isomerisation* 252
9.1. Olefin Dimerisation 252
9.2. Olefin Trimerisation 255
9.3. Oligomerisation and Cyclooligomerisation of Dienes 255
9.4. Oligomerisation of Acetylenes 257
9.5. Polymerisation of Olefins 258
9.5.1. Inorganic Oxide Supported Olefin Polymerisation Catalysts 259
9.5.2. Polymer Supported Olefin Polymerisation Catalysts 261
9.6. Diene Polymerisation 262
9.7. Acetylene Polymerisation 264
9.8. Copolymerisation of Propylene Oxide with Carbon Dioxide 264
9.9. Olefin Metathesis 265
9.10. Olefin Isomerisation 268
9.10.1. Zirconium Complexes 269
9.10.2. Iron, Ruthenium and Osmium Carbonyl Complexes 269
9.10.3. Ruthenium(II) and Rhodium(I) Carbonyl and Carboxylate Complexes 270
9.10.4. Silica Supported Rhodium Catalysts 273
9.10.5. Nickel Catalysts 273
9.10.6. Palladium Catalysts 274

9.11.   Quadricyclane-Norbornadiene Isomerisation   274
            9.11.1. Quadricyclane to Norbornadiene Isomerisation   274
            9.11.2. Norbornadiene to Quadricyclane Isomerisation   276
9.12.   Grignard Cross-Coupling Reactions   276
References   277

CHAPTER 10. *Oxidation and Hydrolysis*   285
10.1.   Hydrocarbon Oxidation   285
10.2.   Decomposition of Peroxides   288
10.3.   Oxidation of Organic Compounds   289
10.4.   Oxidation of Inorganic Compounds   291
10.5.   Chlorination   292
10.6.   Ammoxidation   293
10.7.   Hydroxylation of Aromatic Compounds   293
10.8.   Hydroxylation of Olefins   293
10.9.   Carboxylation of Olefins and Aromatic Compounds   294
10.10. Vinyl Ester and Ether Exchange   295
10.11. Nitrile Hydrolysis   295
10.12. Nucleophilic Substitution of Acetate Groups   295
10.13. Stereoselective Hydrolysis of Esters   296
References   296

CHAPTER 11. *Conclusions And Future Possibilities*   299
11.1.   Sequential Multistep Reactions   299
11.2.   Selectivity Enhancement   304
11.3.   Activity   306
11.4.   Organic versus Inorganic Supports   308
11.5   Future Developments   309
References   310

INDEX   313

PREFACE

It is now 15 years since the first patents in polymer supported metal complex catalysts were taken out. In the early days ion-exchange resins were used to support ionic metal complexes. Soon covalent links were developed, and after an initially slow start there was a period of explosive growth in the mid to late 1970s during which virtually every homogeneous metal complex catalyst ever reported was also studied bound to a support. Both polymers and inorganic oxides were studied as supports, although the great preponderance of workers studied polymeric supports, and of these polystyrene was by far the commonest used. This period served to show that by very careful design polymer-supported metal complex catalysts could have specific advantages over homogeneous metal complex catalysts. However the subject was a complicated one. Merely immobilising a successful metal complex catalyst to a functionalised support rarely yielded other than an inferior version of the catalyst.

Amongst the many discouraging results of the 1970s, there were more than enough results that were sufficiently encouraging to demonstrate that, by careful design, supported metal complex catalysts could be prepared in which both the metal complex and the support combined together to produce an active catalyst which, due to the combination of support and complex, had advantages of activity, selectivity and specificity not found in homogeneous catalysts. Thus a new generation of catalysts was being developed. If heterogeneous and homogeneous catalysts are regarded as the first and second generations of man-made catalysts then supported metal complex catalysts are the third generation. Hence the title of this book *Supported Metal Complexes: A New Generation of Catalysts.*

This book is in many ways an update of the review I wrote between 1975 and 1976 (published in 1977). It describes both how supported metal complex catalysts have been prepared and used and suggests ideas for future developments. To date no supported metal complex catalysts have been used in large scale commercial processes; however I believe that this situation will change gradually, particularly as research into mechanically robust supports increases at the expense of investigation into mechanically weak supports such as polystyrene. The considerable opportunities that supported metal complex catalysts

xiii

offer for combining both the support and the metal centre in such a way as to obtain high activity, selectivity and specificity combined with easy separation of the catalyst from the reaction products will, I am sure, lead to commercial successes within the next decade. By attempting to summarise the current state of the art I hope that this book will hasten that event.

A book such as the present one is never merely the product of one man's pen alone. I owe a great deal, firstly to that international company of chemists who have provided the raw material for this work, and secondly to my own research students both at Southampton and Shrivenham. I should particularly like to thank Mr P. N. Vezey, Dr D. J. A. McCaffrey, Dr P. N. Nicholson, Professor J. A. Davies and Mr A. T. Sayer, all of whom have worked with me in this field. Not only has Dr S. G. Murray worked closely with me in this field, but this book would probably never have appeared without his encouragement of me to get on and write it. Mrs P. Trembath and Mrs K. J. Hunt helped with some of the typing. I am most grateful to Professors Brian James and Renato Ugo for inviting me to contribute this book to the *Catalysis by Metal Complexes* series and to Mr Ian Priestnall of Reidel's for so sympathetically living with the inevitable delays in writing a book during the period of major turbulence that has beset tertiary education in the United Kingdom in the 1980s. Finally I acknowledge with a deep sense of gratitude the understanding of my wife and daughters during the long hours of preparation. It is to them that this book is dedicated.

FRANK HARTLEY   *Shrivenham*
                                                                1984

ABBREVIATIONS

Ac	acetyl	EXAFS	extended X-ray absorption fine structure
acac	acetylacetonate		
aibn	azobisisobutyronitrile		
Ar	aryl	γ-	gamma-
bipy	bipyridyl	hfacac	hexafluoroacetylacetone
Bu	butyl		
Bz	benzyl	μ-	bridging ligand
		MAS	magic angle spinning
cod	1,5-cyclooctadiene	Me	methyl
cot	1,3,5-cyclooctatriene		
Cp	cyclopentadienyl	nbd	norbornadiene
CP	cross-polarisation	NMR	nuclear magnetic resonance
Cys	cystine		
		Ⓟ	polymer (usually polystyrene)
dba	dibenzylideneacetone	Ph	phenyl
diop	2,3-o-isopropylidene-2,3-dihydroxy-1,4-bis(diphenylphosphino)butane	Pr	propyl
		pvc	polyvinylchloride
		py	pyridyl
dma	dimethylacetamide		
dmbq	2,6-dimethyl-1,4-benzoquinone	R	alkyl group
dme	dimethoxyethane	rad	unit of radiation
dmf	dimethylformamide		
dmso	dimethylsulphoxide	{Si}	silica
dppe	bis(1,2-diphenylphosphino)ethane		
dpq	3,3′,5,5′-tetramethyldiphenylquinone	thf	tetrahydrofuran
		tmeda	tetramethylethylenediamine
		Tos	tosyl
η-	hapto-	TPP	tetraphenylporphyrin
ESCA	electron spectroscopy for chemical analysis		
		UV	ultra-violet
ESR	electron spin resonance		
Et	ethyl	XPS	X-ray photoelectron spectroscopy
eV	electron volt		

CHAPTER 1

INTRODUCTION

1.1. Catalysis

Although the subject of catalysis has been around ever since the first enzyme 'catalysed' an organic reaction it first began to develop into a science in 1836 when Berzelius used the word 'catalysis' to describe a number of previous experimental observations including Thenard's (1813) that ammonia was decomposed by metals and Dobereiner's (1825) that manganese dioxide modifies the rate of decomposition of potassium chlorate. Berzelius developed the word κατάλυσισ from two Greek words κατα- a prefix meaning down and λυσειν a verb meaning to split or break. Presumably Berzelius saw a catalyst as something that broke down the inhibitions of molecules towards reaction. Indeed this is consistent with the alternative Greek meaning of κατάλυσισ which is variously described as meaning either an inn or tavern [1] or the consequences of a failure of social or ethical restraints, such as a riot [2]. The popular press now uses the word catalysis in its alcoholic sense of 'bringing together' rather than its more literal sense of 'breaking down'. A catalyst may be defined as 'a substance which increases the rate at which a chemical reaction approaches equilibrium without being consumed in the process'. As befits a committee a longer, although more precise, description has been suggested by the UK Science Research Council [3]: "A system is said to be 'catalysed' when the rate of change from state I, to state II, is increased by contact with a specific material agent which is not a component of the system in either state, and when the magnitude of the effect is such as to correspond to one or more of the following descriptions:

(a) Essentially, measurable change from state I to state II occurs only in the presence of the agent.

(b) A similarly enhanced rate of change is found with the same sample of agent in repeated experiments using fresh reactants.

(c) The quantity of matter changed is many times greater than that of the agent."

Man's earliest attempts to emulate the enzymes generally led to catalysts

1

that existed in a separate phase to the reactants with a distinct interface between them; such catalysts are described as 'heterogeneous'. However more recently a number of catalysts have been developed which operate in the same phase as the reactants and are consequently known as 'homogeneous'. Of these, for our present purposes, by far the most important are the metal complex catalysts that dissolve in a solution containing the reactants. Although such homogeneous catalysts were first used for acetylene reactions as early as 1910 (e.g. reaction (1)),

$$HC{\equiv}CH + H_2O \xrightarrow{\;Hg^{2+}\;} CH_3CHO \tag{1}$$

their use only really began to be developed in the 1940s when wartime restrictions on raw materials forced the Germans to investigate both carbon monoxide and acetylene based processes for the production of fuels and plastics. The inherently greater difficulty of separating a homogeneous catalyst from the products at the conclusion of a reaction led industry to adopt a rather conservative attitude to the introduction of homogeneously catalysed processes in place of heterogeneous processes in which the catalyst could often be separated by some form of coarse filtration such as a cyclone filter. However the greater selectivity of homogeneous catalysts, arising out of their molecular nature, which ensures that only one type of active site is present, conveys an ability to produce pure products in high yield that the heterogeneous catalysts, whose activity is often based on 'defect sites' on their surface, of which there may be several different types, are hard pressed to match. Accordingly a significant number of homogeneously catalysed industrial processes have been brought on stream in recent years [4–5a].

The high selectivity of homogeneous catalysts [5b] arising from their molecular nature has led industrial and academic chemists into a search for catalysts that combine such 'molecular sites' with the ease of separation of heterogeneous catalysts. The first approach used was to support the molecular catalyst on an insoluble support. This followed from the work of Merrifield in supporting enzymes for polymer synthesis and degradation on polystyrene resin [6, 7]. The first published work on supported metal complex catalysts, involving the support of cationic metal complexes such as $[Pt(NH_3)_4]^{2+}$ on sulphonated polystyrene, was published in 1969 [8–11]. After a slow initial start, work has expanded to the extent that several hundreds of papers are published annually on supported metal complex catalysts. A number of good reviews have been written [12–33], of which 18, 23, 28, 30 and 33 are particularly recommended. To date only one book has appeared in the West [34]. Reviews in Japanese [35–37] and Russian [38, 39] reflect the worldwide importance of the subject. Except for the specialised field of Ziegler–Natta catalysis (Section 9.5) no commercial processes involving supported metal complex catalysts have yet been developed although it is the present author's view that this is only a few years away because of the enhanced selectivity that can be achieved when both support and metal complex combine together. Indeed it is reported [40] that Mobil did take a polymer-supported rhodium hydroformylation process up to pilot plant scale in the mid-1970s. Further development of a commercial-scale plant was only halted because the existing and projected markets were not large enough to justify the construction of a new plant. Had they been so, then the supported catalyst would have been the catalyst of choice.

Supported metal complex catalysts form the subject of this book. In the present chapter we shall examine the pros and cons of homogeneous and heterogeneous catalysts as a way of identifying the advantages and disadvantages we may expect for supported metal complex catalysts. We shall then examine other areas of chemistry where supported reagents are used, in order to promote an awareness of fields in which those wishing to develop supported metal complex catalysts should look for relevant developments. This chapter concludes with a brief account of alternative approaches to the present one for combining the advantages of homogeneous catalysts, accruing from their molecular nature, with the ease of separation of heterogeneous catalysts. In the remainder of the book we consider first the preparation of the supports (Chapter 2), then the introduction of metal complexes on to them (Chapter 3) and then the characterisation of the products (Chapter 4). After a brief look at the chemical engineering implications of supported metal complex catalysts in Chapter 5, the use of these catalysts in a number of reactions is considered (Chapters 6–10). The book concludes with some suggestions for the future in Chapter 11.

1.2. Reasons for Supporting Metal Complexes

In order to appreciate the reasons for the present interest in supported complex catalysts, which may be called the 'third generation' catalysts, it is useful to look at the advantages and disadvantages of the heterogeneous and homogeneous catalysts under a number of headings:

(a) *Separation of the catalyst*. The major disadvantage of homogeneous catalysts is the problem of separating the very expensive catalyst from the products at the end of the reaction. With heterogeneous catalysts this can be achieved by some kind of coarse filtration whereas with homogeneous catalysts a very efficient distillation or ion-exchange process is required. Distillation is inevitably an endothermic process and is therefore expensive and unless it is efficient it will result in small catalyst losses which may (a) render the process uneconomic and/or (b) contaminate the product which, in the case of a foodstuff, for example, would be unacceptable. It should be pointed out that two of the commercially successful homogeneously catalysed processes, namely the Wacker process for oxidation of ethylene to acetaldehyde [41] and the Monsanto process for the carbonylation of methanol to yield acetic acid [42], depend in part for their success on the relatively low boiling-points of the products (acetaldehyde 20.8 °C; acetic acid 117.9 °C). Of course, in some cases such as the hydrogenation of soft oils to yield components suitable for incorporation into margarine, the products decompose before their boiling-points (even under reduced pressure) so that distillation is not a practical method for separation of the catalyst in such cases. Distillation is also impossible in the case of reactions where there are high-boiling side-products which would steadily build up in concentration if they were not removed.

(b) *Efficiency*. In a heterogeneous system, the catalytic reaction must necessarily take place on the surface of the catalyst so that all atoms or molecules of the catalyst not present at the surface remain unused. By contrast all the molecules in a homogeneous catalyst are theoretically available as catalytic centres so that these catalysts are potentially more efficient in terms of the amount of catalyst needed to catalyse a given amount of reaction.

(c) *Reproducibility*. Homogeneous catalysts have the advantage over hetero-geneous catalysts of being totally reproducible because they have a definite stoichiometry and structure; by contrast the structure of the surface of a hetero-geneous catalyst is heavily dependent on both its method of preparation and its history subsequent to preparation.

(d) *Specificity*. A given homogeneous catalyst will generally have only one type of active site and therefore will often be more specific than a heterogeneous catalyst where several types of active site may be present in the form of different surface defects. These defects are extremely difficult to control. The specificity of a homogeneous catalyst can often be selectively modified by altering the other ligands present in such a way as to alter either the electronic nature or the steric requirements of the site.

(e) *Controllability*. Closely related to specificity is the fact that because a homogeneous catalyst has a definite structure it is much easier to modify it in order to control a reaction. Thus when the homogeneous catalyst [Rh(acac)(CO)$_2$] is altered to [Rh(acac)(CO)(PPh$_3$)] the ratio of normal to branched aldehydes obtained when 1-hexene is hydroformylated is altered from

1.2 : 1 to 2.9 : 1 [43]. Substitution of PPh_3 by $PPh_2(p$-styryl) further enhances this ratio to 3.9 : 1 [44]. By contrast the ill-defined active sites of a heterogeneous catalyst make systematic design and improvement very difficult.

(f) *Thermal Stability*. The thermal stability of heteorgeneous catalysts, such as pure metals and metal oxides, is often much higher than that of homogeneous catalysts. Since the rate of most reactions increases with temperature, a high operating temperature may be an advantage. It should be noted that high temperatures are not always ideal because some reactions involve a pre-equilibrium step which may be disfavoured by increasing the temperature. This is particularly true of reactions involving olefins where the entropy change on metal-olefin complex formation is almost invariably negative so that the stability of metal-olefin complexes decreases with increasing temperature [45]. The lower thermal stability of homogeneous catalysts is often compensated for by their significantly higher activities at lower temperatures and pressures.

(g) *Oxygen and moisture sensitivity*. Homogeneous catalysts are often organo-metallic compounds with metals in low oxidation states. Accordingly many of them are sensitive to oxygen and moisture. However heterogeneous catalysts are frequently subject to poisoning by 'soft' ligands to a much greater degree than homogeneous catalysts.

(h) *Solvent*. Whereas the range of suitable solvents for a homogeneous catalyst is often limited by the solubility characteristics of the catalyst, this clearly presents no problem for a heterogeneous catalyst.

(i) *Corrosion and plating out*. The use of some homogeneous catalysts on a commercial scale has led to a number of practical problems such as corrosion and plating out on the reactor walls that are not immediately obvious when the reaction is carried out in all-glass apparatus on the laboratory scale. The oxidative acetylation of ethylene to vinyl acetate catalysed by palladium(II) (reaction (2)) is an example of a process that suffers severe corrosion problems.

$$C_2H_4 + NaOAc \quad \xrightarrow[HOAc]{Pd^{II}, Cu^{II}, O_2} \quad CH_2CHOAc \qquad (2)$$

Although the original driving force for supporting homogeneous catalysts was to combine most of the advantages of homogeneous catalysts that accrue from their molecular character, particularly their selectivity and controllability, with the paramount virtue of ease of separation of the heterogeneous catalyst, experience has shown that the presence of both support and catalyst can have synergically beneficial effects. Thus the process of attachment to a support may have the following effects:

(i) The support may not be merely an inert back bone. It may take a positive role leading to preferred orientations of the substrate at the catalytic site so promoting selectivity [45a]. This, of course, is what the supporting back bones in many enzymes have been doing in nature for many millenia. It is this effect that is believed to be largely responsible for the $3\frac{1}{2}$-fold enhancement of the normal : branched aldehyde selectivity when 1-hexene is hydroformylated over polypropylene supported [Rh(acac)(Ph$_2$PC$_6$H$_4$CH$=$CH$_2$-p)CO] as compared to catalysis by the same complex unsupported [44].

(ii) Organic functional groups covalently bound to the surface of crystalline solids or polymers are subject to special constraints which may alter their chemical reactivity relative to the analogous small molecules [45b]. Thus the chemical properties of supported metal complexes can be different to their homogeneous analogues.

(iii) Supporting a metal complex on what is effectively a multidentate ligand may alter the stereochemistry around the metal ion in a beneficial way. This is well illustrated by the selectivity of platinum on nylon catalysts in the hydrogenation of benzene [46]. Thus platinum anchored on Nylon 66, 6 and 610 catalyses the formation of cyclohexene whereas platinum on Nylon 3, although an active hydrogenation catalyst, yields cyclohexane exclusively.

(iv) Supporting a metal complex may alter the position of equilibrium between metal ions and their surrounding ligands. This undoubtedly occurs when rhodium(I) complexes are bound on phosphinated supports and accounts for the fact that much lower phosphorus : rhodium ratios give greater enhancements of selectivities in olefin hydroformylation in the case of supported catalysts than with homogeneous catalysts [44, 47].

(v) By supporting a complex it is sometimes possible to stabilise catalytically active but normally unstable structures. Often this arises through site separation that prevents dimerisation to form stable inactive species, a process that is an important deactivation mechanism for many rhodium(I) systems [48]. Activation through site isolation is very important in hydrogenations catalysed by titanocene hydrides [49] and olefin polymerisations catalysed by chromium oxides [50] which are considered in more detail in Chapters 6 and 9. Although there has been a lot of interest in developing supported catalysts in which the metal centres are isolated, these catalysts are often used in swelling solvents. An important paper examined the effect of swelling solvents on site isolation by studying reaction (3) using infrared spectroscopy [50a].

$$\text{(P)}—PPh_2 + [Co(NO)(CO)_3] \xrightarrow{25\,°C} [Co(\text{(P)}—PPh_2)(NO)(CO)_2]$$

$$\nu_{NO} = 1755 \text{ cm}^{-1}$$

$$\Big\downarrow 70\,°C$$

$$[Co(\text{(P)}—PPh_2)_2(NO)(CO)] \qquad (3)$$

$$\nu_{NO} = 1710 \text{ cm}^{-1}$$

When $[Co(NO)(CO)_3]$ was reacted with a low cross-linked and a 20% cross-linked phosphinated polystyrene in the absence of solvent or in the presence of n-hexadecane, which is a poor swelling solvent, only the 1 : 1 complex was formed, whereas on adding m-xylene, which is a good swelling solvent, there was a rapid increase in site-site interaction and formation of a 1 : 2 complex. Thus the presence of cross-links are not in themselves sufficient to maintain site isolation in swelling solvents [50a]. A number of other reactions have been used to study the kinetics of site-site reactions and their dependence on both the degree of cross-linking and the nature of the solvent [50b–50e]. These include reaction (4), where the product is the monomer $[(\text{(P)}—CH_2C_5H_4)FeH(CO)_2]$ rather than the dimer formed by the unsupported analogue (reaction (5), and reaction (6)).

$$2\,Cp + [Fe_2(CO)_9] \longrightarrow 2\,[FeCpH(CO)_2] + 5\,CO$$

$$\Big\downarrow$$

$$[FeCp(CO)_2]_2 + H_2 \qquad (5)$$

$$\underset{\text{CH}_2\text{Cl}}{\overset{\text{CH}_2\overset{+}{\text{N}}\text{Me}_2(^n\text{Bu})\text{OAc}^-}{\boxed{\text{P}}}} \xrightarrow[4\,^\circ\text{C}]{\text{dioxane}} \underset{\text{CH}_2\text{OAc}}{\overset{\text{CH}_2\overset{+}{\text{N}}\text{Me}_2(^n\text{Bu})\text{Cl}^-}{\boxed{\text{P}}}} \qquad (6)$$

When homogeneous catalysts are photoactivated by the light-induced extrusion of a ligand or by metal–metal bond cleavage, recombination of the cleaved components can readily lead to deactivation. However, supports can be used to prevent this and to stabilise the coordinatively unsaturated products, as has been demonstrated by supporting $[\text{RuL(CO)}_4]$ and $[\text{Ru}_3\text{L}_3(\text{CO})_9]$ on high-surface-area silica [50f].

(vi) Enzymes often achieve their remarkably high selectivities through the simultaneous action of more than one type of catalytic site [51]. An excellent example of site cooperation is found in the catalysis of the hydrolysis of un-saturated esters by acid ion-exchange resins in which some of the protons on the acid sites had been replaced by silver(I) ions [52]. This substitution increases the rate of unsaturated ester hydrolyses, due to silver(I) binding the olefinic side chain so tying down the ester while hydrolysis is catalysed at nearby acidic sites.

(vii) Attachment of a metal complex to a support can sometimes provide some protection for the catalytic species against poisons such as water or atmospheric oxygen. Thus water is a poison for such Lewis acid catalysts as aluminium(III) chloride. However supporting aluminium(III) chloride on poly-styrene results in almost complete insensitivity to moisture during manipulation in air [53]. Similarly the instability of soluble rhodium(I) phosphine catalysts in the presence of oxygen is well known. However, there are now several examples of analogous supported rhodium(I) complexes that are resistant to oxygen degradation and can be filtered and recycled in air without the need to take special precautions [54–56]. Similarly reduced oxygen sensitivity has been found with supported palladium(0) phosphine complexes [57].

Clearly the long term goal of supported metal complex catalysts is to mimic nature and use both the support and the catalytic sites to unite to produce a highly selective catalyst in the same way as enzymes do. Some progress has been made in this direction. However it has been emphasised [58] that if the full benefits of supporting metal complexes are to be realised it may not necessarily be best to choose the same ligands that have proved so effective in homogeneous catalysis. For example, when a ligand is selected for a homogeneous catalyst it should coordinate fairly strongly to the metal in order to prevent ligand loss and subsequent reduction of the metal ions to the free metal. However on a support ligand dissociation is more spacially restricted so that weaker bonding ligands may be used. These may have electronic and steric advantages that are not realisable in homogeneous situations.

One way of promoting high selectivity has been to support the metal complex within the interstices of a cross-linked polymer. Diffusion of the reactants into the polymer and up to the active site then provides for selectivity when the reactant is a mixture of compounds. However such selectivity is necessarily achieved at the expense of catalyst activity. If activity is all important then it is necessary not only to ensure that all the catalytic sites are on the surface of the support, but better still that they hang off the surface and are effectively "dissolved" in the reaction medium. Some work along these lines has been done by insertion of a thirteen atom chain between polystyrene and pendant diphenylphosphine groups. Rhodium(I) complexes linked to these supports have comparable or superior hydrogenation ability to that observed with their homogeneous counterparts [59].

1.3. Catalyst Requirements

If a catalyst is to be commercially useful it must possess several desirable features. It must exhibit high selectivity, otherwise the cost of separating the products will be prohibitive. The catalyst must have a reasonable activity per unit volume of reactor space and the cost of the catalyst per unit of the product being produced should be low. This in turn implies a high turnover number. If the metal complex is attached solely to the surface of the support the activity per unit volume of reactor will be low. Accordingly, porous supports are normally used, although even here pellet size needs to be taken into account because diffusion and mass transport will be lower inside the support than in the bulk solution.

Most reactors involve considerable agitation of the catalyst. If degradation is to be avoided the catalyst must be mechanically strong. Furthermore, since many reactions are exothermic, the supported catalyst must not only be stable to reasonable temperatures, it must also have adequate heat transfer properties to disperse the heat generated at the active site. Above all, of course, the support must be chemically inert to the reactants, products and solvent of the reaction to be catalysed at the temperatures generated.

1.4. Types of Support

Two broad classes of supported complex catalyst have been developed. In the first class, the metal complex is linked to the support through attachment to one of its ligands, as in $[(ʃ–PPh_2)Rh(PPh_3)_2Cl]$ where $ʃ–PPh_2$ = supported diphenylphosphine; the environment of the metal ion is unaltered on supporting. In the second class, reaction of a metal complex with the support results in displacement of ligands attached to the metal and their substitution by groups that form an essential part of the support as in reactions (7) and (8). In both classes two broad types of support are used, organic polymers and inorganic oxides.

$$
\text{Zr(CH}_2\text{Ph)}_4 + \begin{array}{l} \text{Si—OH} \\ \text{Si—OH} \end{array} \longrightarrow \begin{array}{l} \text{Si—O} \\ \text{Si—O} \end{array} \!\!\!\!\!\!\!\!\!\! \underset{}{\overset{\text{CH}_2\text{Ph}}{\diagdown\!\!\!\underset{}{Zr}\!\!\!\diagup}}\!\!\!\!\!\!\!\!\!\! \overset{}{\underset{\text{CH}_2\text{Ph}}{}} + 2\,\text{PhCH}_3 \qquad (7)
$$

$$
(-\text{CH}_2-\text{CH}-)_n + \text{Mo(CO)}_6 \longrightarrow (-\text{CH}_2-\text{CH}-)_n \quad + 3\,\text{CO} \quad (8)
$$

with the pendant phenyl bearing —Mo(CO)$_3$.

Both have advantages as well as disadvantages. A significant difference between them is the degree to which they can be functionalised. Inorganic matrices have an upper limit of monofunctional groups of 1–2 meq/g of matrix. Organic matrices can carry up to 10 meq/g of matrix [30].

At least three analytical factors are important in order to characterise a supported catalyst: the percentage substitution, the distribution of any linking agent in the support and the structure of the linking agent on the support. The percentage substitution is usually determined by microanalysis. However it is important to determine the uniformity of substitution. Thus functionalisation of the support may or may not be even. If the functionalisation reaction is rapid relative to the rate of diffusion of the reagents then a shell of progressive introduction of functionality from the exterior of the particle towards the centre will result. If diffusion is more rapid, then a more uniform functionalisation will result. Very often, as we shall see in Chapter 2, functionalisation will involve several steps, and any unreacted intermediate functional groups that do not react in subsequent steps will remain on the support. In addition, any side reactions that occur will also leave their products on the support. When the metal complex is introduced it may react with undesirable functional groups as well as with the intended ones. A good example of the problems that may arise occurs in the introduction of phosphine and amine functional groups into polystyrene by the chloromethylation route (see Section 2.2.1.b). If the initial numbers of chloromethyl groups are high then quaternisation of the newly introduced phosphine or amine groups may occur (reaction 9).

$$
\text{P}\!-\!\text{CH}_2\text{Cl} + \text{P}\!-\!\text{PR}_2 \longrightarrow \text{P}\!-\!\text{CH}_2\overset{+}{\text{P}}\text{R}_2\text{Cl}^- \qquad (9)
$$
$$
\underset{\text{P}}{\big|}
$$

Even in commercial tertiary amine resins between 10 and 15% of the nitrogen atoms are present as quaternary ammonium salts [33]. The uniformity of functionalisation can be determined by sectioning the support and carrying out an electron microprobe analysis (Section 4.4) [60].

The structure of the linking group is normally assumed from its mode of preparation and subsequent reactions. However the advent of solid-state NMR will enable this aspect of functionalisation to be studied in more detail (see Section 4.3.5). This has already proved particularly valuable in the case of phosphinated supports where it has been found that the introduction of even one alkyl group in the *para*-position of one phenyl ring on triphenylphosphine renders the phosphine very sensitive to air oxidation [61, 62]. Thus one commercially available diphenylphosphinated polystyrene support has been found to contain a very large amount of phosphine oxide on delivery [62]. Infrared is sometimes helpful in characterising the nature of the link although it is not a sensitive technique and is therefore unable to help at low functionalisations.

There are two main types of support, organic polymers and inorganic supports, mainly metal oxides. We shall consider each in turn.

1.4.1. ORGANIC POLYMERS

Organic polymers that have been used as supports include polystyrene, polypropylene, polyacrylates and polyvinyl chloride. Polymers offer several advantages over other supports [62a]:

(i) They are easily functionalised; this is particularly true of polymers containing aryl groups.

(ii) Unlike metal oxide surfaces, most hydrocarbon polymers are chemically inert. As a result the support does not interfere with the catalytic group.

(iii) Polymers, particularly poly(styrene-divinylbenzene) can be prepared with a wide range of physical properties. As a result their porosity, surface area and solution characteristics can be altered by varying the degree of cross-linking [63—63b]. Thus with polystyrene, which is by far the most widely studied polymer, variation in the degree of cross-linking allows an almost continual change from virtually a soluble material (2% cross-linked polystyrene can be studied by solution NMR) to a completely insoluble material at 20% cross-linking.

The principal disadvantages of polymers are their poor heat transfer ability and in many cases their poor mechanical properties which prevent them from being used in stirred reactors in which they are pulverised. Recently polymers such as polypropylene [44, 63c, 63d] and poly(phenylene oxides) [63e] have been studied specifically because they are mechanically strong. Although there are many attractions to using commercial polymers, these are not always very well defined and often contain unknown impurities. It is important in developing polymer supported catalysts to have as well-defined and pure a support as possible.

The physical properties of polymers vary widely depending on the molecular weight, the chemical nature of the monomer or combinations of monomers, and the conditions of polymerisation which affect the arrangement of the polymer molecules and their interactions with one another. There are essentially three major types of polymers, although in practice a continuous range of material is available between these three extremes:

(i) Gels, or gellular, or microporous polymers are a particularly simple type of polymer in which the long strands of polymer molecules are either lightly cross-linked or merely entangled randomly. A gel of a hydrocarbon molecule such as polystyrene is very much like a hydrocarbon liquid. The molecules are in constant random motion and can accommodate high concentrations of chemically similar molecules such as other hydrocarbons, although they repel polar molecules such as water. Small molecules like benzene can diffuse through the gel rapidly, encountering almost as little resistance to diffusion as they would through a solution of long hydrocarbon molecules. Not only does this free diffusion promote transport of reactants and products to and from the catalytic site it also facilitates the removal of heat. Gellular polymers can be used in hydrocarbon solvents as essentially soluble supports that may be separated at the end of the reaction either by precipitation by changing the solvent, or by osmotic procedures such as membrane filtration [64–67]. However in polar solvents, which are often used for certain catalytic reactions, such as hydrogenation, gel type polymers are not swollen but tend to close up their pores [63a].

(ii) Macroreticular or macroporous polymers have a carefully controlled regular cross-linking (reticulation) which allows a high internal surface area to the polymer [30, 63]. They can be formed either from styrene, using divinylbenzene as the copolymer, or from acrylates. In addition to their permanent, or dry state porosity, macroreticular polymers can often be swollen. Thus cross-linked polystyrene is swollen by benzene. Donor ligands can be supported within, as well as on the surface of the polymer. Clearly reagents do not have the access to internal catalytic sites that they have in the case of gels; however, by carefully controlling the degree of cross-linking it is possible to control access to the sites of a macroreticular polymer-supported catalyst, and so enhance the selectivity of the catalyst. Thus the rates of hydrogenation of a series of olefins in benzene solution in the presence of $[\text{Rh}(\text{\textcircled{P}}-\text{PPh}_2)(\text{PPh}_3)_2\text{Cl}]$, where $\text{\textcircled{P}}-\text{PPh}_2$ is 100–200 mesh 2% cross-linked phosphinated polystyrene, decreased as the steric bulk of the olefin increased in the order 1-hexene $\gg$ cyclohexene $\gg$ cyclooctene $>$ cyclododecene $\gg$ Δ^2-cholestene [68]. This same effect can be used to effect a regioselective reduction of the side-chain double bond of steroid **1** without reduction of the double bond in the large steroid nucleus (reaction 10), whereas homogeneous catalysts promote the reduction of both double-bonds [68a].

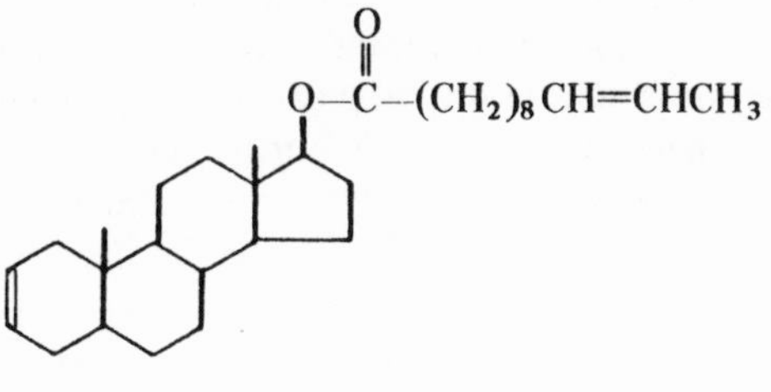

$$O-\overset{\overset{\displaystyle O}{\|}}{C}-(CH_2)_8\,CH{=}CHCH_3$$

1

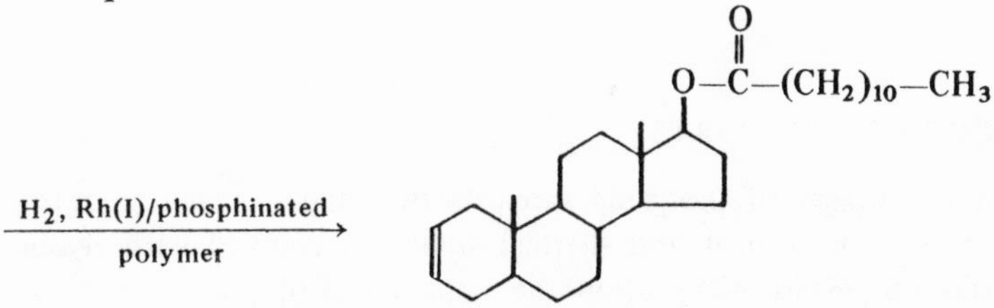

$$O-\overset{\overset{\displaystyle O}{\|}}{C}-(CH_2)_{10}-CH_3$$

$$\xrightarrow[\text{polymer}]{H_2,\ Rh(I)/\text{phosphinated}}$$

(10)

(iii) Proliferous or 'popcorn' polymers [69, 70] are formed spontaneously in butadiene and butadiene copolymer plants, for example butadiene/styrene plants are particularly susceptible to the formation of proliferous polymers. They are hard, porous, opaque materials which swell somewhat in benzene and carbon tetrachloride but are insoluble in all common solvents.

As already mentioned, one of the highly attractive features of organic polymer supports is the opportunity they offer to introduce extra selectivity into the catalyst through the control of diffusion of reactants within the polymer. This, however, is necessarily at the expense of activity. Diffusion rates through liquids are generally of the order of 10^{-5} cm sec^{-1}, whereas diffusion rates within polymer matrices are typically an order of magnitude lower (*ca* 10^{-6} cm sec^{-1}) [30]. If the actual reaction is fast, relative to diffusion, then the potential activiy of the catalyst will not be realised and the supported catalyst will be less active than its homogeneous counterpart. Common methods used to reduce the diffusion limitation of supported catalysts are to reduce the particle size, to increase the overall surface area of the catalyst, for example by using a gellular polymer, and to introduce a bimodal pore size distribution allowing rapid diffusion through a portion of the catalyst. Since diffusion within a polymer depends on the nature of the porous structure of the polymer, it is governed not only by the degree of cross-linking but also by the extent of swelling in the presence of the reactants and solvents present. For example with polystyrene, solvents that are more polar than benzene (1) decrease the pore width due to decreased resin swelling and (2) create polar gradients between the bulk solvent and the local environment of the active site [68]. The first of these decreases the diffusion rate of large, bulky reactants whilst the second selectively enhances the diffusion rate of non-polar reactants. Thus as the ethanol content of a benzene-ethanol solvent system is increased the rate of reduction of non-polar olefins in the presence of a polystyrene-immobilised hydrogenation catalyst increases until the ethanol content becomes sufficiently high as to shrink the pores so much that molecular size dominates selectivity. Thus the proper choice of both solvent and polymer can result in some control of the porosity and hence yield a catalyst that has molecular shape selectivity [16, 68].

1.4.2. INORGANIC SUPPORTS

The major advantages of inorganic supports over their organic counterparts are their better mechanical and thermal stabilities coupled with reasonable heat transfer properties. Very often the upper limit of the thermal stability of organic polymer supported complex catalysts is set by the polymer rather than the metal complex [28]. Inorganic supports that have been used include silica, alumina, glasses, clays and zeolites. Some inorganic supports such as silica gel lack the chemical stability and flexibility of organic polymers. Clays, on the other hand, have great potential. Although often highly desirable, flexibility is not always an asset; in particular where catalyst deactivation occurs by dimerisation flexibility should be avoided either by using an inorganic support or a highly cross-linked organic polymer. Conversely inflexibility can also introduce complications in that the distance between the surface and the functional group can sometimes be critical to catalytic performance [71]. Rhodium(I) hydroformylation catalysts supported on phosphinated silica with short chains between the silica and the phosphine are less active than those with longer links. A similar phenomenon has been observed with heterogenised enzymes [51].

Although there are situations in which variations in diffusion rates can be disadvantageous there are plenty of situations in which the ability to control diffusional factors is an advantage. This is often difficult with organic polymers due to the variation of swelling as a function of both temperature and solvent. In contrast inorganic substrates can be selected for stable diffusional characteristics at most reaction conditions [20]. Zeolites, with their well-understood controlled pore sizes, should be of particular value [72–76], although they have been studied relatively little. If diffusion control through swelling is desirable then this can be achieved using naturally occurring silicates known as smectites. These relatively abundant minerals [77], which include hectorite and montmoril-lonite, possess mica-like structures in which the crystallites are made up of alternating layers of cations and negatively charged silicate sheets (Figure 1). However unlike micas the cations can be readily exchanged, for example for cationic rhodium complexes [77–81a]. The intracrystal space can be swelled by water, alcohols and other organic solvents, the degree of swelling depending on the intralayer cations, the substrate and the negative charge density on the silicate sheet.

Most inorganic oxide supports possess surface hydroxyl-groups which are used to effect attachment. In some cases these surface hydroxyl-groups, which give rise to a very polar surface, are undesirable in the final catalyst and where that is so they can be removed, usually by reaction with a chlorosilane to give a non-polar lipophilic surface. This technique is commonly used in gas chromato-graphy. t-Butyldimethylchlorosilane is a particularly effective silylation reagent

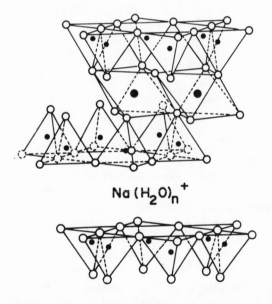

$$Na\,(H_2O)_n^{+}$$

Fig. 1. Structure of Hectorite. Oxygen atoms of silicate sheet are represented by open circles. The upper and lower layers of tetrahedral holes are occupied mainly by silicon. Magnesium and lithium occupy the central layer of octahedral holes. The layer exchange ions are mainly solvated Na^+ ions (reproduced with permission from [78]).

that is 10 000 times more stable to hydrolysis than other silylation reagents [82].

1.5. Chemically Modified Electrodes

The chemical attachment of metal complexes on to the surface of electrodes and photoelectrodes is currently a subject of considerable interest. The object is to immobilise a redox compound on the surface of the electrode to act as an electron transfer mediator between the electrode surface and the solution surrounding it [83—95]. Although we shall not be concerned further with such systems in this book, the reader should be aware of their existence because some of the techniques used to prepare and characterise the electrodes are of relevance in the preparation and characterisation of supported catalysts.

1.6. Immobilised Enzymes and Reagents for Organic Syntheses

As already mentioned in Section 1.1, the idea of supporting metal complexes developed from Merrifield's pioneering work on supporting enzymes. Since some of the approaches used to support enzymes are relevant in the present context, the interested reader is referred to [96—101] for more recent developments in the field. Similarly [102—115] should provide the reader with an entrée into polymer and silica supported organic reagents, which will not be considered further in the present book.

1.7. Triphase Catalysis

Phase transfer catalysis [116—120] has proved to be a very valuable technique for promoting the reaction of a water-soluble compound with a water insoluble compound. For example the reaction between sodium cyanide and an alkyl bromide can be promoted by the addition of a crown ether to a suspension of aqueous sodium cyanide in a benzene solution of the alkyl bromide. Alternatively, a soap could disperse the organic phase in the aqueous phase and promote micelle formation, so increasing the reaction rate. If a quaternary salt such as R_4N^+ with sufficiently long alkyl chains is used, the cation will be sufficiently soluble to take the cyanide into the organic phase. A recent development of this is to attach the quaternary ammonium or phosphonium salt to a polymer [121—126]. In this way a third phase is created within the polymer wherein the components of the organic and aqueous phases can be brought together to react (Scheme 1).

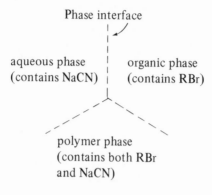

Phase interface

aqueous phase | organic phase
(contains NaCN) | (contains RBr)

polymer phase
(contains both RBr
and NaCN)

Scheme 1

This is known as *triphase catalysis*. Some clustering of the quaternary sites occurs so that on stirring some aqueous phase is imbibed into the resin along with the organic phase so promoting the reaction. After reaction the liquid is filtered from the resin and the alkyl cyanide extracted from the benzene solution. The polymer can then be recycled. The techniques used to synthesise the polymer phase are essentially those described in Chapter 2. The effectiveness of the polymer catalyst depends on the quaternary group used, phosphonium salts being more effective than ammonium salts [123–125], the length of the backbone between the polymer and the quaternary group, longer backbones being more effective so long as vigorous stirring is maintained [122, 124], and the nature of the organic solvent [122]. The support for the quaternary salt need not be an organic polymer. Silica has been used, and indeed has the advantage that in the reaction of alkyl bromides with potassium iodide the organic solvent can be dispensed with and the reaction run in aqueous solution alone [127–129]. Non-ionic phase transfer catalysts involving linear (reaction (11)) and crown ethers (reaction (12)) have also been described [130–133].

$$\text{P}\!-\!\langle\text{C}_6\text{H}_4\rangle\!-\!CH_2Cl + HO(CH_2CH_2O)_nCH_3$$

$$\xrightarrow{(n\ averages\ 13)} \text{P}\!-\!\langle\text{C}_6\text{H}_4\rangle\!-\!CH_2O(CH_2CH_2O)_nCH_3 + HCl \qquad (11)$$

$$(12)$$

Montmorillonite clay has been used to support phase transfer catalysts by exchanging some of the cations in the clay for benzyltributylammonium ions. This is very effective as a triphase catalyst for reactions (13–15) due to the large surface area of the clay which is able to promote a high degree of surface contact between the two solvent phases [134].

$$(13)$$

$$(14)$$

$$[RhCl(PPh_3)_3] + NaOAc$$

$$\xrightarrow[\text{benzene/water}]{\text{Montmorillonite}/(PhCH_2)^n Bu_3 NBr} [Rh(OAc)(PPh_3)_3] + NaCl \qquad (15)$$

1.8. Heterogenisation of Metal Complex Catalysts

As we have already discussed in Section 1.2, there would be many advantages to be gained if the ease of separation of a heterogeneous catalyst could be combined with the catalytic advantages stemming from the molecular character of a metal complex catalyst. Supporting the metal complex, which forms the subject matter of this book, is only one way of 'heterogenising' a metal complex. In this section we briefly look at some of the other approaches that have been developed to tackle this problem.

1.8.1. PHASE TRANSFER

The use of a phase transfer agent to transfer water-soluble $[RhCl_4]^-$ ions into an organic medium where they catalyse the isomerisation of alcohols to ketones (reaction (16)) or olefins (reaction (17)) [135] is but one example of many phase transfer transition metal catalysed reactions [116, 136].

$$CH_3(CH_2)_n \overset{\overset{\displaystyle OH}{|}}{CH}CH{=}CH_2 \xrightarrow[n\,=\,2-4]{R_4N^+RhCl_4^-\ or\ R_4P^+RhCl_4^-} CH_3(CH_2)_nCOC_2H_5 \quad (16)$$

$$cis\text{-}PhCH{=}CHCH_3 \xrightarrow{R_4N^+RhCl_4^-\ or\ R_4P^+RhCl_4^-} trans\text{-}PhCH{=}CHCH_3 \quad (17)$$

The metal complex is added in aqueous solution, carried into the organic phase using a quaternary ammonium or phosphonium salt and then recovered into the aqueous phase at the end of the reaction using a lipophilic anion such as perchlorate (reaction (18)).

$$R_4E^+RhCl_4^-(org.) + NaClO_4(aq.) \xrightarrow{E\,=\,N,\,P} R_4EClO_4(org.) + Na^+RhCl_4^-(aq.) \quad (18)$$

A very important development in phase transfer catalysis is the reporting of an aqueous rhodium system for the highly selective hydroformylation of higher olefins such as 1-decene [135a, 135b]. $p\text{-}Ph_2PC_6H_4CO_2H$ ligands (P : Rh = 10 : 1) and $C_{12}H_{25}NMe_3^+Br^-$ as a phase transfer reagent enable the hydroformylation to take place in neat olefin, with the rhodium catalyst remaining in the aqueous layer, so that product-catalyst separation is easy. Normal : branched aldehyde selectivities of between 10 : 1 and 120 : 1 can be achieved, depending upon the precise conditions. Sulphonated phosphines give lower selectivities [135a, 135c, 135d]. Phase transfer catalysts, many of which involve crown ethers, have been immobilized on both polymeric [136a–136k] and silica supports [136ℓ–136p].

1.8.2. SUPPORTED LIQUID AND GAS PHASE CATALYSIS

Supported liquid and gas phase catalysts are prepared by absorbing a catalyst within the pores of an inert carrier such as kieselguhr, silica or alumina. The resulting product has the properties of a molecular complex catalyst but can be handled like a heterogeneous catalyst. Much of the experimental study on these catalysts has been carried out by loading the catalyst into gas chromatographic columns, but commercially they could be used in both fixed [137] and fluidised [138, 139] bed reactors. Although the idea of supporting metal complex catalysts dates only from 1966 [140], the method has been used since the 1930s for the polymerisation of mixtures of paraffins and olefins in contact with supported sulphuric acid [141–144]. Supported metal complex catalysts have been used to study the hydrogenation [145–150], isomerisation [145, 147, 148] and hydroformylation [145, 147, 148, 150–163b] of olefins. They have the important advantages of eliminating the severe reactor corrosion problems associated with many homogeneous catalysts. Supported catalysts can be used either in the liquid phase or in the gas phase.

When used in the liquid phase, a solution of the metal complex in a non-volatile solvent is soaked into a porous mineral support. The activity, stability and selectivity of the catalyst can be varied by altering the polarity of the solvent. For hydroformylation in the presence of $[RhH(CO)(PPh_3)_3]$, triphenyl-phosphine has commonly been used as the solvent because it promotes the formation of linear as opposed to branched aldehydes [154–162]. However, more traditional gas-liquid chromatography solvents such as benzyl butyl phthalate [152], diphenyl ethyl hexyl phosphate [164] and diphenyl ether [164] have also been used. The activity and selectivity of supported liquid phase catalysts is highly dependent on the type of support, the degree of liquid loading, the adsorption of the catalyst on to the support surface and the concentration of the gases in the high-boiling solvent [147, 148, 150, 152, 156a–156c, 157, 158]. Too much liquid phase can decrease the activity by blocking the pores in the catalyst preventing the reactants, which are in the gas phase, from reaching the catalyst except by a long diffusion path through the liquid phase [147, 148, 152].

Supported catalysts for use in the gas phase are prepared by impregnating the support with a solution of the catalyst in a volatile solvent such as chloroform [163, 165–167] which is subsequently volatilised off leaving the metal complex held by van der Waals' forces. Both metal oxides and activated charcoal have been used as the support material [166, 168]. Activated charcoal was used to absorb an aqueous-acetic acid solution of rhodium trichloride that had been treated with carbon monoxide and hydriodic acid before absorption. After removal of the solvent the resulting catalyst was effective in promoting the vapour phase carbonylation of methanol to acetic acid at low pressure [168].

Since the contact time in the presence of supported liquid and gas phase catalysts is very short, the catalysts must have very high activities to be of any use in this way [154]. For this reason a very popular system for study has been the hydroformylation of olefins in the presence of $[RhCl(CO)(PPh_3)_2]$ and $[RhH(CO)(PPh_3)_3]$. In preparing these catalyst it is important to note that a theoretical study [153] predicting the existence of an optimum loading of the catalyst on the support has been confirmed experimentally [152], although it has practically no effect on the ratio of normal : branched aldehydes formed. The great advantage of using triphenylphosphine as the solvent for $[RhH(CO)(PPh_3)_3]$ is that it enables very high ratios of normal : branched aldehydes to be obtained (47 : 1) so long as only low pressures (e.g. 2 bar) of hydrogen and carbon monoxide are used, without severely depressing the activity of the catalyst due to the presence of very large amounts of phosphine [154, 155]. This effect arises because only the rhodium(I) complexes at the gas-triphenylphosphine boundary are active [158]. Although the support is often inert, this is not always so [159]. For example in the hydroformylation of allyl alcohol the undesirable side reaction of isomerisation of allyl alcohol to propionaldehyde is caused by interaction between aluminium in supports such as kieselguhr and alumina with $[RhH(CO)(PPh_3)_3]$. When silica with a very low aluminium content was used in conjunction with excess triphenylphosphine this side reaction was eliminated [156].

Although most liquid and gas phase catalysts have been supported on metal oxides this is not exclusively so. Thus phosphinated styrene-divinylbenzene resins have been formed over silica gel and used to support rhodium(I) (reaction (19)) to obtain polymer bound catalysts in which the reactants do not have to diffuse through the polymer [168a, 168b].

The catalysts were more active in the gas phase hydroformylation of olefins at $100-150\,°C$ and atmospheric pressure than their homogeneous counterparts, even when these latter were dispersed on silica gel; for propene a normal : branched selectivity of $6 : 1$ was achieved. Secondary olefins were not hydroformylated under these conditions. $[RhH(CO)(PPh_3)_3]$ supported on phosphinated highly cross-linked polystyrene has been used in the gas phase hydroformylation of propene [168c, 168d]. The performance of these catalysts depends closely on their method of preparation; in particular all chlorine introduced prior to phosphination must be rigorously removed. Catalysts prepared by chloromethylation and subsequent reaction with $LiPPh_2$ (see Sections 2.2.1.b and 2.2.1.c) deactivate slightly during propene hydroformylation but give a higher normal : branched selectivity than those prepared by chlorophosphination (reaction (20)).

$$\text{(P)}-\langle\rangle + PCl_3 \xrightarrow{\ BF_3\cdot Et_2O\ } \text{(P)}-\langle\rangle-PCl_2$$

$$\xrightarrow{\ LiPh\ } \text{(P)}-\langle\rangle-PPh_2 \qquad (20)$$

The hydroformylation activity of these latter decreases with increasing phosphine content. Typical normal : branched selectivities of these systems $(0.62-1.0)$ are very low, their real advantage being their high activity [168c, 168d].

Since triphenylphosphine is a good ligand for rhodium(I), its presence necessarily depresses the activity of the catalyst. Accordingly, $[Rh(\eta^3\text{-}C_3H_5)(CO)(PPh_3)_2]$ supported on alumina has been evaluated, since the η^3-allyl ligand is readily lost, creating a vacant coordination site. The catalyst is an active olefin hydrogenation and hydroformylation catalyst [169, 170]. $[RhCl(PPh_3)_3]$ supported on silica gel is more active as an olefin hydrogenation catalyst when used in the gas phase than when supported in an involatile solvent such as 1,2,4-trichlorobenzene [147, 148]. $[RuCl_2(CO)_2(PPh_3)_2]$ is active for the hydrogenation of aldehydes and ketones at 15 bar and $160\,°C$ both in solution in carbowax and in the gas phase [150]. In the gas phase it is possible to hydrogenate reactants at temperatures very close to their boiling points.

In summary, supported liquid and gas phase catalysts have a number of important advantages over their homogeneous counterparts. They can be as active as the homogeneous catalysts, but more selective. They are easy to separate from the products and give rise to less corrosion. However, optimisation of the reaction conditions is more complex because there are more variables, although this can be advantageous in enhancing selectivity. The fact that the link between support and metal complex is due to van der Waals' interactions has been said to be a disadvantage in that, in the liquid phase, leaching may be a problem [171], although some catalysts have been used for very long periods without signs of deterioration in their activity [156, 157].

1.8.3. USE OF MELTS

A method for overcoming the problem of separating the reaction products from the catalyst solution that has only rarely been used is to dissolve the catalyst in an involatile molten salt [172–174]. The products can then be readily separated, either at the end of the reaction or continuously by distillation. One of the problems of using molten salts is that they are often corrosive and present problems of mass transfer. However the tetraalkylammonium salts of $GeCl_3^-$ and $SnCl_3^-$ are stable, low melting solids (m.p. less than 100 °C) which are good solvents for olefins as well as many Group VIII metal complexes. Thus, a solution of $PtCl_2$ in such salts has been shown to catalyse the hydrogenation, isomerisation, hydroformylation and alkoxycarbonylation of olefins [172]. Molten ammonium and phosphonium salts, such as $[P^nBu_4]Br$, which melts at 100 °C, containing either ruthenium(IV) oxide or ruthenium(III) acetylacetonate yield highly active catalysts for the synthesis of ethylene glycol from synthesis gas (reaction (21)) [173, 174].

$$2\,CO + 3\,H_2 \xrightarrow[\text{Ru catalyst}]{220\,°C;\,430\,atm.\,1:1\,CO:H_2} \begin{array}{c} CH_2OH \\ | \\ CH_2OH \end{array} \qquad (21)$$

The reaction is, of course, accompanied by the formation of considerable amounts of methanol and ethanol as well as the monoether formed by condensation of these alcohols with ethylene glycol.

1.8.4. LATTICE METAL COMPLEXES

A potential route for the immobilisation of metal complexes is to link them into lattices in which the metal atoms themselves are integral components of the lattices. An example of the preparation of such a material is given in reaction (22) [175, 176].

CN—⟨benzene⟩—NC + [Rh(CO)$_2$Cl]$_2$

$$\longrightarrow \left[\cdots \right]^{2n+} (Cl^-)_{2n} \quad (22)$$

Although these complexes have been developed in the hope that they may have valuable catalytic properties, these have still to be demonstrated.

1.8.5. WATER SOLUBLE COMPLEXES

A possible approach to facilitating the separation of a molecular complex catalyst from the products of a reaction on an olefinic substrate is to prepare a water-soluble catalyst and use a two-phase system. The catalyst is then dissolved in water and the olefin substrate and its products are held in an immiscible organic phase. Rhodium(I) complexes of sulphonated arylphosphines such as $Ph_{3-n}P(C_6H_4SO_3Na)_n$ [135b, 177] have been shown to be effective olefin hydrogenation and hydroformylation catalysts [177–180]. In practice however these complexes may become distributed between the two phases, making clean separations impossible, as in the hydroformylation of olefins [178]. $[Rh(Ph_2PCH_2CH_2NMe_2)_2(MeOH)_2]^{2+}(NO_3^-)_2$ is an active olefin hydrogenation catalyst that can be used in one phase (water soluble olefins) or two phases with water insoluble olefins [181]. In contrast to the sulphonated phosphine system this system is stable to short (one hour) exposures to air.

An alternative approach would be to use the metal complex in a homogeneous solution in an organic solvent and then subsequently extract the metal complex into aqueous solution. This approach has been used to hydroformylate olefins in the presence of a cobalt carbonyl complex containing the amine substituted phosphine $P(CH_2CH_2NEt_2)_3$. The cobalt complex is extracted into dilute mineral acid after completion of the reaction. It can then be re-extracted back into the organic solvent by the addition of base [182]. A similar approach for the hydroformylation of olefins in the presence of the tris(2-pyridyl)phosphine complex of rhodium, $[RhH(CO)(PPh_3)\{P(2\text{-py})_3\}_2]$ in the presence of excess tris(2-pyridyl)phosphine, gave successful hydroformylation of 1-hexene in acetophenone as solvent. However attempts to recover the rhodium-phosphine complex by extraction with water were unsuccessful. When dilute hydrochloric or hydrofluoroboric acids were used, the complexes decomposed with liberation of hydrogen. Attempts to adsorb the complex from the hydroformylation solution on to an ion-exchange resin were also unsuccessful. Ironically, the most successful approach to catalyst recovery was to precipitate the complex with petroleum, although recoveries were never greater than 95% [183].

The rhodium(I) complex 2 as well as the iron, molybdenum and tungsten complexes $[Fe(CO)_4L]$, $[Mo(CO)_5L]^+$ and $[W(CO)_5L]^+$, where $L = Ph_2P\text{-}CH_2CH_2NMe_3^+I^-$, catalyse the water gas shift reaction slowly in aqueous solution [184, 184a].

2

$[RhCl(Ph_2PC_6H_4SO_3Na\text{-}m)_3]$ catalyses the hydrogenation of unsaturated triglycerides at an oil-water interface and when codispersed in bilayers of saturated phospholipids [185]. Other studies have shown that unsaturated phospholipids dispersed as bilayers in water can be successfully hydrogenated in the presence of the water-soluble catalyst as well as with $[RhCl(PPh_3)_3]$, which partitions from the aqueous phase into the hydrocarbon interior of the phospholipid bilayer [186]. The ordered arrangement of phospholipids in lamellar structures enables highly selective hydroformylation reactions to be achieved by interpolating substrates into bilayers of dipalmitoylphosphatidyl-choline in the presence of water soluble, but not lipid soluble, homogeneous catalysts such as $[RhH(CO)(Ph_2PC_6H_4SO_3Na\text{-}m)_3]$ [187].

1.9. Polymer Supported Metal Catalysts

In Chapter 3 we shall repeatedly note that following support of a metal complex, that complex is reduced to the free metal as a means of producing highly dispersed active metal catalysts. An alternative method of depositing metals on polymeric supports, that may one day find application in the preparation of supported metal complexes, involves cocondensing a metal vapour on to a polymer [188]. In this way very small clusters of metal atoms can be widely dispersed throughout a polymer.

References

1. G. Wilke in: *Fundamental Research in Homogeneous Catalysis* (Ed. M. Tsutsui) Plenum Press, New York. Vol. 3, p. 1 (1979).
2. G. C. Bond: *Heterogeneous Catalysis*, Oxford (1974).
3. *Catalysis*, U.K. Science Research Council Report (1975).
4. G. W. Parshall: *Homogeneous Catalysis*, John Wiley, New York (1980).
5. C. Masters: *Homogeneous Transition-Metal Catalysis*, Chapman and Hall, London (1981).
5a. D. T. Thompson: *Chem. in Brit.* **20**, 333 (1984).
5b. B. M. Trost: *Chem. in Brit.* **20**, 315 (1984).
6. R. B. Merrifield: *Federation Proc.* **21**, 412 (1962).
7. R. B. Merrifield: *Science* **150**, 178 (1965).
8. W. O. Haag and D. D. Whitehurst: *German Patent* 1,800,371 (1969); *Chem. Abs.* **71**, 114951 (1969).
9. W. O. Haag and D. D. Whitehurst: *German Patent* 1,800,379 (1969); *Chem. Abs.* **72**, 31192 (1970).
10. W. O. Haag and D. D. Whitehurst: *German Patent* 1,800,380 (1969); *Chem. Abs.* **71**, 33823 (1969).
11. W. O. Haag and D. D. Whitehurst: *Belgian Patent* 721,686 (1969).
12. J. Manassen: *Platinum Metals Review* **15**, 142 (1971).
13. C. U. Pittman: *ChemTech* **1**, 416 (1971).
14. N. Kohler and F. Dawans: *Revue Inst. Franc. Pétrole* **27**, 105 (1972).
15. Z. M. Michalska and D. E. Webster: *Platinum Metals Review* **18**, 65 (1974).
16. J. C. Bailar: *Cat. Rev. – Sci. Eng.* **10**, 17 (1974).
17. Z. M. Michalska and D. E. Webster: *ChemTech* **5**, 117 (1975).
18. F. R. Hartley and P. N. Vezey: *Adv. Organometal. Chem.* **15**, 189 (1977).
19. D. Commereuc and B. Martino: *Revue Inst. Franc. Pétrole* **30**, 89 (1975).
20. C. H. Brubaker in: *Encyclopaedia of Polymer Science and Technology*, Wiley, New York. Vol. suppl. I, p. 166 (1976).
21. L. L. Murrell in: *Advanced Materials in Catalysis* (Ed. J. L. Burton and R. L. Garten) Academic Press, New York, p. 235 (1977).
22. L. Dawans: *Inf. Chim.* **163**, 191 and 201 (1977); *Chem. Abs.* **87**, 91253 (1977).
23. Y. Chauvin, D. Commereuc and F. Dawans: *Prog. Polymer Sci.* **5**, 95 (1977).
24. G. Webb: *Catalysis*, Chem. Soc. Spec. Per. Reports **2**, 163 (1978).
24a. M. S. Scurrell: *Catalysis*, Chem. Soc. Spec. Per. Reports **2**, 215 (1978).
24b. V. Davydov: *Acta Polym.* **30**, 119 (1979).
24c. B. Sedlacek, G. C. Overberger and H. F. Mark (Eds.): *Polymer Catalysts and Affinants – Polymers in Chromatography*, Wiley (1981).
25. R. H. Grubbs: *ChemTech* **7**, 512 (1977).
26. C. U. Pittman: *Polymer News* **4**, 5 (1977) and subsequent articles by the same author in *Polymer News*.

26a. G. Jannes in: *Catalysis, Heterogeneous and Homogeneous* (Ed. B. Delmon and G. Jannes) Elsevier, Amsterdam, p. 83 (1975).

27. C. U. Pittman and G. O. Evans: *ChemTech* **3**, 560 (1973).

28. D. C. Bailey and S. H. Langer: *Chem. Rev.* **81**, 109 (1981).

29. C. U. Pittman in: *Polymer-Supported Reactions in Organic Synthesis* (Ed. P. Hodge and D. C. Sherrington) Wiley, Chichester, p. 249 (1980).

30. D. D. Whitehurst: *ChemTech* **10**, 44 (1980).

30a. M. Kaneko and E. Tsuchida: *Macromol. Rev.* **16**, 397 (1981).

31. F. Ciardelli, G. Braca, C. Carlini, G. Sbrana and G. Valentini: *J. Mol. Cat.* **14**, 1 (1982).

32. T. Imanaka and K. Kaneda: *Kagaku Sosetsu* **34**, 176 (1982); *Chem. Abs.* **97**, 161871 (1982).

33. C. U. Pittman in: *Comprehensive Organometallic Chemistry* (Ed. G. Wilkinson, F. G. A. Stone and E. W. Abel) Pergamon Press, Oxford. Vol. 8, chapter 55 (1982).

34. Yu. I. Ermakov, B. N. Kuznetsov and V A. Zakharov: *Catalysis by Supported Complexes*, Elsevier, Amsterdam (1981).

35. T. Mizoroki: *Sekiya Gakkai Shi* **19**, 455 (1976); *Chem. Abs.* **87**, 118095 (1977).

36. T. Imanaka: *Hyomen* **16**, 333 (1978); *Chem. Abs.* **89**, 186522 (1978).

37. N. Toshima: *Yuki Gosei Kagaku Kyokaishi* **36**, 909 (1978); *Chem. Abs.*. **90**, 104033 (1979).

38. V. M. Akhmedov, A. A. Medzhidov and A. G. Azizov: *Azerb. Khim. Zh.* 122 (1979); *Chem. Abs.* **92**, 11692 (1980).

39. Yu. I. Ermakov (Ed.): *Catalysts Containing Supported Complexes*, publ. by Akad. Nauk SSSR, Sib. Otd. Inst. Katal. Novosibirsk, USSR (1980); *Chem. Abs.* **94**, 163356 (1981).

40. A. L. Robinson: *Science* **194**, 1261 (1976).

41. J. Smidt: *Chem. Ind. (London)* 54 (1962).

42. J. F. Roth, J. H. Craddock, A. Hershman and F. E. Paulik: *ChemTech* **1**, 600 (1971).

43. K. G. Allum, R. D. Hancock, S. McKenzie and R. C. Pitkethly: *Catalysis — Volume 1* (Ed. J. W. Hightower) North-Holland, Amsterdam, p. 477 (1973).

44. F. R. Hartley, S. G. Murray and P. N. Nicholson: *J. Mol. Catal.* **16**, 363 (1982).

45. F. R. Hartley: *Chem. Rev.* **73**, 163 (1973).

45a. W. Heitz: *Adv. Polymer Sci.* **23**, 1 (1977).

45b. S. Mazur, P. Jayalekshmy, J. T. Anderson and T. Matusinovic: *Amer. Chem. Soc., Symp. Ser.* **192**, 43 (1982).

46. D. P. Harrison and H. F. Rase: *Ind. Eng. Chem. Fund.* **6**, 161 (1967).

47. A. T. Jurewicz, L. D. Rollmann and D. D. Whitehurst: *Adv. Chem. Ser.* **132**, 240 (1974).

48. F. H. Jardine: *Prog. Inorg. Chem.* **28**, 63 (1981).

49. R. H. Grubbs, C. Gibbons, L. C. Kroll, W. D. Bonds and C. H. Brubaker: *J. Amer. Chem. Soc.* **95**, 2373 (1973).

50. H. L. Krauss in: *Catalysis — Volume 1* (Ed. J. W. Hightower) North-Holland, Amsterdam, p. 207 (1973).

50a. S. L. Regen and D. P. Lee: *Macromolecules* **10**, 1418 (1977).

50b. G. Gubitosa, M. Boldt and H. H. Brintzinger: *J. Amer. Chem. Soc.* **99**, 5174 (1977).

50c. S. L. Regen and D. Bolikal: *J. Amer. Chem. Soc.* **103**, 5248 (1981).

50d. J. Rebek and J. E. Trend: *J. Amer. Chem. Soc.* **101**, 737 (1979).

50e. S. Mazur and P. Jayalekshmy: *J. Amer. Chem. Soc.* **101**, 677 (1979).

50f. D. K. Liu and M. S. Wrighton: AD-A110082 Report (1982); *Chem. Abs.* 97, 64064 (1982).

51. J. Manassen in *Catalysis, Progress in Research* (Ed. F. Basolo and R. L. Burwell) Plenum, New York, p. 177 (1977).

52. S. Affrossman and J. P. Murray: *J. Chem. Soc.* (*B*) 1015 (1966).

53. D. C. Neckers, D. A. Kooistra and G. W. Green: *J. Amer. Chem. Soc.* 94, 9284 (1972).

54. G. Bernard, Y. Chauvin and D. Commereuc: *Bull. Soc. Chim. France* 1163 (1976).

55. G. Bernard, Y. Chauvin and D. Commereuc: *Bull. Soc. Chim. France* 1168 (1976).

56. W. Dumont, J.-C. Poulin, T.-P. Dang and H. B. Kagan: *J. Amer. Chem. Soc.* 95, 8295 (1973).

57. B. M. Trost and E. Keinan: *J. Amer. Chem. Soc.* 100, 7779 (1978).

58. N. L. Holy in: *Fundamental Research in Homogeneous Catalysis* (Ed. M. Tsutsui) Plenum Press, New York 3, 691 (1979).

59. J. M. Brown and H. Molinari: *Tetrahedron Lett.* 2933 (1979).

60. R. H. Grubbs and E. M. Sweet: *Macromolecules* 8, 241 (1975).

61. S. Franks and F. R. Hartley: *J. Chem. Soc. Perkin I* 2233 (1980).

62. L. Bemi, H. C. Clark, J. A. Davies, D. Drexler, C. A. Fyfe and R. Wasylishen: *J. Organometal. Chem.* 224, C5 (1982).

62a. J. Lieto, D. Milstein, R. L. Albright, J. V. Minkiewitz and B. C. Gates: *ChemTech* 13, 46 (1983).

63. W. Heitz: *Adv. Poly. Sci.* 23, 1 (1977).

63a. D. E. Bergbreiter: *Amer. Chem. Soc., Symp. Ser.* 192, 1 (1982).

63b. A. Guyot and M. Bartholin: *Prog. Poly. Sci.* 8, 277 (1982).

63c. F. R. Hartley, D. J. A. McCaffrey, S. G. Murray and P. N. Nicholson: *J. Organometal. Chem.* 206, 347 (1981).

63d. F. R. Hartley, S. G. Murray and P. N. Nicholson: *J. Poly. Sci., Poly. Chem. Ed.* 20, 2395 (1982).

63e. L. Verdet and J. K. Stille: *Organometallics* 1, 380 (1982).

64. B. C. Gates: *Nato Adv. Study Inst., Ser. E.* 39, 437 (1980).

65. E. Bayer and V. Schurig: *Angew. Chem. Int. Ed.* 14, 493 (1975).

66. V. Schurig and E. Bayer: *ChemTech* 6, 212 (1976).

67. V. A. Kabanov and V. I. Smetanyuk: *Macromol. Chem. Phys. Suppl.* 5, 121 (1981).

68. R. H. Grubbs, L. C. Kroll and E. M. Sweet: *J. Macromol. Sci.* A7, 1047 (1973).

68a. R. H. Grubbs, E. M. Sweet and S. Phisabut in: *Catalysis in Organic Syntheses 1976* (Ed. P. N. Rylander and H. Greenfield) Academic Press, New York, p. 153 (1976).

69. H. Staudinger and E. Husemann: *Ber.* 68, 1618 (1935).

70. M. S. Karasch, W. Nudenberg, E. V. Jensen, P. E. Fischer and D. L. Mayfield: *Ind. Eng. Chem.* 39, 830 (1947).

71. L. J. Boucher, A. A. Oswald and L. L. Murell: *Preprints* Petrol. Div., ACS Meeting, Los Angeles, p. 162 (March 1974).

72. J. A. Rabo: *Zeolite Chemistry and Catalysis*, ACS Monograph 171 (1976).

72a. R. Le Van Mao: *Rev. Inst. Franc. Pétrole* 34, 429 (1979).

73. L. V. C. Rees (Ed.): *Proc. 5th Int. Conf. on Zeolites*, Heyden, London (1980).

74. L. B. Sand and F. A. Mumpton (Eds.): *Natural Zeolites: Occurrence, Properties, Use*, Pergamon, Oxford (1978).

75. D. W. Breck: *Zeolite Molecular Sieves: Structure, Chemistry and Use*, Wiley, London (1974).

76. See also *Zeolites*, an international journal first published in 1981.
77. T. J. Pinnavaia and P. K. Welty; *J. Amer. Chem. Soc.* **97**, 3819 (1975).
78. T. J. Pinnavaia, P. Raythatha, J. G. S. Lee, L. J. Halloran and J. F. Hoffman: *J. Amer. Chem. Soc.* **101**, 6891 (1979).
79. W. H. Quayle and T. J. Pinnavaia: *Inorg. Chem.* **18**, 2840 (1979).
80. M. Mazzei, M. Riocci and W. Marconi: *German Patent* 2,845,216 (1979); *Chem. Abs.* **91**, 56329 (1979).
81. J. F. Hoffman: *Diss. Abs.* **B37**, 4454 (1977).
81a. T. J. Pinnavaia: *Amer. Chem. Soc., Symp. Ser.* **192**, 241 (1982).
82. E. J. Corey and A. J. Venkateswarlu: *J. Amer. Chem. Soc.* **94**, 6190 (1972).
83. R. W. Murray: *Phil. Trans. Roy. Soc. London* **A302**, 253 (1981).
84. E. Tsuchida, E. Hasegawa and K. Honda: *Biochim. Biophys. Acta* **427**, 520 (1976).
85. K. Ichimura and S. Watanabe: *Chem. Lett.* 1289 (1978).
86. H. R. Allcock, P. P. Greigger, J. E. Gardner and J. L. Schmutz: *J. Amer. Chem. Soc.* **101**, 606 (1979).
87. J. P. Collman and C. A. Reed: *J. Amer. Chem. Soc.* **95**, 2048 (1973).
88. S. Tazuke, H. Tomono, N. Kitamura, K. Sato and N. Hayashi: *Chem. Lett.* 85 (1979).
89. D. J. Hucknall and B. M. Willatt: *Brit. Patent* 2,013,901 (1979); *Chem. Abs.* **92**, 79280 (1980).
90. N. Oyama and F. C. Anson: *J. Amer. Chem. Soc.* **101**, 3450 (1979).
91. H. D. Abruña, J. L. Walsh, T. J. Meyer and R. W. Murray: *Inorg. Chem.* **20**, 1481 (1981).
92. H. D. Abruña, P. Denisevich, M. Umaña, T. J. Meyer and R. W. Murray: *J. Amer. Chem. Soc.* **103**, 1 (1981).
93. T. P. Henning, H. S. White and A. J. Bard: *J. Amer. Chem. Soc.* **103**, 3937 (1981).
94. C. A. Melendres and F. A. Cafasso: *J. Electrochem. Soc.* **128**, 755 (1981).
94a. J. M. Calvert and T. J. Meyer: *Inorg. Chem.* **20**, 27 (1981).
94b. M. Salmon, A. Diaz and J. Goitia: *Amer. Chem. Soc., Symp. Ser.* **192**, 65 (1982).
94c. M. S. Wrighton: *Amer. Chem. Soc., Symp. Ser.* **192**, 99 (1982).
94d. J. A. R. van Veen, J. F. van Baar and K. J. Kroese: *J. C. S. Faraday I* **77**, 2827 (1981).
94e. O. Hirabaru, T. Nakase, K. Hanabusa, H. Shirai, K. Takemoto and N. Hojo: *J. C. S. Chem. Commun.* 481 (1983).
94f. N. R. Armstrong, T. Mezza, C. L. Linkous, B. Thacker, T. Klofta and R. Cieslinski: *Amer. Chem. Soc., Symp. Ser.* **192**, 205 (1982).
95. C. A. Melendres and X. Feng: *J. Electrochem. Soc.* **130**, 811 (1983).
96. A. Ledwith and D. C. Sherrington in: *Molecular Behaviour and the Development of Polymeric Materials* (Ed. A. Ledwith and A. M. North) Chapman and Hall, London, Chapter 9 (1975).
97. C. J. Suckling: *Chem. Soc. Rev.* **6**, 215 (1977).
98. I. Chibata (Ed.): *Immobilised Enzymes*, John Wiley, New York (1978).
99. G. Manecke, H. G. Vogt and D. Polakowski: *Makromol. Chem. Suppl.* **3**, 107 (1979).
100. M. D. Trevan: *Immobilised Enzymes*, John Wiley, New York (1980).
100a. L. L. Wood, F. J. Hartdegen and P. A. Hahn: *Swiss Patent*, 624,715 (1981); *Chem. Abs.* **95**, 146163 (1981).
100b. R. C. Sheppard: *Chem. in Brit.* **19**, 402 (1983).
100c. J. Klein: *Nachr. Chem., Tech. Lab.* **29**, 850 (1981); *Chem. Abs.* **96**, 67175 (1982).
100d. S. Fukui and A. Tanaka: *Kagaku to Seibutsu* **19**, 620 (1981); *Chem. Abs.* **96**, 18586 (1982).

100e. V. N. R. Pillai and M. Mutter: *Naturwissenschaften* **68**, 558 (1981).
100f. G. P. Royer: *Enzyme Eng.* **6**, 117 (1982).
100g. I. Chibata and T. Tosa: *Kagaku Zokan* 175 (1982); *Chem. Abs.* **98**, 32948 (1983).
101. A. M. Klibanov: *Science* **219**, 722 (1983).
102. G. G. Overberger and K. N. Sannes: *Angew. Chem. Int. Ed.* **13**, 99 (1974).
103. C. C. Leznoff: *Chem. Soc. Rev.* **3**, 65 (1974).
104. E. Cernia and M. Graziani: *J. Appl. Polym. Sci.* **18**, 2725 (1974).
105. N. K. Mathur and R. E. Williams: *J. Macromol. Sci. – Rev. Macromol. Chem.* **C15**, 117 (1976).
106. J. I. Crowley and H. Rapoport: *Acc. Chem. Res.* **9**, 135 (1976).
107. C. R. Harrison and P. Hodge: *JCS Chem. Commun.* 813 (1978).
108. C. U. Pittman and G. A. Stahl: *Polymer News* **4**, 280 (1978).
109. G. Manecke and W. Storck: *Angew. Chem. Int. Ed.* **17**, 657 (1978).
110. G. Manecke and P. Reuter: *Pure Appl. Chem.* **51**, 2313 (1979).
111. A. McKillop and D. W. Young: *Synthesis* 401 (1979).
112. A. McKillop and D. W. Young: *Synthesis* 481 (1979).
113. J. P. Tam, F. J. Tjoeng and R. B. Merrifield: *J. Amer. Chem. Soc.* **102**, 6117 (1980).
113a. M. A. Kraus and A. Patchornik: *Macromol. Rev.* **15**, 55 (1980).
114. P. Hodge and D. C. Sherrington (Eds.): *Polymer Supported Reactions in Organic Synthesis*, John Wiley, Chichester (1980).
114a. A. Akelah: *Br. Poly. J.* **13**, 107 (1981).
114b. S. Gestrelius: *Appl. Biochem. Biotechnol.* **7**, 19 (1982).
115. J. M. J. Fréchet, P. Darling and M. J. Farrall: *J. Org. Chem.* **46**, 1728 (1981).
116. H. Alper: *Adv. Organometal. Chem.* **19**, 183 (1981).
117. W. P. Weber and G. W. Gokel: *Phase Transfer Catalysis in Organic Synthesis*, Springer-Verlag, Berlin and Heidelberg (1977).
118. C. M. Starks and C. Liotta (Eds.): *Phase Transfer Catalysis: Principles and Techniques*, Academic Press, New York (1979).
119. E. V. Dehmlow and S. S. Dehmlow: *Phase Transfer Catalysis*, Verlag Chemie, Weinheim (1980).
120. K. Kondo and K. Takemoto: *Kagaku* **35**, 1013 (1980); *Chem. Abs.* **94**, 181360 (1981).
121. S. L. Regen: *J. Amer. Chem. Soc.* **97**, 5956 (1976).
122. H. Molinari, F. Montanari, S. Quici and P. Tundo: *J. Amer. Chem. Soc.* **101**, 3920 (1979).
123. S. L. Regen, J. C. K. Heh and J. McLick: *J. Org. Chem.* **44**, 1961 (1979).
124. M. S. Chiles, D. D. Jackson and P. C. Reeves: *J. Org. Chem.* **45**, 2915 (1980).
125. M. Tomori and W. T. Ford: *J. Amer. Chem. Soc.* **103**, 3821 (1981).
125a. M. Tomori and W. T. Ford: *J. Amer. Chem. Soc.* **103**, 3828 (1981).
126. F. Montanari, S. Quici and P. Tundo: *J. Org. Chem.* **48**, 199 (1983).
127. P. Tundo: *JCS Chem. Commun.* 641 (1977).
128. P. Tundo and P. Venturello: *J. Amer. Chem. Soc.* **101**, 6606 (1979).
129. P. Tundo and P. Venturello: *J. Amer. Chem. Soc.* **103**, 856 (1981).
130. S. L. Regen and L. Dulak, *J. Amer. Chem. Soc.* **99**, 623 (1977).
131. M. Tomoi, O. Abe, M. Ikeda, K. Kikara and H. Kakinchi: *Tetrahedron Lett.* 3031 (1978).
132. M. Tomoi, K. Kikara and H. Kakinchi: *Tetrahedron Lett.* 3485 (1979).
132a. G. Manecke and P. Reuter: *J. Mol. Catal.* **13**, 355 (1981).

132b. E. Blasius, K. P. Janzen, H. Klotz and A. Toussaint: *Makromol. Chem.* **183**, 1401 (1982).

132c. F. L. Cook, J. R. Robertson and W. A. Ernst: *Polym. Preprints, Amer. Chem. Soc., Div. Petrol. Chem.* **22**, 161 (1981).

132d. W. M. MacKenzie and D. C. Sherrington: *Polymer* **22**, 431 (1981).

133. B. Arkles, W. R. Peterson and K. King: *Amer. Chem. Soc., Symp. Ser.* **192**, 281 (1982).

134. P. Monsef-Mirzai and W. R. McWhinnie: *Inorg. Chim. Acta* **52**, 211 (1981).

135. Y. Sasson, A. Zoran and J. Blum: *J. Mol. Catal.* **11**, 293 (1981).

135a. M. J. H. Russell and B. A. Murrer: reported at 12th Sheffield-Leeds International Organometallic Chemistry Conference (1983).

135b. M. J. H. Russell and B. A. Murrer: *Brit. Patent Appl.* 2,085,874 (1981).

135c. J. Jenck and D. Morel: *US Patent* 4,248,802 (1981).

135d. E. Kuntz: *German Patent* 2,627,354 (1976); *Chem. Abs.* **87**, 101944 (1977).

136. H. Alper: *Pure Appl. Chem.* **52**, 607 (1980).

136a. S. L. Regen: *J. Amer. Chem. Soc.* **97**, 5695 (1975).

136b. S. L. Regen: *J. Amer. Chem. Soc.* **98**, 6720 (1976).

136c. S. L. Regen: *J. Org. Chem.* **42**, 875 (1977).

136d. M. Tomoi and W. Ford: *J. Amer. Chem. Soc.* **103**, 3821 and 3829 (1981).

136e. M. Cinquini, S. Collons, H. Molinari, F. Montanari and F. Tundo: *J.C.S. Chem. Commun.* 394 (1976).

136f. A. Warshawsky: *Talanta* **21**, 962 (1974).

136g. H. Molinari, F. Montanari and P. Tundo: *J.C.S. Chem. Commun.* 639 (1977).

136h. F. Montanari and P. Tundo: *Tetrahedron Lett.* 5055 (1979).

136i. K. Kikukawa, S. Takamura, H. Hirayama, H. Namiki, F. Wada and T. Matsuda: *Chem. Lett.* 511 (1980).

136j. Y. Y. Hsu, H.-L. Chin, K.-H. Huang, C. C. War and Y.-C. Huang, *K'o Hsueh Pao* **26**, 408 (1981); *Chem. Abs.* **95**, 43735 (1981).

136k. F. Montanari and P. Tundo: *J. Org. Chem.* **46**, 2125 (1981).

136l. P. Tundo and P. Ventarello: *J. Amer. Chem. Soc.* **101**, 5505 (1979).

136m. T. Waddell, D. Leyden and D. Hercules in: *Silylated Surfaces* (Ed. D. Leyden and W. Collins) Gordon and Breach, New York (1980).

136n. T. Waddell and D. Leyden: *J. Org. Chem.* **46**, 2105 (1981).

136o. B. Arkles, W. R. Peterson and K. King: *Amer. Chem. Soc., Symp. Ser.* **192**, 281 (1982).

136p. P. Tundo, *J.C.S. Chem. Commun.* 641 (1977).

137. B. C. Gates, J. R. Katzer and G. C. A. Schuit: *Chemistry of Catalytic Processes*, McGraw-Hill, New York (1979).

138. H. W. Flood and B. S. Lee: *Scientific American* 219 (July), 94 (1968).

139. D. Kunii and O. Levenspiel: *Fluidization Engineering*, Wiley, New York (1969).

140. G. J. K. Acres, G. C. Bond, B. J. Cooper and J. A. Dawson: *J. Catal.* **6**, 139 (1966).

141. R. Z. Moravec, W. T. Schelling and C. F. Oldershaw: *Brit. Patent* 511,556 (1939); *Chem. Abs.* **34**, 7102(6) (1940).

142. R. Z. Moravec, W. T. Schelling and C. F. Oldershaw: *Canadian Patent*, 396,994 (1941); *Chem. Abs.* **35**, 6103(1) (1941).

143. F. G. Ciapetta: *US Patent* 2,430,803 (1947); *Chem. Abs.* **42**, 1398h (1948).

144. F. G. Ciapetta: *US Patent* 2,434,833 (1948); *Chem. Abs.* **42**, 2983c (1948).

145. G. C. Bond: *German Patent* 2,047,748 (1971); *Chem. Abs.* **75**, 25968 (1971).

146. Monsanto: *Brit. Patent* 1,185,453 (1967); *Chem. Abs.* **72**, 132059 (1970).

147. P. R. Rony and J. F. Roth in: *Catalysis, Heterogeneous and Homogeneous*, (Ed. B. Delmon and G. Jannes) Elsevier, 373 (1975).

148. P. R. Rony and J. F. Roth: *J. Mol. Catal.* 1, 13 (1975).

149. H. Pscheidl, E. Moeller, H. U. Juergens and D. Haberland: *East German Patent* 138,153 (1979); *Chem. Abs.* 92, 65370 (1980).

149a. D. Haberland, E. Moeller and H. Pscheidl: *East German Patent* 147,914 (1981); *Chem. Abs.* 96, 92496 (1982).

150. W. Strohmeier, B. Graser, R. Marćec and K. Holke: *J. Mol. Catal.* 11, 257 (1981).

151. G. C. Bond: *German Patent* 2,055,539 (1971); *Chem. Abs.* 75, 48429 (1971).

152. P. R. Rony: *J. Catal.* 14, 142 (1969).

153. P. R. Rony: *Chem. Eng. Sci.* 23, 1021 (1968).

154. W. Strohmeier and M. Michel: *J. Catal.* 69, 209 (1981).

155. W. Strohmeier, R. Marćec and B. Graser: *J. Organometal. Chem.* 221, 361 (1981).

156. N. A. De Munck, J. P. A. Notenboom, J. E. De Leur and J. J. F. Scholten: *J. Mol. Catal.* 11, 233 (1981).

156a. D. M. Hercules: *Report* ARO-14839.3-C (1980); *Chem. Abs.* 94, 163248 (1981).

156b. A. Luchetti and D. M. Hercules: *J. Mol. Catal.* 16, 95 (1982).

156c. D. Hesse: *Ber. Bunsenges. Phys. Chem.* 86, 746 (1982).

157. L. A. Gerritsen, A. van Meerkerk, M. H. Vreugdenhil and J. J. F. Scholten: *J. Mol. Catal.* 9, 139 (1980).

158. L. A. Gerritsen, J. M. Herman, W. Klut and J. J. F. Scholten: *J. Mol. Catal.* 9, 157 (1981).

159. L. A. Gerritsen, J. M. Herman and J. J. F. Scholten: *J. Mol. Catal.* 9, 241 (1980).

160. L. A. Gerritsen, W. Klut, M. H. Vreugdenhil and J. J. F. Scholten: *J. Mol. Catal.* 9, 257 (1980).

161. L. A. Gerritsen, W. Klut, M. H. Vreugdenhil and J. J. F. Scholten: *J. Mol. Catal.* 9, 265 (1980).

162. J. Hjortkjaer, M. S. Scurrell, P. Simonsen and H. Svendsen: *J. Mol. Catal.* 12, 179 (1981).

162a. J. Hjortkjaer, M. S. Scurrell and P. Simonsen: *J. Mol. Catal.* 10, 127 (1980).

163. J. Hjortkjaer, M. S. Scurrell and P. Simonsen: *J. Mol. Catal.* 6, 405 (1979).

163a. J. Villadsen and H. Livberg: *Cat. Rev. Sci. Eng.* 17, 203 (1978).

163b. J. Villadsen and H. Livberg: *NATO Adv. Study Inst. Ser. E.* 51, 541 (1981).

164. Monsanto: *Belgian Patent* 711,042 (1968); *Platinum Metals Rev.* 11, 147 (1967).

165. F. E. Paulik, K. K. Robinson and J. F. Roth: *US Patent* 3,487,112 (1969); *Chem. Abs.* 72, 68984 (1970).

166. K. K. Robinson, F. E. Paulik, A. Hershman and J. F. Roth: *J. Catal.* 15, 245 (1969).

167. G. Biale: *US Patent* 3,733,362 (1973).

168. K. K. Robinson, A. Hershman, J. H. Craddock and J. F. Roth: *J. Catal.* 27, 389 (1972).

169. Th. G. Spek and J. J. F. Scholten: *J. Mol. Catal.* 3, 81 (1977).

170. P. W. H. L. Tjan and J. J. F. Scholten: 6th Internat. Congr. Catal. (London) (1967).

171. G. Jannes, in: *Catalysis, Heterogeneous and Homogeneous* (Ed. B. Delmon and G. Jannes) Elsevier, 83 (1975).

172. G. W. Parshall: *J. Amer. Chem. Soc.* 94, 8716 (1972).

173. J. F. Knifton: *US Patent* 4,265,828 (1981).

174. J. F. Knifton: *J. Amer. Chem. Soc.* 103, 3959 (1981).

175. A. Efraty, I. Feinstein, F. Frolow and L. Wackerle: *J. Amer. Chem. Soc.* 102, 6341 (1980).

176. A. Efraty, I. Feinstein, F. Frolow and A. Goldman: *JCS Chem. Commun.* 864 (1980).
177. K. Emile: *German Patent* 2,627,354 (1976); *Chem. Abs.* 87, 101944 (1977).
178. A. F. Borowski, D. J. Cole-Hamilton and G. Wilkinson: *Nouv. J. Chim.* 2, 137 (1978).
179. F. Joó, Z. Tóth and M. T. Beck: *Inorg. Chim. Acta* 25, L61 (1977).
179a. M. E. Wilson, R. G. Nuzzo and G. M. Whitesides: *J. Amer. Chem. Soc.* 100, 2269 (1978).
180. Y. Dror and J. Manassen: *J. Mol. Catal.* 2, 219 (1977).
181. R. T. Smith, R. K. Ungar and M. C. Baird: *Trans. Met. Chem.* 7, 288 (1982).
182. G. Gugliemo and A. Andreetta: *German Patent* 2,313,102 (1973); *Chem. Abs.* 79, 140101 (1973).
183. K. Kurtev, D. Ribola, R. A. Jones, D. J. Cole-Hamilton and G. Wilkinson: *J. Chem. Soc. Dalton* 55 (1980).
184. R. G. Nuzzo, D. Feitler and G. M. Whitesides: *J. Amer. Chem. Soc.* 101, 3683 (1979).
184a. R. T. Smith and M. C. Baird: *Inorg. Chim. Acta* 62, 135 (1982).
185. T. D. Madden, W. E. Peel, P. J. Quinn and D. Chapman: *J. Biochem. Biophys. Meth.* 2, 19 (1980).
186. C. Vigo, F. M. Goni, P. J. Quinn and D. Chapman: *Biochim. Biophys. Acta* 508, 1 (1978).
187. P. J. Quinn and C. E. Taylor: *J. Mol. Catal.* 13, 389 (1981).
188. G. A. Ozin, C. G. Francis and H. X. Huber: *US Patent* 4,292,253 (1981); *Chem. Abs.* 95, 210494 (1981).

CHAPTER 2

PREPARATION OF THE SUPPORTS

Although the requirements of a material to act as a support for a metal catalyst may be readily defined, it is a difficult task to produce such a support on demand. It may not be difficult to arrange for a given particle size to be produced, but once we begin to consider factors such as the distribution of donor sites, then the characterisation becomes much more vague. Thus, when considering the preparation of functionalised supports a great deal of thought should be given to the structure at a molecular level since it is a mistake to assume that functionalised sites are evenly distributed and do not affect each other. Equally it should not be assumed that the support is a simple carrier, particularly for resin-type materials where matrix effects and solvation may play an important role in their subsequent use.

Having noted that it is not straightforward to employ polymer supports for catalysts a considerable number of systems have been studied either using commercially available materials or by synthesising polymers in the laboratory. This chapter will consider the preparation of functionalised supports by modification of pre-formed polymers and inorganic materials and by polymerisation of functionalised monomers capable of polymerisation. The properties of these materials will also be considered where it is believed that they may have some bearing on their use as catalyst supports.

2.1. General Considerations for Organic Polymers

For the present purposes organic polymers can be divided into two groups, linear and cross-linked. It is not the purpose of this book to consider in detail the chemistry involved in the formation of polymers as this subject is covered in a wealth of texts [1—4]. However, it is pertinent to consider the types of polymers which can be readily prepared or purchased from the point of view of their properties related to their use as support materials.

(a) *Linear organic polymers* are prepared by one of two methods: a polycondensation or addition polymerisation. Polycondensation usually requires reaction between two difunctional reagents with chain growth at both ends of the forming polymer. Examples of this type of reaction include a diacid reacting with either a diol or a diamine to give a polyester or a polyamide, respectively. It is often advantageous to use the diacid chloride or a diisocyanate in place of the diacid. These reactions are capable of forming materials with very high average molecular weights and are produced commercially in many forms, for instance, the range of nylons (polyamides). As supports for catalysts their major drawback is the difficulty in modifying them to give donor sites to support transition metal complexes particularly for the polyesters. Polyamides can be modified using the amide N—H as the reactive site [5].

Addition polymerisation is the reaction of alkene monomers undergoing self-polymerisation catalysed by either Ziegler-Natta, anionic, cationic or free radical catalysts. It is relatively easy to control chain length and, if no cross-linking is introduced, this factor will be important in determining the behaviour of linear polymers in contact with organic solvents, as will be discussed below. Suspension polymerisation is probably the most relevant method of achieving addition polymerisation as this produces approximately spherical solid particles called 'beads'. This process employs a solvent in which the monomer is insoluble and in a suitable stirred reactor the liquid monomer will form spherical droplets the size of which can be controlled by experience. The catalyst precursor is dissolved in the monomer and is activated once the stirring and droplet formation are established. The usual catalyst is the free radical formed by thermal cleavage or ultraviolet irradiation of azobis (*iso*-butyronitrile). Problems may arise in suspension polymerisation when this method is used to form a copolymer. Copolymerisation is an addition polymerisation in which two or more different monomer types are to be incorporated into the polymer. It may be extremely difficult to find a suitable solvent, particularly if one monomer is hydrophobic and the other hydrophilic. The structures of copolymers will only be discussed where relevant in later sections.

The conformation of linear polymers in a solvent is dependent mainly on the interactions between the solvent and each polymer chain but other factors are pertinent such as chain length and concentration of the polymer 'solution'. Using a good solvent for that particular polymer, for example polystyrene in benzene, it is possible to obtain a true molecular solution with the chains having a random coil configuration of an expanded nature. In poorer solvents the coil contracts on itself until, in the limit, the interactions between polymer and solvent are unfavourable and the polymer precipitates. To obtain a solution of a polymer in which there is effectively no chain interaction, i.e., the chains are separated, very dilute systems are required where the polymer is present in concentrations of less than two per cent by weight. Increased concentrations lead to increasing chain entanglement even in a good solvent. Thus, in a practical sense, if linear polymers are used as catalyst supports both intra-chain and inter-chain interactions are to be expected.

(b) *Cross-linked polymers* may remove some of the uncertainty of what occurs structurally in a polymer-solvent system; however, new factors must now be considered. Cross-linked polymers consist of linear polymer chains joined together either in a random manner or in a highly symmetrical pattern, thus forming an infinite three-dimensional network. Again, as for linear polymers, it is possible to use either a condensation reaction or an addition reaction to obtain a cross-linked polymer.

Condensation cross-linked polymers may be formed by the addition of a trifunctional species to the difunctional reagents which would be employed to form a linear polymer. Thus, in the preparation of a polyester, inclusion of the triol glycerol will yield a cross-linked polyester. Similarly, diol plus triol when reacted with a diisocyanate yields a polyurethane. These polymers, together with the phenolformaldehyde resins, are of considerable commercial importance; however, to our knowledge these types of cross-linked polymers have yet to be studied as supports for metal complex catalysts.

Addition polymers which are cross-linked have, to date, been the most used supports for transition metal catalysts. This has been due to one particular system which has lent itself to ready functionalisation, namely cross-linked polystyrene-based materials, which will be considered in detail in Section 2.2.

This type of polymer is formed by the copolymerisation of a monoene and a diene, as shown in Scheme 1.

$$CH_2{=}CHR + CH_2{=}CH{\sim\sim\sim}CH{=}CH_2$$

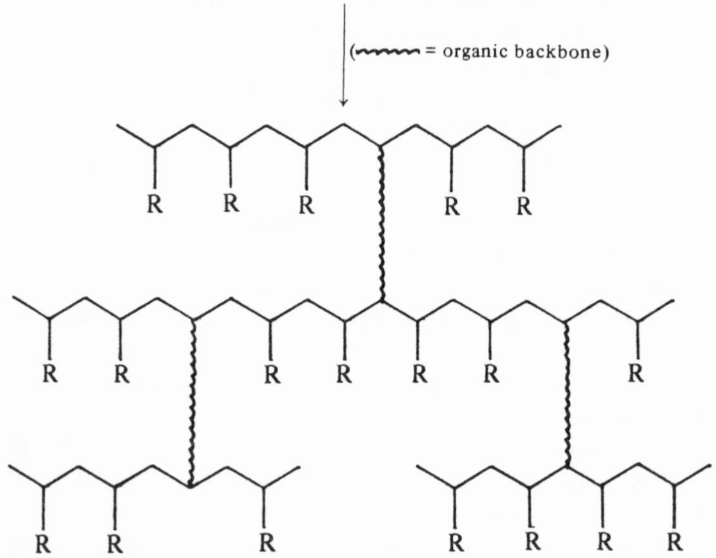

($\sim\sim\sim$ = organic backbone)

Scheme 1

It is evident that the ratio of monoene to diene will affect the amount of cross-linking and thus the rigidity of the product. However, it is a somewhat simplistic approach, as other factors have a large effect on the final macrostructure of the polymer. A knowledge of the macrostructure is important, since it determines the accessibility of both solvents and reagents to the interior of the polymer. The important factor is that of pore size, which will vary for a particular monoene-diene system, with the ratio of monoene to diene, with the method of copolymerisation, with the solvents used and also with the solvent used to swell the polymer in subsequent use [6, 7]. The latter point, concerning which solvents may be used to swell the polymer, is a secondary effect since the macrostructure which regulates this swellability is determined by the ratio of monoene to diene and the manufacturing process. This leads to two basic types of polymer: microporous and macroporous or macroreticular [7—9]. These structural types are often quoted when authors are describing the support materials used in their supported catalyst studies.

Microporous resins are produced when no inert solvent is used in the copolymerisation reaction. As the polymer chains grow they are solvated by unreacted monomer which is used up as the polymerisation proceeds causing the chains to aggregate, finally yielding a glassy product. In the absence of solvent the pore size is approximately the distance between polymer chains, thus giving the title microporous. It should be noted, however, that addition of a good solvent, i.e. one with a high affinity for the polymer, will cause a large degree of swelling, giving increased porosity. These swollen polymers are often called 'gels' and the degree of swelling will be dependent not only on the solvent used but also on the amount of cross-linking. The higher the amount of cross-linking the more difficult it will be for the polymer to swell. Thus for this type of resin the pore size will be limited to a large degree by the amount of diene used [6], and this is often quoted when describing a resin material. A figure of 2% cross-linking is common and it is usually less than 10%. For very low degrees of cross-linking (<0.5%) the resin will behave more like a linear polymer and is unlikely to retain its structure in the presence of a good solvent.

To give increased porosity another technique must be used employing inert solvents in the copolymerisation process [6, 7]. Generally a large amount of cross-linking agent is used (up to 60%) and a solvent in which the monomer and the cross-linking agent are both soluble. Due to the dilution effect and the large amount of cross-linking agent the initial copolymer is highly cross-linked and forms precipitated particles. These are then connected by polymer with a decreasing amount of cross-linking that is formed from voids filled with monomer solution which become increasingly poor in monomer. When the solvent is removed these resins may well retain some proposity due to their heterogeneous nature and are sometimes described as heterogeneously cross-linked. They readily take up good solvents, but will also accomodate poor solvents due to their macroporous nature. If a solvent is used which precipitates the forming polymer then large permanent pores will form. These resins have a pore structure in the dry solventless state and will accommodate both good and bad solvents alike. These macroreticular resins have large pores and a high internal surface area. The copolymers are highly cross-linked and only swell a little in good solvents.

All of the processes described above to form cross-linked resins can be per-formed using an emulsion technique yielding beads of different sizes. The physical properties of the final resin are of importance since their subsequent use as supports requires ease of handling. One problem which occurs for highly cross-linked resins is that of brittleness resulting in their mechanical degradation. Low cross-linking may give resins which form intractable gums in the presence of solvents.

One other method of forming a cross-linked resin is that of 'popcorn' polymer-isation. This process occurs when a monoene and a diene are warmed together in the absence of an initiator. A spontaneous reaction occurs which may form a typical glassy microporous resin or, depending on the conditions used, can produce a granular material which rapidly expands, filling the space occupied by the original liquid reactants, and then pushing up into the space above the liquid. This resin production is similar in appearance to popcorn thus giving it the somewhat unusual name. The actual polymeric material swells little even in good solvents, but the resin will absorb large amounts of solvent, probably into large pores in the macrostructure.

2.2. Styrene Based Systems

Two broad types of styrene based polymers have been used. In the first a commercial polymer is purchased and then functionalised, whereas in the second the monomers styrene, divinylbenzene and a functionalised styrene are copolymerised.

2.2.1. FUNCTIONALISATION OF PREFORMED POLYSTYRENE

Undoubtedly the most popular route to supported catalysts has been to start with commercial polystyrene and introduce functional groups on to it. For those wishing to make their own polystyrene, detailed routes for the preparation of polystyrene membranes, gel form (microporous) and macroporous beads have been described [8, 9]. If commercially produced cross-linked polystyrene beads are used, it is essential that they are pretreated so as to remove the impurities that remain on the surface after the emulsion polymerisation reactions used to produce them [10]. The commonest impurities remaining after emulsion polymerisation are hydrated colloidal alumina, fuller's earth, carboxymethyl cellulose, stearic acid, sodium lauryl sulphate and sodium polyacrylamide [11] which, if they are left on the surface of the beads, result in the formation of surface ions which prevent penetration of ionic reagents such as butyl lithium or lithium diphenylphosphide. A satisfactory procedure [10] for removing the surface impurities involves successive washings in 1N NaOH (60 °C), 1N HCl (60 °C), 1N NaOH (60 °C), 1N HCl (60 °C), H_2O (25 °C), dmf (40 °C), 1N HCl (60 °C), H_2O (60 °C), CH_3OH (20 °C), 3 : 2 (v/v) $CH_3OH : CH_2Cl_2$, 1 : 3 (v/v) $CH_3OH : CH_2Cl_2$, 1 : 9 (v/v) $CH_3OH : CH_2Cl_2$, pure CH_2Cl_2, followed by drying to constant weight at 100 °C *in vacuo* (10 torr overnight, then 0.1 torr for several hours).

Polystyrene has been functionalised with a very wide range of groups. The principle routes for initial activation involve lithiation and chloromethylation.

2.2.1.a. *Lithiation* can be effected either by the direct action of *n*-butyllithium complexed with *N,N,N',N'*-tetramethylethylenediamine, tmeda, (reaction (1)) [12–17], or by metal-halogen exchange on a ring-halogenated polymer (reaction (2)) [17–22].

$$\text{(1)}$$

$$\text{(2)}$$

Reaction 1 involves the abstraction of a proton from the aromatic ring by direct reaction of the *n*-butyllithium–tmeda complex; the tmeda considerably enhances the basicity of the *n*-butyllithium. Lithiation always occurs in the aromatic ring, suggesting that either equilibration to form the thermodynamically more stable backbone-lithiated species either takes place very slowly or is totally inhibited by steric constraints. The position of lithiation is complex, the majority of opinion favours largely *meta*-substitution (~66%) [14, 15, 23]; the earlier suggestion [12] that about 80% of the lithiation was in the *para*-position arose because the experimental evidence could not at that time distinguish between the *meta*- and *para*-products. The degree of lithiation possible prior to precipitation depends on the solvent: in cyclohexane precipitation occurs after between 8 and

20% lithiation of the aromatic rings [12, 13]; with a tetrahydrofuran/benzene mixture about 22% lithiation causes precipitation, whereas with pure benzene greater degrees of lithiation are possible [24].

The metal-halogen exchange reaction (reaction 2) is valuable because it allows greater degrees of lithiation. It depends upon the formation of a more stable lithium-carbon bond than that in the simple alkyl lithium compound. The efficiency of the reaction depends on the halogen in the order $Cl < Br < I$ [18]. The effectiveness of the overall reaction depends on the cleanliness of the original halogenation reaction, for which iodination of the polystyrene using either I_2/H_2SO_4 [25] or I^-/IO_3^- [26] was originally the recommended route [19–21], since it takes place largely in the *para*-position and the subsequent lithium exchange occurs smoothly and without side reactions [27]. Care is needed to avoid Wurtz coupling to form a cross-linked product, but by adding a benzene solution of the iodinated polymer slowly to an excess of *n*-butyllithium in benzene this is readily avoided.

An alternative, now commoner, halogenation route uses bromination [17], which can be carried out using bromine in carbon tetrachloride in the presence or absence of a Lewis acid catalyst, and by using positive bromine reagents in the presence of thallic acetate to promote specifically *para*-substitution [28, 29]. Bromination allows high loadings to be achieved [17]. When a lightly cross-linked resin is stirred in a solution of bromine in carbon tetrachloride in the dark in the absence of a catalyst only small degrees of bromination are achieved, whereas with a more highly cross-linked polystyrene much greater amounts of halogen substitution can occur [30]. This results from bromine addition to residual double bonds, as well as radical bromination of ethylstyryl residues, which form a significant proportion of the aromatic groups in these resins. Most brominations are carried out in the presence of ferric chloride [17, 25, 31], aluminium trichloride [32] or thallic acetate [17, 33–36] Lewis acid catalysts, of which the last gives a much cleaner product and is now the recommended route.

In general, resins with higher degrees of halogenation usually require more than one treatment with *n*-butyllithium to achieve high levels of exchange. The necessity for this depends to some extent on the nature of the resin. Thus macroreticular resins tend to exchange completely in tetrahydrofuran even with high degrees of bromination [17]. By contrast, lightly cross-linked gels give varying results in tetrahydrofuran and cyclohexane. However in benzene or toluene, facile and complete lithiation occurs [17]. In spite of this however, halogen-lithium exchange is notoriously difficult to reproduce, and each system must be investigated in detail to optimise the results.

2.2.1.b. *Chloromethylation* of polystyrene can be effected by the Friedel-Crafts reaction of chloromethylmethylether in the presence of tin(IV) chloride (reaction (3)) [37, 38]. Full preparative details have been given [8, 9, 37]; in particular it

$$\text{\textcircled{P}}\!-\!\!\langle\underline{}\rangle \; + \; ClCH_2OCH_3 \; \xrightarrow{\;SnCl_4\;} \; \text{\textcircled{P}}\!-\!\!\langle\underline{}\rangle_{CH_2Cl} \tag{3}$$

is important to handle chloromethylmethylether carefully as it is carcinogenic. The product is often known as Merrifield's resin because of its extensive use by Merrifield in the synthesis of peptides [39]. When cumene is treated in a similar manner the ratio of $o:m:p$ isomers formed is $12:3:85$ [40]. Thus with polystyrene the majority of the substitution is expected to be in the *para*-position and the infrared spectrum seems to largely confirm this at high degrees of chloromethylation [41]. Merrifield showed that the chloromethyl groups are distributed right throughout the polymer, a considerable proportion of them being well within the interstices of the cross-linked resins [42]. The resulting benzylic halide can itself also take part in further electrophilic aromatic substitution producing methylene cross-links (reaction (4)).

$$\text{\textcircled{P}}\!-\!\!\langle\underline{}\rangle\!-\!CH_2Cl \; + \; \text{\textcircled{P}}\!-\!\!\langle\underline{}\rangle$$
$$\xrightarrow{\;SnCl_4\;} \; \text{\textcircled{P}}\!-\!\!\langle\underline{}\rangle\!-\!CH_2\!-\!\langle\underline{}\rangle\!-\!\text{\textcircled{P}} \tag{4}$$

The reaction obviously becomes more significant at higher degrees of chloromethylation. It can be minimized by keeping the solutions relatively dilute and reaction times fairly short [43]. Carbon tetrachloride [43], tetrachloroethylene [44] and dioxane [45] have all been found to act as successful diluents. Weaker Lewis acids, such as zinc chloride, are also effective at promoting reaction (3) whilst minimizing reaction (4) [45, 46].

2.2.1.c. *Phosphination* can be effected to yield either $-PPh_2$ or $-CH_2PPh_2$ groups bound to the aromatic residues of polystyrene. $-PPh_2$ groups can be introduced either by reacting the brominated or iodinated polymers with $MPPh_2$ [26, 47–56], M = Li, Na or K, or by reacting the lithiated polymer with Ph_2PCl [16, 53, 57–59] (reactions (5) and (6)).

$$\text{P}\!\!-\!\!\langle\text{Ph}\rangle\!\!-\!\!Br \xrightarrow[\text{(M = Li, Na, K)}]{MPPh_2} \text{P}\!\!-\!\!\langle\text{Ph}\rangle\!\!-\!\!PPh_2 \qquad (5)$$

$$\text{P}\!\!-\!\!\langle\text{Ph}\rangle\!\!-\!\!Li \xrightarrow{Ph_2PCl} \text{P}\!\!-\!\!\langle\text{Ph}\rangle\!\!-\!\!PPh_2 \qquad (6)$$

Although formally the two products appear the same the rates of reaction of the various steps are different and so the detailed nature of the end-product including both the nature and amount of the unreacted intermediates and the detailed distribution of the phosphine groups within the polymer differ. Reaction (5) has on occasions been reported as being difficult to control and reproduce [16]. Reaction (6) using $ClP(menthyl)_2$ has been used to prepare a supported asymmetric catalyst [60, 61].

Reaction of Merrifield's resin with $MPPh_2$, M = Li, Na or K yields $\text{P}\text{-}(p\text{-}C_6H_4)CH_2PPh_2$ [26, 48, 53, 62–66]. The reaction occurs relatively slowly and since the phosphine reagents are relatively bulky it is sometimes found that evenly chloromethylated polystyrene beads are not evenly phosphinated; this is especially obvious in samples containing narrow pores [67]. The use of LiPPhR where R = methyl or menthyl, or $LiP(menthyl)_2$ enables optical activity to be introduced into the polymeric support [60, 68, 68a]. One of the difficulties with $-CH_2PPh_2$ groups is that they can interact with neighbouring chloromethyl groups leading to quaternarisation (reaction (7)); the resulting

$$\begin{array}{c}\text{CH}_2\text{PPh}_2 \\ \text{P} \\ \text{CH}_2\text{Cl}\end{array} \longrightarrow \begin{array}{c}\text{CH}_2 \quad\quad \text{Ph} \\ \text{P} \qquad \overset{+}{\text{P}} \qquad \text{Cl}^- \\ \text{CH}_2 \quad\quad \text{Ph}\end{array} \qquad (7)$$

phosphonium complexes yield very poor rhodium(I) hydroformylation catalysts [69]. Reaction of $Li(CH_2)_3PR_2$ with brominated polystyrene has been used to prepare polystyrene functionalised with $-(CH_2)_3PR_2$ groups [70]; by using $Li(CH_2)_nPRR'$ asymmetric phosphines have been supported [70]. Very long alkyl chains have been used to link phosphines to polystyrene in order to ensure that the resulting phosphine metal complexes are suspended from the resin surface into the bulk solution. These have been prepared as in reaction (8) [71].

$$\text{(P)}-CH_2NH_2 + CH_2=CH(CH_2)_8COCl \xrightarrow[12\,hrs]{py,\,60^\circ} \text{(P)}-CH_2NHCO(CH_2)_8CH=CH_2$$

1. BH_3, SMe_2, 0°, 1 hr
2. H_2O_2

$$\text{(P)}-CH_2NHCO(CH_2)_{10}OTos \xleftarrow{TosCl} \text{(P)}-CH_2NHCO(CH_2)_{10}OH \qquad (8)$$

LiPPh$_2$
in thf

$$\text{(P)}-CH_2NHCO(CH_2)_{10}PPh_2$$

A completely different route for the phosphination of polystyrene involves treating the polymer with phosphorus trichloride in the presence of $BF_3 \cdot Et_2O$ to introduce $-PCl_2$ groups directly onto the aromatic rings [72]. On reaction with an alkyl or aryl lithium these phosphine groups are alkylated or arylated [72].

In addition to the introduction of unidentate phosphine ligands, chelating phosphines have also been introduced on to polystyrene. The DIOP-type of phosphine has been introduced by first oxidising the $-CH_2Cl$ side groups of Merrifield's resin to $-CHO$ groups using dimethylsulphoxide followed by functionalisation as in reaction (9) [73–75]. Reaction (10) has been used to prepare an asymmetric supported aminophosphine [76]. Simple bidentate phosphines have been introduced as in reactions (10) and (11) [77, 78].

$$R \overset{H}{\underset{H_2N}{\text{—}C\text{—}}} COOH \rightarrow \rightarrow \rightarrow \quad R \overset{H}{\underset{Me_2NHCl}{\text{—}C\text{—}}} CH_2Cl \quad \xrightarrow[\text{thf}]{H_2PPh,\ KO^t Bu,} \quad R \overset{H}{\underset{Me_2N}{\text{—}C\text{—}}} CH_2PHPh$$

$$\begin{array}{c} 1.\ n\text{-BuLi} \\ 2.\ \textcircled{P}\text{—CH}_2\text{Cl} \end{array} \quad (10)$$

$$\textcircled{P}\text{—CH}_2\text{P} \overset{CH_2\text{—}C\overset{H}{\underset{NMe_2}{\text{—}R}}}{\underset{Ph}{\big\backslash}}$$

$$\triangle\!\!-Cl + 2KPPh_2 \longrightarrow Ph_2P \overset{OK}{\diagup\!\!\diagdown} PPh_2 + KCl$$

$$\downarrow \textcircled{P}\text{—CH}_2\text{Cl} \qquad (11)$$

$$\textcircled{P}\text{-}CH_2\text{-} \begin{array}{c} \text{—}PPh_2 \\ \text{—}PPh_2 \end{array}$$

$$\textcircled{P}\text{—Br} + KPhP(CH_2)_nPPh_2 \xrightarrow[(n = 2, 3, 4)]{} \textcircled{P}\text{—P(Ph)}(CH_2)_nPPh_2 \qquad (12)$$

2.2.1.d. *Other Functional Groups.* Although phosphines are by far the commonest functional groups that have been introduced on to polystyrene, a wide range of other groups have been introduced. These are summarised in Table I. For detailed preparative procedures readers should consult the original literature which is referenced in the Table.

There is some evidence that polystyrene containing cyclopentadienyl groups attached directly to the phenyl ring prepared by reaction (13) [123, 124] does

$$\textcircled{P}\text{—Li} + \overset{O}{\underset{}{\bigcirc}} \longrightarrow \textcircled{P}\overset{HO}{\diagdown\!\!\bigcirc} \xrightarrow{heat} \textcircled{P}\diagdown\!\!\bigcirc \qquad (13)$$

not have the integrity of structure expected, and behaves anomalously when used as a support for transition metal complexes [129]. Further evidence in

TABLE I

Functional groups introduced into polystyrene

Functional Group	Starting Polymer[a]	Reagent	References
$-SO_3H$	(P)	conc. H_2SO_4	[8,9,79,80]
$-CHO$	(P)$-CH_2Cl$	dmso, $NaHCO_3$	[37,81−86]
$-COOH$	(P)$-Li$	CO_2	[17,87]
	(P)$-CH_2Cl$	1. dmso, $NaHCO_3$ 2. $Na_2Cr_2O_7$, H_2SO_4, HOAc	[84,85]
$-CH_2OH$	(P)$-CH_2Cl$	NaOH	[88]
$-CH_2CH_2OH$	(P)$-Li$		[8,9.17]
$-CH_2CH_2OP(OEt)_2$	(P)$-CH_2CH_2OH$	$ClP(OEt)_2$	[8,9]
$-CH_2O$crown ether	(P)$-CH_2Cl$	Na salt of crown ether	[89−92]
$-CH_2NHCO(CH_2)_nBr$, used to support crown ethers	(P)$-CH_2NH_2$	$ClCO(CH_2)_nBr$	[96]
	(P)$-CH_2Cl$	either Hacac + trace of NaOEt in thf	[93]
		or Na acac + NaI	[94]
$-CH(COOEt)_2$	(P)$-CH_2Cl$	$CH_2(COOEt)_2$ + $NaHCO_3$	[95]
	(P)$-CH_2Cl$		[97−99]
$-NH_2$	(P)	1. HNO_3, Ac_2O, HOAc 2. $SnCl_2$, HCl, HOAc 3. KOH, MeOH	[100]
$-CH_2NH_2$	(P)$-CH_2Cl$	1. K phthalimide, dmf 2. Ethanolic hydrazine	[96]
$-CH_2NR_3{}^+Cl^-$	(P)$-CH_2Cl$	NR_3	[37,101,102]

TABLE I (*continued*)

Functional Group	Starting Polymer[a]	Reagent	References
−CH₂N{(CH₂)₃NH₂}₂	(P)−CH₂Cl	1. NaI, Me₂Co 2. HN(CH₂CH₂CN)₂, thf 3. BH₃, thf	[103,104]
−CH₂NHCH₂CH₂NH₂ −CH₂NHCH₂CH₂NHCH₂CH₂NH₂ −CH₂NH(CH₂CH₂NH)₂CH₂CH₂NH₂	(P)−CH₂Cl	{ en dien trien }	[105]
−CH₂OCONH(CH₂)ₙNH₂	(P)−CH₂OH	1. ClCOOC₆H₄NO₂-*p* 2. H₂N(CH₂)ₙNH₂	[106]
(2-substituted pyridine ring)	(P)−Li	pyridine	[107]
(2,2′-bipyridine ring)	(P)−Li	bipyridine	[32,108–110]
−CH₂N(imidazol-1-yl)	(P)−CH₂Cl	Li imidazolate in thf Na imidazolate in dmf	[111] [112]
−SO₂NH−(1,10-phenanthrolin-5-yl)	(P)−SO₂Cl	H₂N−(1,10-phenanthrolin-5-yl)	[113]
−SO₂Cl	(P)	ClSO₃H, CCl₄	[113]
−CH₂O−⟨C₆H₄⟩−NRPPh₂	(P)−CH₂Cl	1. NaO−⟨C₆H₄⟩−NHR 2. ClPPh₂	[114]
−CH₂XTPP (X = NH, COO, CO; TPP = tetraphenylporphyrin)	(P)−CH₂Cl	TPPXH (X = NH, COO) or TPPCOCl (X = CO)	[115]

TABLE I (*continued*)

Functional Group	Starting Polymer[a]	Reagent	References
—NHYTPP (Y = CO or SO$_2$)	(P)—NH$_2$	TPPH$_2$(YCl)$_4$	[100]
—CH$_2$CN	(P)—CH$_2$Cl	NaCN, dmso	[116]
—CH$_2$CH$_2$NC	(P)—CH$_2$Cl	LiCH$_2$NC	[117]
—AsPh$_2$	(P)—Li	ClAsPh$_2$	[118]
	(P)—Br	LiAsPh$_2$	[118]
—SH	(P)—Li	1. sulphur 2. LiAlH$_4$	[17,119]
—SMe	(P)—Li	MeSSMe	[31,33]
—CH$_2$SH	(P)—CH$_2$Cl	S=C(NH$_2$)$_2$	[120]
—CH$_2$SMe	(P)—CH$_2$Cl	KSMe	[121]
—CH$_2$S—C$_6$Cl$_4$(Cl)(SH) (tetrachloro-mercaptophenyl)	(P)—CH$_2$Cl	1. C$_6$Cl$_5$SNa 2. NaSH	[122]
cyclopentadienyl[b]	(P)—Li	1. (cyclopent-2-enone) 2. H$^+$	[123,124,129]
—CH$_2$-cyclopentadienyl	(P)—CH$_2$Cl	NaCp	[125,126]
—CH$_2$C$_2$B$_{10}$H$_{11}$	(P)—CH$_2$Cl	LiC$_2$B$_{10}$H$_{11}$	[127]
—CH$_2$C$_2$B$_9$H$_{10}^{-}$	(P)—CH$_2$Cl	Na$_2$C$_2$B$_9$H$_{10}$	[128]

[a] (P) = polystyrene; (P)—Li and (P)—CH$_2$Cl = functionalised polystyrenes as described in Sections 2.2.1.a and 2.2.1.b; (P)—NH$_2$, (P)—CH$_2$NH$_2$, (P)—CH$_2$OH, (P)—SO$_2$Cl = functionalised polystyrenes prepared according to references in this Table.

[b] Material does not have the structural integrity once thought; see discussion in Section 2.2.1.d.

support of the problem is obtained from the model reaction (reaction (14)) which gives only 10–20% of the desired product. An alternative route for the formation of the cyclopentadienyl substituted polystyrene is given in Section 2.2.2.

$$\text{(14)}$$

2.2.2. COPOLYMERISATION OF FUNCTIONALISED STYRENES

An alternative approach to functionalising preformed polystyrene is to polymerise a functionalised monomer. Homopolymerisation of the monomer produces a linear product with every segment carrying a functional group, while copolymerisation either with styrene or another monomer also produces a linear polymer, but now with only a proportion of the monomer groups functionalised [18]. This readily allows the degree of functionalisation to be controlled although there is, of course, no automatic guarantee that the functional groups will be evenly distributed throughout the polymer. If a divinyl monomer such as divinylbenzene is incorporated in the polymerisation then a cross-linked resin will result. Copolymerisation has been used successfully in the synthesis of polystyrene-supported phosphines [58, 70, 130–136] and pyridines [110, 111, 130, 137, 138].

The suspension polymerisation technique is a valuable approach to forming the polymers in bead form (see Section 2.1.b). However the use of functionalised co-monomers can give special problems. For example if the co-monomer has a significant water solubility then at best some of this co-monomer will be lost into the aqueous phase, and at worst it may function as a surfactant and adversely affect the suspension conditions. With 4-vinylpyridine as co-monomer, bead polymerisation can still be achieved successfully, although the degree of pyridine incorporation is usually significantly less than theoretical [137]. The relative reactivities of the separate monomers is a vital factor that influences both the proportion and distribution of each monomer unit along the polymer chain. Industrially this difference in reactivity is less of a problem than in the laboratory, since on an industrial scale it can be overcome by feeding the two monomers into the reaction mixture at such a rate as to maintain the monomer ratio at a fixed level; the same could be done with rather greater difficulty in the laboratory. Instead the initial mixture of monomers is usually polymerised to high conversion and then a statistical analysis is made of the fluctuations in the composition of the resulting copolymer [140, 141].

p-Styryldiphenylphosphine is considerably more reactive than styrene itself so that unless appropriate concentration corrections are made as the copolymerisation proceeds there is a tendency for multiple units of phosphinated monomers to be formed in the early stages of the polymerisation, whereas as its concentration becomes depleted isolated phosphinated segments would become more common [132]. It is very difficult to prepare a copolymer in this way with all the phosphinated residues significantly isolated from each other.

By contrast the reactivities of styrene and 4-vinylpyridine are very similar, so that these two copolymerise more or less randomly [142, 143]. As a result the initial molar ratio of the monomers provides a good guide to the resulting backbone composition and the distribution of monomer units along the chain.

In addition to p-styryldiphenylphosphine and 4-vinylpyridine a number of other monomers have been copolymerised with styrene. These include $Ph_2P(CH_2)_nCH=CH_2$ where $n \geqslant 1$ [144] and the optically active monomer **1** prepared as in reaction (15) [145, 146]. **1** Was copolymerised with 2-hydroxyethyl methacrylate (reaction (16)).

$$\tag{15}$$

$$1 + CH_2=C(CH_3)COOCH_2CH_2OH$$

$$\tag{16}$$

It is not, of course, necessary to introduce the final functional groups into the initial monomers. Thus styrene copolymers containing vinylphenols or glycidyl ethers of vinylphenols on treatment with phosphorus(III) compounds yield phosphinated supports that have high swellability in hydrocarbon solvents, low hygroscopicity and good stability [147].

By copolymerising styrene and divinylbenzene in the presence of silica either in aqueous emulsion or in methanolic solution it is possible to coat the silica with cross-linked polystyrene. Such materials have a very high active surface area and obviate the need for the reactants to diffuse through the polymer to the catalytically active sites [148].

As already mentioned in Section 2.2.1.d there is evidence that cylcopentadienyl-substituted polystyrene prepared by reaction (13) may not be as simple as was once thought [129]. A route that does lead to the desired product is shown in Scheme 2. Although this route is complex it does have the advantages

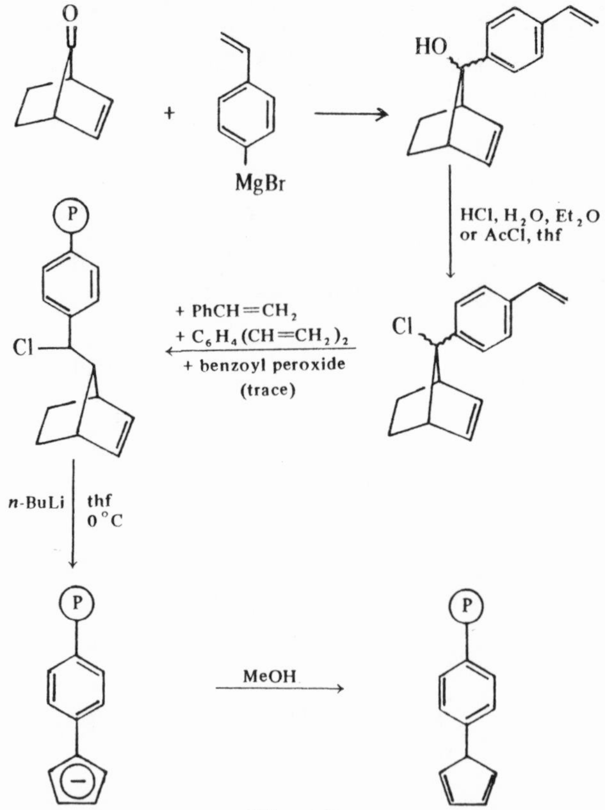

Scheme 2

that: (i) the structure and purity of the ligand is assured; (ii) the concentration of cyclopentadienyl groups within the polymer can be regulated and hence the cyclopentadienyl groups can be isolated from each other; (iii) the nature of the backbone and its polarity can be altered by varying the co-monomer; and (iv) varying degrees of cross-linking can be introduced [129].

2.3. Non-Styrene Polymers

Poly(phenylene oxide) polymers have been used to support metal complex catalysts because they are thermally more stable and mechanically stronger than polystyrene. Poly(2,6-dimethylphenylene oxide) and poly(2,6-diphenylphenylene oxide) have been functionalised by introducing cyclopentadienyl groups through reactions (17) and (18) [149]. The resulting polymers have been used to support titanium hydrogenation catalysts and cobalt and rhodium hydroformylation catalysts.

There has been some interest in the use of polyvinyl chloride as a catalyst support. Pvc is potentially simple to use because of the reactive chlorine atoms present on the free polymer [50, 52, 53]. However attempts to replace these by phosphine groups using reaction (19) have been unsuccessful because it leads to

$$-(CH_2-\underset{\underset{Cl}{|}}{CH})_n + LiPPh_2 \xrightarrow{\text{thf}} -(CH_2-\underset{\underset{Cl}{|}}{CH})_{n-x}(CH_2-\underset{\underset{PPh_2}{|}}{CH})_x \quad (19)$$

breakdown of the polymer chains [150, 151]; complete replacement of all the chlorine atoms required a 40 hour reflux, by which time the molecular weight had dropped to about 1500, corresponding to chains of 10−12 units [152]. Alkali metal diphosphine alcoholates have been used to functionalise halogenated vinyl polymers [153]. Hydroxyl-containing polymers such as polyvinyl alcohol, polyallyl alcohol, polyethylene glycol, polyesters and polystyrene substituted with $-CH_2OH$ groups can be phosphinated through reaction (20) [154].

$$\text{(P)}-OH + XPR^1R^2 \longrightarrow \text{(P)}-OPR^1R^2 \quad (20)$$

(X = halogen or OR^3)

A series of hydrocarbon-supported polymers containing phosphine groups, either pendant or as integral components of the polymer backbone, have been prepared from dienes [155], polybutadiene [156] and di-Grignard reagents [155] (reactions (21−23)).

$$\text{\Large\wedge\wedge\hspace{-2pt}/} + RPH_2 \longrightarrow -(\underset{\underset{R}{|}}{P}-(CH_2)_6)_n \quad (21)$$

$$-(CH_2CH{=}CHCH_2)_n + Ph_2PH \xrightarrow[\text{initiation}]{h\nu \text{ or radical}} -(CH_2CH_2\underset{\underset{PPh_2}{|}}{CH}CH_2)_n \quad (22)$$

$$ClMg(CH_2)_2MgCl + RPCl_2 \longrightarrow -(\underset{\underset{R}{|}}{P}-CH_2)_2)_n \quad (23)$$

Polyamides such as poly(*m*-phenyleneisophthalimide) have been phosphinated with Ph_2PCl as in reaction (24) [5, 157]. The hydrogen chloride liberated is

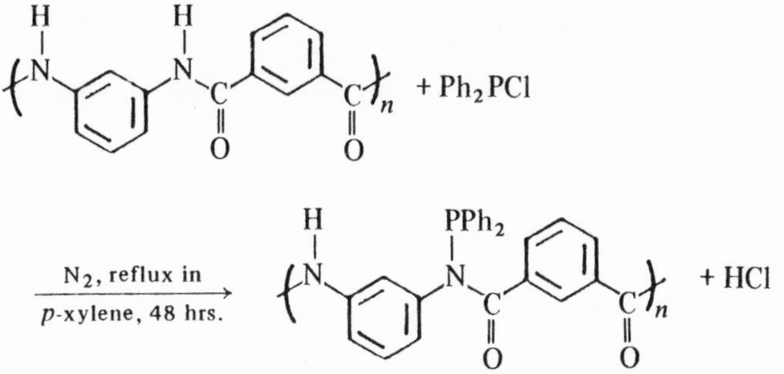

(24)

swept out with the nitrogen purge gas. Phosphinated polydiacetylenes have been prepared using reaction (25) [158, 159].

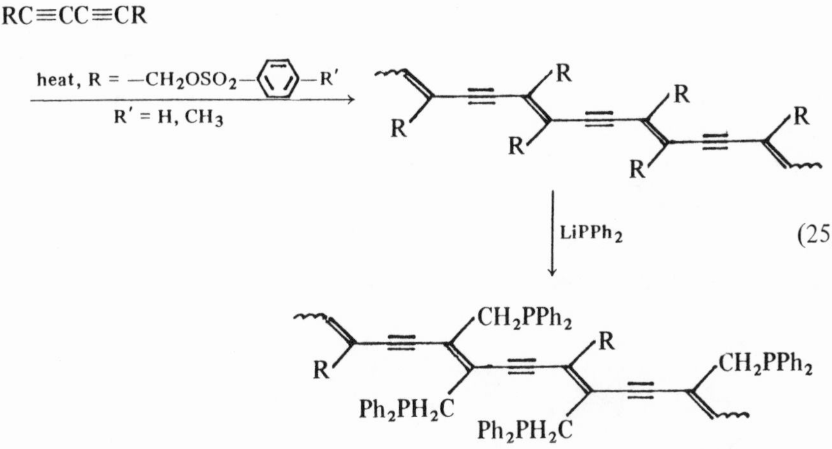

$$RC{\equiv}CC{\equiv}CR$$

(25)

$(Ph_2PCH_2CH_2)_2NH^+Cl^-$, prepared as in reaction (26), has been used to introduce a chelating diphosphine on to a series of polymers functionalised with acyl chloride groups (reaction (27)) [160]. $Li(tmeda)^+(Ph_2P)_2CH^-$ reacts

$$(ClCH_2CH_2)_2NH_2^+Cl^- + 2HPPh_2$$

$$(Ph_2PCH_2CH_2)_2NH_2^+Cl^-$$

(26)

$$\text{(P)}-COX + (Ph_2PCH_2CH_2)_2NH_2^+Cl^-$$

$$\xrightarrow{\text{base}} \quad \text{(P)}-\overset{\overset{O}{\|}}{C}-N(CH_2CH_2PPh_2)_2 \qquad (27)$$

with polymers functionalised with $-SiMe_2Cl$ groups to introduce a multidentate phosphine on to the support (reaction (28)) [161].

$$\text{(P)}-\langle\bigcirc\rangle-SiMe_2Cl + Li(tmeda)^+(Ph_2P)_2CH^-$$

$$\longrightarrow \quad \text{(P)}-\langle\bigcirc\rangle-SiMe_2CH(PPh_2)_2 \qquad (28)$$

Polymeric supports functionalised with chiral phosphines have been prepared by binding diop to styrene and then copolymerising the product with hydroxyethyl methacrylate (reaction (29)) [162, 163]. The resulting polymers have been used

$$(29)$$

to support rhodium(I) asymmetric hydrogenation catalysts (see Section 6.24). Other copolymers have been used in order to alter the local environment to the active site [164].

Copolymerisation of $CH_2{=}CHPO(OEt)_2$ with acrylic acid yields **2** which has

$$-CH_2CH-CH_2-CH-$$
$$(EtO)_2PO \qquad COOH$$

2

been used to support cobalt liquid-phase oxidation catalysts [165]. The phosphine-bearing poly(aryloxyphosphazenes), $\{NP(OC_6H_4PPh_2\text{-}p)(OPh)_{2-x}\}_n$, have been used to support metal complex catalysts [166].

A number of 2- and 4-vinylpyridine catalysts have been prepared, either by polymerising 2- or 4-vinylpyridine themselves [167–169] or by copolymerising 4-vinylpyridine with styrene [170] or by radiation grafting 4-vinylpyridine on to polypropylene (see Section 2.4 below) [171, 172]. Poly(4-vinylpyridine) can be cross-linked with 1,2-dibromoethane (reaction (30)), although this introduces

positive charges on the cross-linked nitrogen atoms which can inhibit metal ion uptake [167].

2.4. Radiation Grafting

The polymeric supports used in Sections 2.2 and 2.3 have been limited to those that can be functionalised chemically. An alternative approach has been to graft on to a polymer a material that can itself be used to support metal complexes. In this section we shall be particularly concerned with grafting using radiation, usually UV- or γ-radiation but also including high energy (~3.5 MeV) electrons and gas discharge plasmas, as the means of attachment [173–176]. Examples of monomers that have been radiation grafted on to a range of polymers are given in Table II.

TABLE II
Some examples of monomers that have been radiation grafted on to polymers

Monomer	Polymer	Radiation	References
4-vinylpyridine	polypropylene	γ	[171,172,176,177]
allylX (X = OH, NH$_2$, SH)	polyethylene	γ, 3.5 MeV	
CH$_2$=CHCOOR (R = H, Me)	polypropylene	electrons,	[176, 178–180]
CH$_3$COCH=CH$_2$	polystyrene	plasma	
CH$_2$=CHCN			
p-styryldiphenylphosphine	polypropylene	γ	[177,185]
Ph$_2$PCH=CH$_2$	polypropylene, pvc	γ	[181,182]
cis-Ph$_2$PCH=CHPPh$_2$	polypropylene, pvc	γ	[181,182]
o,m,p-PPh$_2$	polyethylene, polypropylene,	γ	[183,184]
—OCOCH=CH$_2$ o,m,p-PPh$_2$	polystyrene, pvc, cellulose, wool	γ	[183,184]
p-nitrostyrene	polypropylene, polyethylene, pvc	γ	[181,182,186]
X— (X = NH$_2$, Cl, Br)	polypropylene	γ	[181]
styrene	polypropylene	uv	[187]
styrene	pvc	γ	[182]

Irradiation of many organic polymers results in the ejection of electrons and the formation of a radical (reaction 31).

$$\text{\textasciitilde\textasciitilde CHR\textasciitilde\textasciitilde} \xrightarrow{\text{irradiation}} \text{\textasciitilde\textasciitilde}\dot{C}R\text{\textasciitilde\textasciitilde} + \dot{H} \qquad (31)$$

In the presence of an unsaturated monomer this radical site can then undergo carbon-carbon bond formation (reaction (32)). The resulting radical can then undergo one of three reactions:

$$\text{\textasciitilde\textasciitilde}\dot{C}R\text{\textasciitilde\textasciitilde} + CH_2=CHX \xrightarrow{\text{absence of O}_2} \begin{array}{c} \text{\textasciitilde\textasciitilde}CR\text{\textasciitilde\textasciitilde} \\ | \\ CH_2-\dot{C}HX \end{array} \qquad (32)$$

(i) radical recombination with the original hydrogen radical displaced, which is relatively unlikely;

(ii) reaction with a solvent molecule to give chain termination accompanied by formation of a new radical originating from the solvent;

(iii) reaction with a further molecule of the unsaturated monomer to yield a graft copolymer (reaction (33)).

$$\underset{\underset{CH_2-\overset{\centerdot}{C}HX}{|}}{\sim\!\!\sim\!CR\!\sim\!\!\sim} \quad + CH_2\!\!=\!\!CHX \longrightarrow \underset{\underset{CH_2-CHX-CH_2-\overset{\centerdot}{C}HX}{|}}{\sim\!\!\sim\!CR\!\sim\!\!\sim} \qquad\qquad \longrightarrow \text{ etc. } (33)$$

There are four main types of radiation that have been used for radiation grafting:

(i) γ-Radiation, which requires a ^{60}Co radiation facility; although such facilities are not widely available due to both their cost and the safety requirements that they pose, γ-radiation grafting has been the method most extensively used for preparing supported catalysts [171, 172, 177, 181–186, 188, 189].

(ii) Uv-irradiation, which may require the use of photosensitising agents [187].

(iii) Accelerated electrons, which clearly requires high energy (~ 3–4 MeV) generators [176, 178, 179].

(iv) Gas discharge plasmas [176, 178, 179].

2.4.1. TECHNIQUES OF RADIATION GRAFTING

The actual radiation grafting can be done in one of four ways:

(i) *Mutual or Simultaneous Grafting*
In this, the most popular method, the polymer, A_n, is irradiated whilst in direct contact with the vinyl monomer, B, to be grafted, which may be present as a vapour, liquid or in solution. This leads to the formation of a graft copolymer (reaction (34)).

$$-A-A-A-A- \xrightarrow{\ \gamma\text{-ray}\ } -A-A-\overset{\centerdot}{A}-A- \xrightarrow{\ mB\ } \underset{\underset{B-B-B-}{|}}{-A-A-A-A-} \qquad (34)$$

Experimentally this is the simplest and most widely used method of radiation grafting [171, 172, 176–179, 181–189], and it can be the most effective as the constant regeneration of polymer sites during the grafting reaction increases the possibility of obtaining higher grafting yields (the amount of monomer incorporated on to the polymer). Variation of the intensity of the radiation can alter the chain length of the grafted monomer by changing the ratio of the concentrations of radical sites to monomer.

(ii) *Pre-irradiation Grafting*
In pre-irradiation grafting the polymer is irradiated *in vacuo* or in the presence of an inert gas. It is then removed from the radiation facility and reacted with the monomer, which may be in the form of a liquid or a vapour, at elevated temperature to achieve grafting (reaction (35)).

$$-A-A-A-A- \xrightarrow{\gamma\text{-ray}} -A-A-\overset{\cdot}{A}-A- \xrightarrow{m\text{B}} \begin{array}{c} -A-A-A-A- \\ | \\ B-B-B- \end{array} \qquad (35)$$

Irradiation of the polymer produces relatively stable trapped free radicals, and diffusion of the monomer to these active sites can produce a graft copolymer. This method has the advantage that very little homopolymer, B_n, is produced during the grafting process. It does, however, have a number of disadvantages. The process requires the formation of stable radical sites on the polymer backbone, and these are generally formed in the crystalline region of a polymer where reactions such as cross-linking and radical hopping, which can lead to cross-linking, are hindered by the rigidity of the crystalline structure. Diffusion of insoluble or viscous monomers into the crystalline regions is hindered and heating the reaction mixture to increase polymer swelling and decrease monomer viscosity can destroy crystallinity, which in turn can remove the radical sites before grafting occurs. The use of solvents to 'thin out' the monomer and increase polymer swelling decreases grafting efficiency, i.e., the proportion of the monomer which becomes grafted on to the polymer, purely on concentration grounds.

(iii) *Peroxidation Grafting*
In peroxidation grafting the polymer is irradiated in air to give stable diperoxides, **3**, or hydroperoxides, **4**, which may then be removed from the radiation source and stored at room temperature until they are required for reaction with the monomer (reactions (36) and (37)).

$$2 \text{ } -A-A-A-A- \xrightarrow{\gamma\text{-ray}} 2 \text{ } -A-A-\overset{\cdot}{A}-A-$$

$$\xrightarrow{O_2} \begin{array}{c} -A-A-A-A- \\ | \\ O \\ | \\ O \\ | \\ -A-A-A-A- \\ \mathbf{3} \end{array} \xrightarrow{\text{heat}} \begin{array}{c} 2 \text{ } -A-A-A-A- \\ | \\ \overset{\cdot}{O} \\ \downarrow m\text{B} \\ 2 \text{ } -A-A-A-A- \\ | \\ O-B-B-B- \end{array} \qquad (36)$$

$$-A-A-A-A- \xrightarrow{\gamma\text{-ray}} -A-A-\overset{\cdot}{A}-A- \xrightarrow[\text{H abstraction}]{O_2} \begin{array}{c} -A-A-A-A- \\ | \\ O-O-H \\ \mathbf{4} \end{array}$$

$$\downarrow \text{heat}$$

$$\begin{array}{c} -A-A-A-A- \\ | \\ O-B-B- \end{array} \xleftarrow{\quad m\text{B} \quad} \begin{array}{c} -A-A-A-A- \\ | \\ \overset{\cdot}{O} \end{array} \qquad (37)$$

The advantage of this technique is that the peroxy-compounds, **3** and **4**, may be stored for long periods of time before performing the final grafting step. Indeed, it has been found that such compounds prepared from cellulose [190] and polyethylene [191] may be stored for several years after irradiation before being used for grafting. It should be noted, however, that the peroxidation method leads to the formation of graft copolymers in which the constituent polymers are joined by an ether linkage instead of the carbon–carbon bond which links graft copolymers prepared by other methods. Ether linkages are more susceptible to cleavage by chemical processes than are carbon–carbon bonds and thus copolymers prepared by this method may well be unsuitable for use as polymer supports in catalytic reactions. This method is particularly unsuitable when the monomer is an olefinic phosphine as it may well lead to oxidation of the phosphine thus rendering it useless as a supporting ligand for most metal complexes.

(iv) *Cross-Linking of Two Different Polymers*

If two polymers (A_n and B_m) are mixed intimately and subjected to radiation, cross-linking can occur between the two different chains (reactions (38)).

$$-A-A-A-A- \qquad\qquad -A-A-\overset{\cdot}{A}-A \qquad\qquad -A-A-A-A-$$

$$+ \qquad \xrightarrow{\gamma\text{-ray}} \qquad + \qquad \longrightarrow \qquad\qquad\qquad\quad | \qquad\qquad (38)$$

$$-B-B-B-B- \qquad\qquad -B-B-\overset{\cdot}{B}-B- \qquad\qquad -B-B-B-B-$$

In practice the two polymers are usually dissolved in the same solvent, or cast as a film mixture from a common solvent. This method thus involves preformation of a polymer from the monomer to be grafted and it also requires that a common solvent be found for the grafting process.

2.4.2. REACTIONS OCCURRING UNDER THE INFLUENCE OF RADIATION

Although reactions (34) to (38) are the desirable reactions that lead to the formation of copolymers, in reality a number of reactions can and do occur in the presence of radiation. The most common reactions which occur during the simultaneous irradiation method are:—

(i) Direct radiation grafting of the unsaturated monomer (B) on to the polymer backbone (A_n) to give:

(a) a monolayer of grafted molecules giving a copolymer directly analogous to that prepared by chemical functionalisation of a polymer (reaction (39)).

$$-A-A-A-A- \xrightarrow{\gamma\text{-ray}} \quad -A-A-\overset{\cdot}{A}-A- $$
$$+ \overset{\cdot}{R} \text{ (usually } \overset{\cdot}{H} \text{)}$$

$$\xrightarrow[2.\,\overset{\cdot}{R}]{1.\,B} \quad -A-A-A-A- \xrightarrow[2.\,B]{1.\,\gamma\text{-ray}} \quad -A-A-A-A- \quad (39)$$
$$\qquad\qquad\qquad\qquad | \qquad\qquad\qquad\qquad\qquad |\quad\, |$$
$$\qquad\qquad\qquad\qquad B \qquad\qquad\qquad\qquad\qquad B\;\; B$$

(b) a long grafted chain of monomer initiated by one radical site on the polymer chain (reaction (40)).

$$-A-A-A-A- \xrightarrow[2.\,B]{1.\,\gamma\text{-ray}} -A-A-A-A- \xrightarrow{B} -A-A-A-A- \longrightarrow \text{ etc. (40)}$$
$$\qquad\qquad\qquad\qquad\qquad\quad | \qquad\qquad\qquad\qquad |$$
$$\qquad\qquad\qquad\qquad\qquad\; \overset{\cdot}{B} \qquad\qquad\qquad\quad B-\overset{\cdot}{B}$$

(ii) Fission of the polymer backbone and reaction with monomer to give block copolymers (reaction (41)).

$$-A-A-A-A- \xrightarrow{\gamma\text{-ray}} -A-\overset{\cdot}{A} + \overset{\cdot}{A}-A-$$

$$\xrightarrow{m\text{B}} -A-A-B-B-B-B-B-A-A- \qquad\qquad (41)$$

(iii) Homopolymerisation of the monomer (reaction 42).

$$B \xrightarrow{\gamma\text{-ray}} B^{\cdot} + R^{\cdot} \xrightarrow{mB} -B-B-B-B-B-B- \tag{42}$$

(iv) Combination of the homopolymer of monomer B with the polymer backbone (reaction (43)).

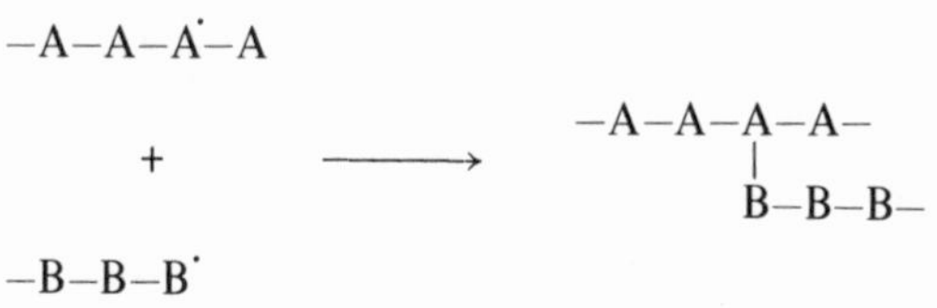

$$
\begin{array}{c}
-A-A-\overset{\cdot}{A}-A \\[2pt]
+ \\[2pt]
-B-B-\overset{\cdot}{B}
\end{array}
\longrightarrow
\begin{array}{c}
-A-A-A-A- \\
| \\
B-B-B-
\end{array}
\tag{43}
$$

(v) Grafting of the monomer onto previously grafted chains of monomer (reaction (44)).

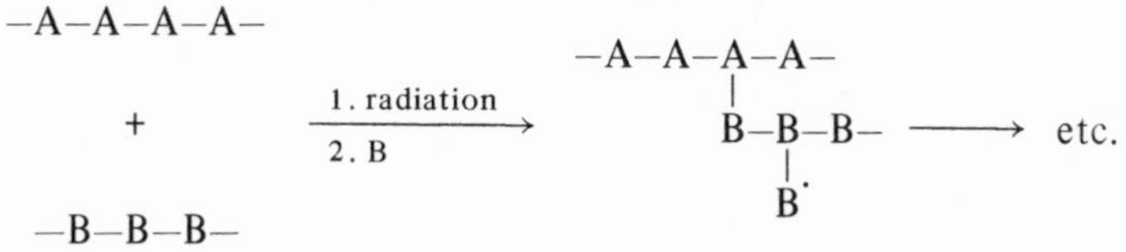

$$
\begin{array}{c}
-A-A-A-A- \\[2pt]
+ \\[2pt]
-B-B-B-
\end{array}
\xrightarrow[\text{2. B}]{\text{1. radiation}}
\begin{array}{c}
-A-A-A-A- \\
| \\
B-B-B- \\
| \\
\overset{\cdot}{B}
\end{array}
\longrightarrow \text{etc.}
\tag{44}
$$

(vi) Cross-linking of the polymer (reaction (45)).

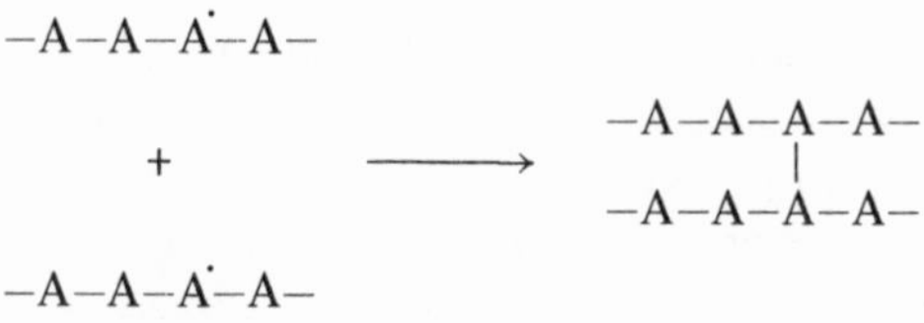

$$
\begin{array}{c}
-A-A-\overset{\cdot}{A}-A- \\[2pt]
+ \\[2pt]
-A-A-\overset{\cdot}{A}-A-
\end{array}
\longrightarrow
\begin{array}{c}
-A-A-A-A- \\
| \\
-A-A-A-A-
\end{array}
\tag{45}
$$

(vii) Degradation of the polymer (reaction (46)).

$$-A-A-A-A- \xrightarrow{\text{radiation}} -A-\overset{\cdot}{A} + \overset{\cdot}{A}-A- \xrightarrow{R^{\cdot}} -A-A-R \tag{46}$$

Looking at the above reactions, (45) and (46) are functions of the polymer alone, and for a given dose of radiation at a given intensity a certain amount of polymer cross-linking or degradation will occur. Degradation of the polymer is undesirable since it affects its solubility characteristics, an important factor in the use of supported catalysts, and thus the degrading types of polymer such as poly(tetrafluoroethylene) and poly(vinyl chloride) [171] are not suitable for radiation grafting. The formation of block copolymers, reaction (41), only occurs with degrading types of polymers and so it should not occur if the polymer employed is of the cross-linking type. Although cross-linking of the polymer, reaction (45), affects the activity and selectivity of supported catalysts it is not such a serious problem and indeed is a potential method by which the characteristics of a supported catalyst may be altered. Thus polymers of the cross-linking type such as polyolefins and polyamides [173] are often the polymers of choice.

Homopolymerisation of the monomer, reaction (42), may be reduced by the presence of a polymerisation inhibitor which inhibits free radical formation in the bulk solution without grossly inhibiting radical formation on the polymer, that is without inhibiting the grafting process.

Reactions (39), (40), (43) and (44) all lead to the formation of graft copolymers which are of interest as catalyst supports. The occurrence of these processes depends upon a large number of inter-related factors which are discussed below.

2.4.3. FACTORS AFFECTING RADIATION GRAFTING

A large number of factors affect whether or not radiation grafting occurs and when it does occur the extent to which it occurs. In each case it is necessary to optimise all the parameters. These are discussed in more detail in Sections 2.4.3.1 to 2.4.3.9.

2.4.3.1. *Monomer Reactivity*

In principle any olefinic monomer may be radiation grafted onto any polymer that produces radicals when subjected to γ-radiation. In practice, however, it appears that only olefins which may be readily homopolymerised using radical initiators will graft onto such polymers, as is illustrated by the numerous radiation grafting studies that have been performed [173−175, 191]. This is perhaps not surprising as the initial step in both reactions is essentially the same, i.e. reaction of the olefin with a radical site. The rate of grafting of monomers has been related to their rate of radical initiated homopolymerisation [192, 193]. It has been shown that the grafted copolymer composition, that is, the ratio of monomers in the grafted side chains, is directly related to the monomer reactivities in radical initiated homopolymerisation, and independent of the polymer. There is, however, no general rule which predicts whether, and under what conditions, a monomer will graft onto a polymer, and it is still necessary to discover this by experiment. It has been found that whilst 4-vinylpyridine [171, 172] and *p*-nitrostyrene [181, 182, 186] are readily radiation grafted, olefinic phosphines are only grafted with considerable difficulty [183, 184, 188].

2.4.3.2. *Total Dose*

Within a given radiation grafting experiment increasing the total dose of radiation received by the reaction mixture increases the grafting yield, i.e. more of the monomer is grafted onto the polymer, until all of the monomer has been converted into either graft copolymer or homopolymer, whereupon the grafting yield levels out to a constant value [193a]. A total dose must therefore be carefully chosen to optimise the grafting yield whilst minimising radiation degradation or cross-linking of the polymer.

2.4.3.3. *Dose Rate*

The rate of formation of radical sites on a polymer is directly related to the incident dose rate, that is, to the intensity of the radiation, and consequently so is the rate of initiation of graft copolymerisation. Higher dose rates thus lead to shorter chain lengths of the grafted monomer under a given set of conditions. It should be noted, however, that increasing the dose rate also increases the rate of formation of monomer radicals as the result of either monomer or solvent interactions with the radiation, and thus the rate of homopolymerisation is also increased. For a given radiation dose the extent of degradation or cross-linking of the polymer is less at lower dose rates [191]. The complex interaction of these factors can be seen by comparing Figures 1 and 2. Figure 1 indicates that

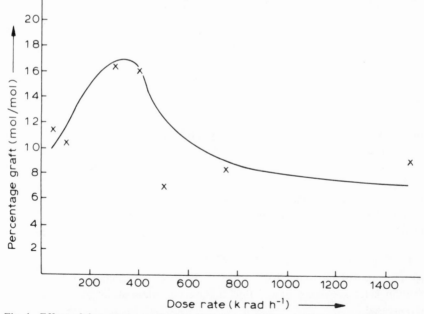

Fig. 1. Effect of dose rate on grafting yield in the γ-radiation grafting of 4-vinylpyridine on to polypropylene in benzene solution (reproduced with permission from [172]).

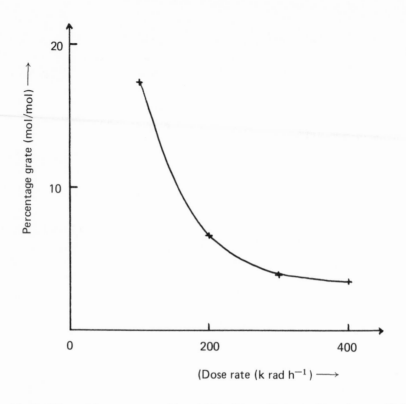

Fig. 2. Effect of dose rate on grafting yield in the γ-radiation grafting of *p*-styryldiphenyl-phosphine on to polypropylene in benzene solution (reproduced with permission from [185]).

in the γ-radiation grafting of 4-vinylpyridine onto polypropylene in benzene solution the optimum dose rate is about 350 krads/hour [172]; lower dose rates probably yield insufficient radical sites, whereas higher dose rates give rise to a great deal of homopolymer. When *p*-styryldiphenylphosphine is γ-radiation grafted onto polypropylene in exactly the same way, Figure 2 indicates that the degree of grafting increases rapidly as the dose rate is decreased from 400 to 100 krads/hour [185].

2.4.3.4. *Monomer-To-Polymer Ratio*
As the ratio of monomer to polymer is increased reactions (40), (42–44) in Section 2.4.2 are enhanced and thus the average length of the grafted monomer is increased.

2.4.3.5. *Solvent and Solvent-To-Monomer Ratio*

The effect of solvent and solvent-to-monomer ratio on radiation grafting depends on many variables, and it is difficult to generalise. It is, however, a general rule that solvents which do not swell the polymer at all give poor grafting yields, whilst solvents that do swell the polymer give higher grafting yields as they assist diffusion of the monomer to the reactive sites on the polymer [194]. However there are some surprising exceptions to this rule. Thus we found that in the γ-radiation grafting of p-styryldiphenylphosphine on to polypropylene, replacing benzene — which is a swelling solvent — by dimethylsulphoxide which is not — gives an eight-fold enhancement in grafting yield [185]. This probably arises because the monomer is more 'soluble' in the polymer than in dimethylsulphoxide, coupled with the insolubility in the solvent of the growing grafted chains which effectively increases the viscosity of the grafting medium.

The use of low solvent-to-monomer ratios, no solvent at all, or of certain solvent mixtures, often leads to an enhanced grafting yield due to the 'gel' or 'Trommsdorff' effect [195]. The appearance of a gel effect thus depends on a number of factors which cannot be easily predicted but must be found by experiment. Its utilisation can maximise the grafting yield obtained under a given set of conditions.

2.4.3.6. *Inhibitor*

An inhibitor such as p-t-butylcatechol may be added to the reaction mixture to decrease the extent of homopolymerisation of the monomer and hence increase the maximum attainable grafting yield. The mode of action of an inhibitor, $X-H$, is outlined in reaction (47).

$$B^{\cdot} + X-H \longrightarrow B-H + X^{\cdot} \tag{47}$$

The inhibitor prevents the homopolymerisation of a monomer, B, by reacting with the monomer radicals, $B^{\cdot}$, as they are formed, giving stable radicals, $X^{\cdot}$, which take no further part in the reaction. Thus the concentration of inhibitor used must be large enough to deal with the concentration of monomer radicals formed during the grafting process.

2.4.3.7. *Chain Transfer Agent*

The chain lengths of grafted monomer may be restricted by the addition of a chain transfer agent whose mode of action is outlined in reaction (48).

$$-A-A-A-A- \ + \ X-H \ \longrightarrow \ -A-A-A-A- \ + \ X^{\cdot} \tag{48}$$
$$\ \ \ \ \ \ \ |\ |$$
$$\ \ \ \ \ \ B^{\cdot} \ B-H$$

In outline this is essentially the same as the mode of action of an inhibitor, reaction (47), the difference being that $X—H$ is chosen so that $X^{\cdot}$ is reactive enough to initiate further grafting reactions by abstraction of a radical from the polymer backbone. It should be noted that a chain transfer agent must be soluble in the polymer and that the radicals, $X^{\cdot}$, produced by its action are also reactive enough to initiate homopolymerisation of the monomer by reaction with an olefinic bond.

2.4.3.8. *Acid*

Mineral acids such as sulphuric, nitric and perchloric acids have been shown to enhance the grafting of some monomers, such as styrene, methyl methacrylate and *p*-nitrostyrene, onto monomers, such as cellulose, polyethylene, pvc and polypropylene, in alcohols [182, 194, 196–199]. This enhancement is thought to be due to a combination of two factors:

(i) Acidification of the solvent increases the concentration of protonated solvent, ROH_2^+, which may react with free electrons produced by irradiation of the polymer to form hydrogen atoms (reaction (49)).

$$ROH_2^+ + e^- \longrightarrow ROH + H^{\cdot} \qquad (49)$$

The hydrogen atoms thus produced may then scavenge radicals from the polymer backbone producing radical sites.

(ii) The hydrogen atoms produced by reaction (49) may react with the monomer and the radicals thus produced may then scavenge radicals from the polymer backone.

In both of these routes the effect of the acid is to increase the number of radical sites produced on the polymer for a given dose of radiation and hence to enhance the grafting process. It should be noted, however, that both of the above processes would also enhance homopolymerisation of the monomer, and an increased amount of homopolymer is observed in the presence of mineral acids [196]. As many monomers, such as amines and phosphines, are bases the addition of mineral acids often fails to enhance the grafting process.

2.4.3.9. *Oxygen*

As mentioned earlier, the irradiation of polymeric materials in the presence of oxygen, an effective free radical scavenger, leads to the production of peroxides and hydroperoxides on the polymer backbone. This leads to attachment of graft copolymers via peroxide or ether linkages which are chemically more reactive than carbon–carbon bonds which are formed under anaerobic conditions. The presence of oxygen on the radiation grafting reaction mixture is particularly undesirable when the monomer is oxygen-sensitive as in the case of phosphines.

2.5. Silica-Based Systems

Silica contains surface silanol groups, $\geqslant Si-OH$, that can either bind metal ions directly or alternatively can be functionalised [200, 201]. Sometimes residual surface silanol groups can be a problem in which case they can be blocked by reaction with Me_2SiCl_2 or Me_3SiCl in toluene followed by thoroughly washing with methanol to decompose any incompletely reacted Me_2SiCl_2 [202, 203].

Several research groups have concentrated on attaching metal complexes to silica by introducing phosphine groups onto the surface. They have used two approaches illustrated in reactions (50) and (51), which differ only in the

$$\{Si\}-OH + X_3SiCH_2CH_2PRR'$$

$$\xrightarrow[\substack{R = R' = Ph \\ R = Ph, R' = menthyl}]{X = EtO, Cl} \{Si\}-O-\overset{|}{\underset{|}{Si}}-CH_2CH_2PRR'$$

$$\downarrow + ML_n$$

$$\{Si\}-O-\overset{|}{\underset{|}{Si}}-CH_2CH_2PRR'ML_{n-1} \qquad (50)$$

$$(EtO)_3SiCH_2CH_2PPh_2 + ML_n \longrightarrow (EtO)_3SiCH_2CH_2PPh_2ML_{n-1}$$

$$\downarrow + \{Si\}-OH$$

$$\{Si\}-O-\overset{|}{\underset{|}{Si}}-CH_2CH_2PPh_2ML_{n-1} \qquad (51)$$

order of the steps [62, 204–215]. The advantages of the first route (reaction (50)) are: (i) the range of bridging groups can readily be varied, for example to $-CH_2CH_2CH_2PPh_2$, $p\text{-}C_6H_4PPh_2$, $-(CH_2)_3C_6H_4PPh_2\text{-}p$ or $4\text{-}(CH_2)_4C_5H_4N$; (ii) metal complexes that are unstable in solution, for example because they readily dimerise or are coordinatively unsaturated, may be prepared because the rigidity of the surface prevents molecular interactions; (iii) the surface of the silica after reaction with the bridging group will still be very polar due to unreacted silanol groups, unless these have been removed by subsequent silylation of the surface. In this way the microenvironment of the catalytic centre can be carefully controlled, as it is of course in a metalloenzyme, with consequent advantages in terms of the activity and selectivity of the catalyst.

The main disadvantage of reaction (50) is the difficulty of determining the precise nature of the catalytic site since this is formed in the support. Reaction (51) was designed to obviate this problem since, in principle, the complex is isolated and may be characterised by standard methods. In practice, however, many complexes of this type cannot be isolated as crystalline solids but only as oils which must be purified chromatographically using non-hydroxylic phases.

A number of routes have been described for the preparation of the functionalised phosphines used in reaction (50); some of these are shown in reactions (52–56) [161, 210, 216–219].

$$Ph_2PCH_2PPh_2 \xrightarrow[\text{2. ClSiMe}_3]{\text{1. }n\text{-BuLi, tmeda}} \underset{\underset{SiMe_3}{|}}{Ph_2PCHPPh_2} \tag{52}$$

$$\text{(EtO)}_3SiCH=CH_2 + Ph_2PH \xrightarrow{{}^t\text{BuOO}{}^t\text{Bu}} \text{(EtO)}_3SiCH_2CH_2PPh_2 \tag{54}$$

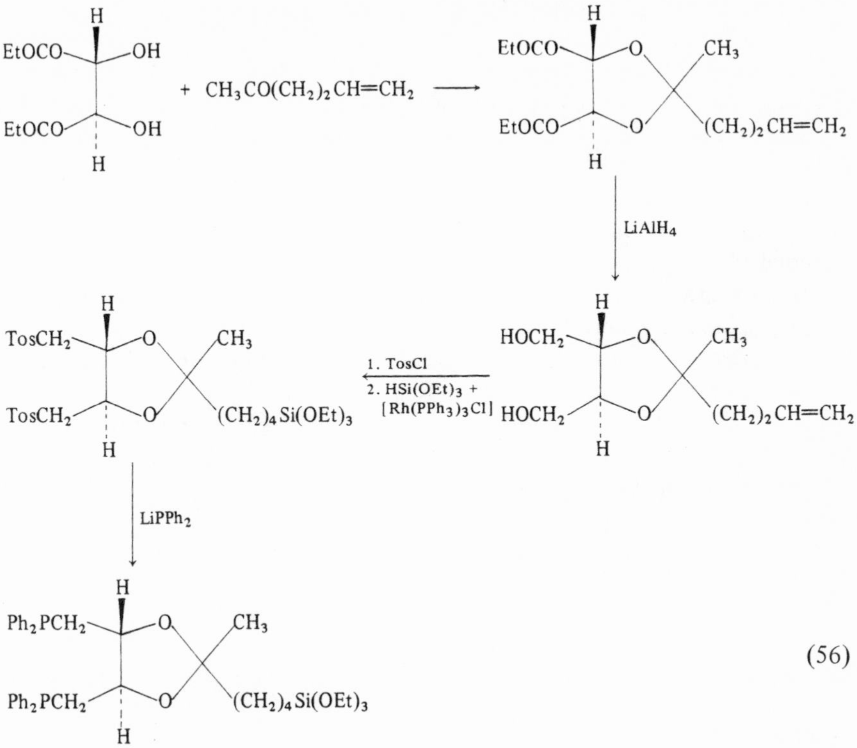

(56)

The difficulty experienced in purifying bidentate phosphines containing —SiR$_3$ functional groups led to the apparently complex route shown in reaction (57) being used to introduce bidentate phosphines on to silica [220]. However

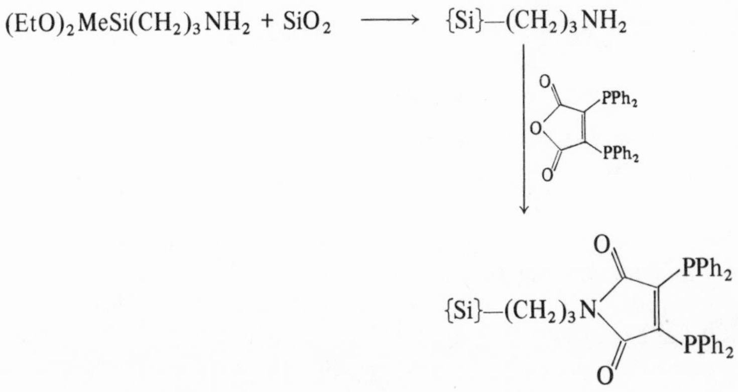

(57)

$$(EtO)_2 Si \begin{cases} (CH_2)_2 PPh_2 \\ \\ CH_2 CH(CH_3)CH_2 PPh_2 \end{cases} \qquad EtOSi\{(CH_2)_3 PPh_2\}_3$$

5   **6**

the bidentate and tridentate functionalised phosphines **5** and **6** have been reported [221].

An alternative approach to the phosphination of silica is to treat the support with phosphorus trichloride in a hydrocarbon medium at 60–80 °C. The resulting phosphinated silica bearing $-PCl_2$ groups can be treated with RXH, where R = alkyl or aryl and X = 0, S or NHR' [222, 223]. Multidentate phosphines such as $Ph_2 P(CH_2)_3 PPh(CH_2)_3 NH_2$ and $(Ph_2 PCH_2 CH_2)_2 P(CH_2)_3 NH_2$ or P_2P-NH_2, have been linked to silica that has been derivatised to contain hydroxyl groups as a glycophase (reaction (58)) [224].

$$\{Si\}-OCH_2 C(OH)_2 CH_2 OH + KIO_3 \longrightarrow \{Si\}-OCH_2 CHO$$

$$\downarrow P_2 P-NH_2$$

$$\{Si\}-OCH_2 CH_2 NH-PP_2 \xleftarrow{NaBH_4} \{Si\}-OCH_2 CH=N-PP_2 \quad (58)$$

Silica has been functionalised by a route analogous to the chloromethylation of polystyrene (reaction (59)) [225]. Chloromethylated silica can then be subjected to most of the reactions applicable to chloromethylated polystyrene. It has been used to support a bidentate phosphine (reaction (60)) [225].

$$Me_2 ClSiCH_2 Cl + SiO_2 \xrightarrow{toluene} \{Si\}-Me_2 SiCH_2 Cl \qquad (59)$$

$$\{Si\}-Me_2 CH_2 Cl + NaCH(CH_2 PPh_2)_2 \longrightarrow \{Si\}-\underset{\underset{Me}{|}}{\overset{\overset{Me}{|}}{Si}}-CH_2 CH \begin{cases} CH_2 PPh_2 \\ \\ CH_2 PPh_2 \end{cases} \quad (60)$$

In addition to phosphine groups, silica has been functionalised with amino, pyridyl, morpholino, piperidino, pyrrolidino, Schiff base, cyano, $-SH$ and $-C_5 H_5$ groups [201, 210, 211, 226–241]. In each case an approach similar to reaction (50) has been adopted using $X_3 Si\sim\sim Y$, $X_2 SiMe\sim\sim Y$ and $XSiMe_2\sim\sim Y$ where X = EtO or Cl and Y = functional group. Acetylacetonate groups have been introduced by using aminofunctionalised silica gel (reaction (61)) [242], which has also been used to prepare an isonitrile functionalised support (reaction (62)) [117].

$$\{Si\}{-}(CH_2)_3NH_2 + CH_3COCHBrCOCH_3$$

$$\longrightarrow \{Si\}{-}CH_2)_3NHCH\underset{\underset{CH_3}{\diagdown C=O}}{\overset{\overset{CH_3}{\diagup C=O}}{}} \qquad (61)$$

$$(EtO)_3Si(CH_2)_3NH_2 \xrightarrow[\text{reflux}]{HCOOEt} (EtO)_3Si(CH_2)_3NHCHO$$

$$\Big\downarrow \begin{array}{l} PPh_3, CCl_4, NEt_3 \\ \text{in } CH_2Cl_2, 65\,^{\circ}C \end{array}$$

$$\{Si\}{-}(CH_2)_3NC \xleftarrow[\text{reflux}]{SiO_2,\text{toluene/}H_2O,} (EtO)_3Si(CH_2)_3NC \qquad (62)$$

A totally different approach to the use of silica as the support is illustrated in reaction (63) in which the functional group, for example a tertiary phosphine, is built into the trialkoxysilane which is then polymerised to produce a non-linear polymer based on an $-Si-O-Si-$ backbone [216].

$$(EtO)_3SiCH_2CH_2PPh_2 + Si(OEt)_4 \xrightarrow[\text{+ trace conc. HCl}]{\text{reflux in HOAc}} \{Si\}{-}CH_2CH_2PPh_2 \quad (63)$$

This reaction demonstrates the limitation of classifying supports as 'organic' or 'inorganic' since it involves an inorganic support prepared from an organic reagent. Another support that is difficult to classify is obtained when silica is covered with a prepolymer of polyphenylsiloxane, which is then chloromethylated and phosphinated to provide an 'organic' support based on an 'inorganic' core [243, 244].

2.6. Other Inorganic Supports

Inorganic materials such as γ-alumina, molecular sieves (zeolites) and glass, although being essentially metal oxides, have hydroxyl groups on the surface that can be used as the point of attachment. Capka has pioneered the use of these materials by attaching groups such as $-PPh_2$, $\geqslant SiCH_2CH_2PPh_2$, $\geqslant SiMe(CH_2)_3CN$, $\geqslant Si(CH_2)_3NMe_2$, $\geqslant Si(CH_2)_3CN$ and $\geqslant Si(CH_2)_2C_5H_4N$ to the surface using reactions (64–66) [245].

$$ \mathfrak{z}{-}OH + PBr_3 \longrightarrow \mathfrak{z}{-}OPBr_2 \xrightarrow{PhMgBr} \mathfrak{z}{-}OPPh_2 \tag{64} $$

($\mathfrak{z}$—OH = 4Å molecular sieve)

$$ \mathfrak{z}{-}OH + CH_2{=}CHSiCl_3 \longrightarrow \mathfrak{z}{-}O\overset{|}{\underset{|}{Si}}CH{=}CH_2 $$

$$ \xrightarrow{HPPh_2 + LiPh} \mathfrak{z}{-}O\overset{|}{\underset{|}{Si}}(CH_2)_2PPh_2 \tag{65} $$

($\mathfrak{z}$—OH = γ-alumina, silica)

$$ \mathfrak{z}{-}OH + X{-}(CH_2)_n{-}SiY_3 \longrightarrow \mathfrak{z}{-}O\overset{|}{\underset{|}{Si}}(CH_2)_nX \tag{66} $$

($\mathfrak{z}$—OH = γ-alumina, silica, glass. $X{-}(CH_2)_n{-}SiY_3$ = $Me_2N(CH_2)_3Si(OEt)_3$, $NC(CH_2)_2SiMe(OEt)_2$, $NC_5H_4(CH_2)_2SiCl_3$, $NC(CH_2)_3SiCl_3$, $NC(CH_2)_3Si(OEt)_3$)

Graphite has been oxidised and then used to support diop. The approach involves the creation of aldehyde functional groups on the surface as in reaction (67) and then functionalisation of these as in reaction (9) [246].

$$ \textcircled{G} \xrightarrow{KMnO_4} \textcircled{G}{-}COOH \xrightarrow{SOCl_2} \textcircled{G}{-}CO\overset{\cdot}{C}l $$

$$ \xrightarrow{HO{-}\langle\!\bigcirc\!\rangle{-}CHO} \textcircled{G}{-}COO{-}\langle\!\bigcirc\!\rangle{-}CHO \tag{67} $$

($\textcircled{G}$ = graphite)

References

1. R. W. Lenz: *Organic Chemistry of Synthetic High Polymers*, Interscience, New York (1967).
2. S. R. Sandler and W. Karo: *Polymer Syntheses*, Academic Press, New York, vol. 1 (1974); vol. 2 (1977); vol. 3 (1980).
3. G. Odian: *Principles of Polymerisation*, Wiley, New York, 2nd Edition (1981).
4. F. W. Billmeyer: *Textbook of Polymer Science*, 2nd Edition, Wiley, London (1971).
5. T. H. Kim and H. F. Rase: *Ind. Eng. Chem., Prod. Res. Dev.* **15**, 249 (1976).
6. W. Heitz: *Adv. Poly. Sci.* **23**, 1 (1977) and refs. therein.
7. D. C. Sherrington in: *Polymer-Supported Reactions in Organic Synthesis*, (Ed. P. Hodge and D. C. Sherrington) Wiley, London (1980) and refs. therein.
8. J. Lieto, D. Milstein, R. L. Albright, J. V. Minkiewicz and B. C. Gates: *ChemTech* **13**, 46 (1983).

9. J. V. Minkiewicz, D. Milstein, J. Lieto, B. C. Gates and R. L. Albright: *Amer. Chem. Soc. Symp. Ser.* **192**, 9 (1982).
10. H. M. Relles and R. W. Schluenz: *J. Amer. Chem. Soc.* **96**, 6469 (1974).
11. G. A. Stark (Ed.): *Biochemical Aspects of Reactions on Solid Supports*, Academic Press, New York, pp. 197–198 (1971).
12. A. J. Chalk: *J. Poly. Sci.* **B 6**, 649 (1968).
13. D. C. Evans, M. H. George and J. A. Barrie: *J. Poly. Sci., Poly. Chem. Ed.* **12**, 247 (1974).
14. N. Plate, M. A. Jampolskaya, S. L. Davydova and V. A. Kargin: *J. Poly. Sci.* **C 22**, 547 (1969).
15. A. T. Bullock, G. G. Cameron and P. M. Smith: *Polymer* **14**, 525 (1974).
16. R. H. Grubbs and S. C. H. Su: *J. Organometal. Chem.* **122**, 151 (1976).
17. M. J. Farrall and J. M. Frechet: *J. Org. Chem.* **41**, 3877 (1976).
18. A. T. Bullock, G. G. Cameron and J. M. Elsom: *Polymer* **18**, 930 (1977).
19. D. Braun: *Makromol. Chem.* **30**, 85 (1959).
20. D. Braun: *Makromol. Chem.* **33**, 181 (1959).
21. D. Braun: *Makromol. Chem.* **44**, 269 (1961).
22. F. C. Leavitt and L. H. Malternas: *J. Poly. Sci.* **45**, 249 (1960).
23. D. C. Evans, L. Phillips, J. A. Barrie and M. H. George: *J. Poly. Sci., Poly. Lett.* **12**, 199 (1974).
24. J. Pannell: *Polymer* **17**, 351 (1976).
25. W. Heitz and R. Michels: *Makromol. Chem.* **148**, 9 (1971).
26. V. Schurig and E. Bayer, *ChemTech.* **6**, 212 (1976).
27. A. T. Bullock, G. G. Cameron and P. M. Smith: *Polymer* **13**, 89 (1972).
28. R. O. C. Norman and R. Taylor: *Electrophilic Substitution in Benzenoid Compounds*, Elsevier, London, pp. 119–134 (1965).
29. A. McKillop and D. Bramley: *Tetrahedron Lett.* 1623 (1969).
30. F. Daly and D. C. Sherrington: unpublished results quoted in [7].
31. E. Seymour and J. M. J. Frechet: *Tetrahedron Lett.* 1149 (1976).
32. R. J. Card and D. C. Neckers: *Inorg. Chem.* **17**, 2345 (1978).
33. G. A. Crosby, N. M. Weinshenker and H. S. Uh: *J. Amer. Chem. Soc.* **97**, 2232 (1975).
34. N. M. Weinshenker, G. A. Crosby and J. Y. Wong: *J. Org. Chem.* **40**, 1966 (1975).
35. F. Camps, J. Castells, M. Ferrando and J. Font: *Tetrahedron Lett.* 1713 (1971).
36. L. T. Scott, J. Rebek, L. Ovsyanko and C. L. Sims: *J. Amer. Chem. Soc.* **99**, 625 (1977).
37. K. W. Pepper, H. M. Paisley and M. A. Young: *J. Chem. Soc.* 4097 (1953).
38. G. D. Jones: *Ind. Eng. Chem.* 44 (1952).
39. R. B. Merrifield: *Fed. Proc.* **21**, 412 (1962).
40. S. K. Freeman, *J. Org. Chem.* **26**, 212 (1960).
41. T. Altares, D. P. Whyman, V. R. Allen and K. Meyersen: *J. Poly. Sci.* **A 3**, 4131 (1965).
42. R. B. Merrifield: *Science* **150**, 178 (1965).
43. B. Green and L. R. Garson: *J. Chem. Soc.* (*C*) 401 (1969).
44. F. Candau and P. Rempp: *Makromol. Chem.* **122**, 15 (1969).
45. Y. A. Ovchinnikov, A. A. Kiryushkin and I. V. Kozhevnikova: *J. Gen. Chem. USSR* **38**, 2546 (1968).
46. R. H. Andreatta and H. Rink: *Helv. Chim. Acta* **56**, 1205 (1973).

47. C. Tamborski, F. E. Ford, W. L. Lehn, G. J. Moore and E. J. Soloski: *J. Org. Chem.* **27**, 619 (1962).
48. G. O. Evans, C. U. Pittman, R. McMillan, R. T. Beach and R. Jones: *J. Organometal. Chem.* **67**, 295 (1974).
49. C. U. Pittman, L. R. Smith and R. M. Hanes: *J. Amer. Chem. Soc.* **97**, 1742 (1975).
50. British Petroleum: *Dutch Patent* 70,06740 (1970).
51. Y. Nonaka, S. Takahashi and N. Hagihara: *Mem. Inst. Sci. Ind. Res., Osaka Univ.* **31**, 23 (1974); *Chem. Abs.* **80**, 140678 (1974).
52. K. G. Allum, R. D. Hancock and P. J. Robinson: *Brit. Patent* 1,277,736 (1972).
53. K. G. Allum, R. D. Hancock, I. V. Howell, R. C. Pitkethly and P. J. Robinson: *J. Organometal. Chem.* **87**, 189 (1975).
54. K. G. Allum, R. D. Hancock and R. C. Pitkethly: *Brit. Patent* 1,295,675 (1972).
55. A. Guyot, C. Graillat and M. Bartholin: *J. Mol. Catal.* **3**, 39 (1977).
56. R. B. King and R. M. Hanes: *J. Org. Chem.* **44**, 1092 (1979).
57. J. P. Collman, L. S. Hegedus, M. P. Cooke, J. R. Norton, G. Dolcetti and D. N. Marquardt: *J. Amer. Chem. Soc.* **94**, 1789 (1972).
58. S. V. McKinley and J. W. Rakshys: *Chem. Commun.* 134 (1972).
59. J. Moreto, J. Albaiges and F. Camps: *Ann. Quim.* **70**, 638 (1974); *Chem. Abs.* **82**, 72485 (1975).
60. H. W. Krause: *React. Kinet. Catal. Lett.* **10**, 243 (1979).
61. H. W. Krause and H. Mix: *East German Patent* 133199 (1978); *Chem. Abs.* **91**, 63336 (1979).
62. K. G. Allum, R. D. Hancock, S. McKenzie and R. C. Pitkethly: *Proc 5th Int. Congr. Catal.* 477 (1972); publ. as *Catalysis* (Ed. J. W. Hightower) North-Holland Publ. Co., Amsterdam, vol. 1 (1973).
63. M. Capka, P. Svoboda, M. Cerny and J. Hetflejs: *Coll. Czech. Chem. Comm.* **38**, 1242 (1973).
64. M. Capka, P. Svoboda, M. Kraus and J. Hetflejs: *Chem. Ind. (London)* 650 (1972).
65. R. H. Grubbs and L. C. Kroll: *J. Amer. Chem. Soc.* **93**, 3062 (1971).
66. R. H. Grubbs, L. C. Kroll and E. M. Sweet: *J. Macromol. Sci.* **7**, 1047 (1973).
67. D. Tatarsky, D. H. Kohn, and M. Cais: *J. Poly. Sci., Poly. Chem. Ed.* **18**, 1387 (1980).
68. P. Svoboda, V. Vaisarova, M. Capka, J. Hetflejs, M. Kraus and V. Bazant: *German Patent* 2,260,260 (1973); *Chem. Abs.* **79**, 78953 (1973).
68a. G. Strukul, M. Bonivento, M. Graziani, E. Cernia and N. Palladino: *Inorg. Chim. Acta* **12**, 15 (1975).
69. T. O. Mitchell and D. D. Whitehurst: *Proc. 3rd North Amer. Conf. Catal. Soc.* paper 27 (1974).
70. S. V. McKinley and J. W. Rakshys: *US Patent* 3,708,462 (1973); *Chem. Abs.* **78**, 85310 (1973).
71. J. M. Brown and H. Molinari: *Tetrahedron Lett.* 2933 (1979).
72. N. A. De Munck, M. W. Verbruggen and J. J. F. Scholten: *J. Mol. Catal.* **10**, 313 (1981).
73. W. Dumont, J. C. Poulin, T. P. Dang and H. B. Kagan: *J. Amer. Chem. Soc.* **95**, 8295 (1973).
74. T. P. Dang and H. B. Kagan: *French Patent* 2,199,756 (1974); *Chem. Abs.* **82**, 22291 (1975).
75. J. C. Poulin, W. Dumont, T. P. Dang and H. B. Kagan: *Compt. Rend. Acad. Sci.* **C277**, 41 (1973).
76. T. Hayashi, N. Nagashima and M. Kumada: *Tetrahedron Lett.* 4623 (1980).

77. I. Tkatchenko: *Compt. Rend. Acad. Sci.* **C282**, 229 (1976).
78. C. U. Pittman and A. Hirao.: *J. Org. Chem.* **43**, 640 (1978).
79. J. R. Millar, D. G. Smith, W. E. Marr and T. R. E. Kressman: *J. Chem. Soc.* 218 (1963).
80. R. W. Wheaton and D. F. Harrington: *Ind. Eng. Chem.* **44**, 1796 (1952).
81. R. L. Letsinger, M. J. Kornet, V. Mahadevan and D. M. Jerina: *J. Amer. Chem. Soc.* **86**, 5163 (1964).
82. J. M. Frechet and C. Schuerch: *J. Amer. Chem, Soc.* **93**, 492 (1971).
83. J. M. Frechet and K. E. Hague: *Macromolecules* **8**, 131 (1975).
84. J. T. Ayres and C. K. Mann: *J. Poly. Sci., Poly. Lett.* **3**, 505 (1965).
85. C. R. Harrison and P. Hodge: *JCS Chem. Commun.* 1009 (1974).
86. C. R. Harrison and P. Hodge: *JCS Parkin I* 605 (1976).
87. T. M. Fyles and C. C. Leznoff: *Can. J. Chem.* **54**, 935 (1976).
88. J. W. Stewart and J. D. Young: *Solid Phase Peptide Synthesis*, W. H. Freeman, San Francisco, pp. 27–28 (1979).
89. M. Tomoi, O. Abe, M. Ikeda, K. Kihara and H. Kakiuchi: *Tetrahedron Lett.* 3031 (1978).
90. A. Warshawsky, R. Kalir, A. Deshe, H. Berkovitz and A. Patchornick: *J. Amer. Chem. Soc.* **101**, 4249 (1979).
91. F. Montanari and P. Tundo: *Tetrahedron Lett.* 5055 (1979).
92. F. Montanari and P. Tundo: *J. Org. Chem.* **46**, 2125 (1981).
93. S. Bhaduri, A. Ghosh and H. Khwaja: *JCS Dalton* 447 (1981).
94. G. L. Linden and M. F. Farona: *J. Catal.* **48**, 284 (1977).
95. A. Ghosh and S. Bhaduri: *Indian Patent* 150,033 (1982); *Chem. Abs.* **98**, 90459 (1983).
96. H. Molinari, F. Montanari and P. Tundo: *JCS Chem. Commun.* 639 (1977).
97. N. L. Holy: *J. Org. Chem.* **43**, 4686 (1978).
98. N. L. Holy in: *Fundamental Research in Homogeneous Catalysis* (Ed. M. Tsutsui) Plenum Press, New York **3**, 691 (1979).
99. E. N. Frankel, J. P. Friedrich, T. R. Bessler, W. F. Kwolek and N. L. Holy: *J. Amer. Oil Chem. Soc.* **57**, 349 (1980).
100. R. B. King and E. M. Sweet: *J. Org. Chem.* **44**, 385 (1979).
101. H. Mueller: *East German Patent* 24,439 (1962).
102. M. Mejstrikova, R. Rericha and M. Kraus: *Coll. Czech. Chem. Comm.* **39**, 135 (1974).
103. R. S. Drago and J. H. Gaul: *JCS Chem. Commun.* 746 (1979).
104. R. S. Drago, J. H. Gaul, A. Zombeck and D. K. Straub: *J. Amer. Chem. Soc.* **102**, 1033 (1980).
105. M. B. Shambhu, M. C. Theodorakis and G. A. Digenis: *J. Poly. Sci., Poly. Chem. Ed.* **15**, 525 (1977).
106. D. M. Dixit and C. C. Leznoff: *Isr. J. Chem.* **17**, 248 (1979).
107. H. C. Meinders, N. Prak and G. Challa: *Makromol. Chem.* **178**, 1019 (1977).
108. R. J. Card and D. C. Neckers: *J. Amer. Chem. Soc.* **99**, 7733 (1977).
109. M. Kaneko, S. Nemoto, A. Yamada and Y. Kurimura: *Inorg. Chim. Acta* **44**, L289 (1980).
110. R. S. Drago, E. D. Nyberg and A. G. El A'mma: *Inorg. Chem.* **20**, 2461 (1981).
111. R. S. Drago and J. H. Gaul: *Inorg. Chem.* **18**, 2019 (1979).
112. F. B. Hulsbergen, J. Manassen, J. Reedijk and J. A. Welleman: *J. Mol. Catal.* **3**, 47 (1977).
113. J. Rebek and F. Gavina: *J. Amer. Chem. Soc.* **97**, 3453 (1975).

114. X. Cochet, A. Mortreux and F. Petit: *Compt. Rend. Acad. Sci.* **C288**, 105 (1979).
115. L. D. Rollmann: *J. Amer. Chem. Soc.* **97**, 2132 (1975).
116. M. Kraus: *Coll. Czech. Chem. Comm.* **39**, 1318 (1974).
117. J. A. S. Howell and M. Berry: *JCS Chem. Commun.* 1039 (1980).
118. W. Tao and X. Hu: *Huaxue Shiji* **4**, 197 (1982); *Chem. Abs.* **98**, 54618 (1983).
119. J. M. Frechet and M. J. Farrall in: *The Chemistry and Properties of Crosslinked Polymers* (Ed. S. S. Labana) Academic Press, New York, p. 59 (1979).
120. J. M. Frechet, M. D. de Smet and M. J. Farrall: *Polymer* **20**, 675 (1979).
121. G. A. Crosby: *US Patent* 3,928,293 (1975); *Chem. Abs.* **84**, 106499 (1976).
122. K. M. Webber, B. C. Gates and W. Drenth: *J. Mol. Catal.* **3**, 1 (1977).
123. W. D. Bonds, C. H. Brubaker, E. S. Chandrasekaran, C. Gibbons, R. H. Grubbs and L. C. Kroll: *J. Amer. Chem. Soc.* **97**, 2128 (1975).
124. J. E. Frommer and R. G. Bergman: *J. Amer. Chem. Soc.* **102**, 5227 (1980).
125. E. S. Chandrasekaran, R. H. Grubbs and C. H. Brubaker: *J. Organometal. Chem.* **120**, 49 (1976).
126. E. S. Chandrasekaran: *Diss. Abs.* **B36**, 6143 (1976).
127. E. S. Chandrasekaran, D. A. Thompson and R. W. Rudolph: *Inorg. Chem.* **17**, 760 (1978).
128. B. A. Sosinsky, W. C. Kalb, R. A. Grey, V. A. Uski and M. F. Hawthorne: *J. Amer. Chem. Soc.* **99**, 6768 (1977).
129. A. Sekiya and J. K. Stille: *J. Amer. Chem. Soc.* **103**, 5096 (1981).
130. D. C. Sherrington, D. J. Craig, J. Dalgleish, G. Domin, J. Taylor and G. U. Meehan: *Europ. Poly. J.* **13**, 73 (1977).
131. J. Manassen: *Israeli Patent* 30505, quoted in J. Manassen: *Plat. Met. Rev.* **15**, 142 (1971).
132. R. Rabinowitz, R. Marcus and J. Pellon: *J. Poly. Sci.* A2, 1241 (1964).
133. R. Rabinowitz and R. Marcus: *J. Org. Chem.* **26**, 4157 (1961).
134. J. Lieto, J. J. Rafalko and B. C. Gates: *J. Catal.* **62**, 149 (1980).
135. W. M. McKenzie and D. C. Sherrington: *J. Poly. Sci., Poly. Chem. Ed.* **20**, 431 (1982).
136. F. Camps, J. Castells, J. Font and F. Velo: *Tetrahedron Lett.* 1715 (1971).
137. J. A. Greig and D. C. Sherrington: *Polymer* **19**, 163 (1978).
138. W. Heitz and K. L. Platt: *Makromol. Chem.* **127**, 113 (1969).
139. R. Arshady, G. W. Kenner and A. Ledwith: *J. Poly. Sci., Poly. Chem. Ed.* **12**, 2017 (1974).
140. C. U. Pittman, L. R. Smith and S. E. Jacobson in: *Catalysis: Heterogeneous and Homogeneous* (Ed. B. Delmon and G. Jannes) Elsevier, Amsterdam, p. 393 (1975).
141. G. E. Ham in: *High Polymers* vol. XVIII, Interscience, New York, Chapter 1 (1964).
142. R. M. Fuoss and G. I. Cathers: *J. Poly. Sci.* **4**, 97 (1949).
143. T. Tamikado: *J. Poly. Sci.* **43**, 489 (1960).
144. K. G. Allum and R. D. Hancock: *Brit. Patent* 1,287,566 (1972).
145. K. Achiwa: *Chem. Lett.* 905 (1978).
146. K. Achiwa: *Heterocycles* **9**, 1539 (1978).
147. V. S. Aliev, S. M. Aliev, A. G. Azizov, B. B. Akhmedov, K. V. Vyrshchikov and G. A. Mamedaliev: *USSR Patent* 687,083 (1979); *Chem. Abs.* **91** 194375 (1979).
148. H. Arai: *J. Catal.* **51**, 135 (1978).
149. L. Verdet and J. K. Stille: *Organometallics* **1**, 380 (1982).
150. N. P. Allen, R. P. Burns, J. Dwyer and C. A. McAuliffe: *J. Mol. Catal.* **3**, 325 (1978).
151. N. P. Allen, J. Dwyer and C. A. McAuliffe: *17th Int. Coord. Chem. Conf., Hamburg* 34 (1976).

152. K. A. Abdulla, N. P. Allen, A. N. Badrun, R. P. Burns, J. Dwyer, C. A. McAuliffe and N. D. A. Toma: *Chem. Ind.* 273 (1976).
153. I. Tkatchenko: *Belgian Patent* 831,084 (1974); *Chem. Abs.* **85**, 94934 (1976).
154. K. G. Allum and R. D. Hancock: *Brit. Patent* 1,291,237 (1972).
155. K. G. Allum and R. D. Hancock: *Brit. Patent* 1,277,737 (1972).
156. K. G. Allum and R. D. Hancock: *Canadian Patent* 903,950 (1972).
157. T. H. Kim: *Diss. Abs.* **B37**, 356 (1976).
158. J. Kiji, S. Kadoi and J. Furukawa: *Angew. Makromol. Chem.* **46**, 163 (1975).
159. J. Kiji and M. Inaba: *Angew. Makromol. Chem.* **65**, 235 (1977).
160. M. E. Wilson, R. D. Nuzzo and G. M. Whitesides: *J. Amer. Chem. Soc.* **100**, 2269 (1978).
161. A. K. Smith: *12th Sheffield-Leeds Int. Symp. on Organometal., Inorg. and Catal. Chem.* (1983).
162. N. Takaishi, H. Imai, C. A. Bertelo and J. K. Stille: *J. Amer. Chem. Soc.* **98**, 5400 (1976).
163. N. Takaishi, H. Imai, C. A. Bertelo and J. K. Stille: *J. Amer. Chem. Soc.* **100**, 264 (1978).
164. T. Masuda and J. K. Stille: *J. Amer. Chem. Soc.* **100**, 268 (1978).
165. A. A. Efendiev, T. N. Shakhtakhtinsky, L. F. Mustafaeva and H. L. Shick: *Ind. Eng. Chem. Prod. Res. Dev.* **19**, 75 (1980).
166. H. R. Allcock, K. D. Lavin, N. M. Tollefson and T. L. Evans: *Organometallics* **2**, 267 (1968).
167. M. Chanda, K. F. O'Driscoll and G. L. Rempel: *J. Mol. Catal.* **12**, 197 (1981).
168. A. J. Moffat: *J. Catal.* **18**, 193 (1970).
169. A. J. Moffat: *J. Catal.* **19**, 322 (1970).
170. R. S. Drago, E. D. Nyberg, A. G. El A'mma and A. Zombeck: *Inorg. Chem.* **20**, 2461 (1981).
171. F. R. Hartley and D. J. A. McCaffrey in: *Fundamental Research in Homogeneous Catalysis* (Ed. M. Tsutsui) Plenum, New York **3**, 707 (1979).
172. F. R. Hartley, D. J. A. McCaffrey, S. G. Murray and P. N. Nicholson: *J. Organometal. Chem.* **206**, 347 (1981).
173. A. Charlesby: *Atomic Radiation and Polymers*, Pergamon, Oxford (1960).
174. A. Chapiro: *Radiation Chemistry of Polymeric Systems*, vol. XV of *High Polymers*, Interscience, New York (1962).
175. H. A. J. Battaerd and G. W. Tregear: *Graft Copolymers*, Interscience, New York (1967).
176. D. A. Kritskaya, A. N. Ponomarev, A. D. Pomogailo and F. S. Dyachovski: *J. Poly. Sci., Poly. Symp.* **68**, 23 (1980).
177. F. R. Hartley, S. G. Murray and P. N. Nicholson: *Brit. Patent Appl.* 8028823 (1980).
178. D. A. Kritskaya, A. D. Pomogailo, A. N. Ponomarev and F. S. Dyaschkovskii: *J. Appl. Poly. Sci.* **25**, 349 (1980).
179. D. A. Kritskaya, A. D. Pomogailo, A. N. Ponomarev and F. S. Dyaschkovskii: *J. Pure Appl. Poly. Sci.* 17 (1978).
180. Z. S. Kiyashkina, A. D. Pomogailo, A. I. Kuzaev, G. V. Lagodzinskaya and F. S. Dyachkovskii: *J. Poly. Sci., Poly. Symp.* **68**, 13 (1980).
181. H. Barker, J. L. Garnett, R. S. Kenyon, R. Levot, M. S. Liddy and M. A. Long: *Proc. 6th Int. Congr. Catal.* 551 (1976).
182. H. Barker, J. L. Garnett, R. Levot and M. A. Long: *J. Macromol. Sci.* **A12**, 261 (1978).

183. Unisearch: *Brit. Patent* 1,519,462 (1978).

184. J. L. Garnett, R. G. Levot and M. A. Long: *European Patent Appl.* 32455 (1981).

185. F. R. Hartley, S. G. Murray and P. N. Nicholson: *J. Poly. Sci., Poly. Chem. Ed.* **20**, 2395 (1982).

186. J. L. Garnett, R. S. Kenyon and M. J. Liddy: *JCS Chem. Commun.* 735 (1974).

187. C. H. Ang, J. L. Garnett, R. Levot, M. A. Long, N. T. Yen and K. J. Nicol: *Stud. Surf. Sci. Catal.* **B7**, 953 (1981).

188. F. R. Hartley, S. G. Murray and P. N. Nicholson: *J. Mol. Catal.* **16**, 363 (1982).

189. A. Rembaum, A. Gupta and W. Volksen: *US Patent* 4,170,685 (1979); *Chem. Abs.* **92**, 7379 (1980).

190. S. Dilli and J. L. Garnett: *Aust. J. Chem.* **24**, 981 (1971).

191. J. E. Wilson: *Radiation Chemistry of Monomers, Polymers and Plastics*, Marcel Dekker, New York, p. 482 (1974).

192. D. S. Ballantine, A. Glines, D. J. Metz, J. Behr, R. B. Mesrobian and A. J. Restaino: *J. Poly. Sci.* **19**, 219 (1956).

193. G. Odian and R. L. Kruse: *Poly. Preprint Amer. Chem. Soc., Div. Poly. Chem.* **9**, 668 (1968).

193a. L. B. Adams, P. J. Fydelor, G. Partridge and R. H. West in: *Power Sources – 4* (Ed. D. H. Collins) Oriel Press, Newcastle-Upon-Tyne, p. 141 (1973).

194. S. Dilli, J. L. Garnett, E. C. Martin and P. H. Phuoc: *J. Poly. Sci., Poly. Symp.* **37**, 59 (1972).

195. E. Trommsdorff, H. Kohle and P. Lagally: *Makromol. Chem.* **1**, 169 (1948).

196. J. L. Garnett and N. T. Yen: *Aust. J. Chem.* **32**, 585 (1979).

197. J. L. Garnett and N. T. Yen: *J. Poly. Sci., Poly. Lett.* **12**, 225 (1974).

198. J. L. Garnett, D. H. Phuoc, P. L. Airey and D. F. Sangster: *Aust. J. Chem.* **29**, 1459 (1976).

199. D. M. Pinkerton and R. H. Stacewicz: *J. Poly. Sci., Poly. Lett.* **14**, 287 (1976).

200. A. Guyot and M. Bartholin: *Inf. Chim.* **166**, 221–4 and 227–8 (1977); *Chem. Abs.* **87**, 91256 (1977).

201. D. C. Locke: *J. Chromatogr. Sci.* **11**, 120 (1973).

202. W. R. Supina in: *The Packed Column in Gas Chromatography* Supelco, Bellefonte, pp. 106–7 (1974).

203. P. H. Bach: *Lab. Practice* **24**, 817 (1975).

204. K. G. Allum, R. D. Hancock, I. V. Howell, S. McKenzie, R. C. Pitkethly and P. J. Robinson: *J. Organometal. Chem.* **87**, 203 (1975).

205. K. G. Allum, S. McKenzie and R. C. Pitkethly: *US Patent* 3,726,809 (1973).

206. A. A. Oswald, L. L. Murrell and L. J. Boucher: *Abstr. 167th Amer. Chem. Soc. Meeting*, PETR 34 and PETR 35 (1974).

207. K. G. Allum, S. McKenzie and R. C. Pitkethly: *German Patent* 2,062,351 (1971); *Chem. Abs.* **75**, 122751 (1971).

208. S. Shinoda, K. Nakamura and Y. Saito: *J. Mol. Catal.* **17**, 77 (1982).

209. V. M. Vdovin, V. E. Fedorev, N. A. Pritula and G. K. Fedorova: *Izv. Akad. Nauk SSSR, Ser. Khim.* 2663 (1981).

210. L. L. Murrell in: *Advanced Materials in Catalysis* (Ed. J. J. Burton and R. L. Garten) Academic Press, New York, chapter 8 (1977).

211. S. C. Brown and J. Evans: *JCS Chem. Commun.* 1063 (1978).

212. J. Pelz, K. Unverferth and K. Schwetlick: *Z. Chem.* **14**, 370 (1974).

213. M. Capka: *Coll. Czech. Chem. Comm.* **42**, 3410 (1977).

214. M. H. J. M. De Croon and J. W. E. Coenen: *J. Mol. Catal.* **11**, 301 (1981).

215. K. Schweltlick, J. Pelz and K. Unverferth: *Proc. 16th Int. Coord. Chem. Conf.,* Dublin, p. 4.4 (1974).
216. F. G. Young: *German Patent* 2,330, 308 (1974); *Chem. Abs.* 80, 121737 (1974).
217. J. M. Moreto, J. Albaiges and P. Camps in: *Catalysis: Homogeneous and Heterogeneous* (Ed. B. Delmon and G. Jannes) Elsevier, Amsterdam, 339 (1975).
218. M. Cerny: *Coll. Czech. Chem. Comm.* 42, 3069 (1977).
219. I. Kolb, M. Cerny and J. Hetflejs: *React. Kinet. Catal. Lett.* 7, 199 (1977).
220. M. K. Neuberg: *Diss. Abs.* B39, 5929 (1979).
221. V. A. Semikolenov, D. Mikhailova and J. W. Sobczak: *React. Kinet. Catal. Lett.* 10, 105 (1979).
222. V. M. Akhmedov, F. R. Sultanova, F. F. Kurbanova, V. S. Aliev and R. Kh. Mamedov: *USSR Patent* 810,261 (1981); *Chem. Abs.* 94, 181541 (1981).
223. V. K. Dudchenko, Z. N. Polyakov, V. A. Zakharov, Yu. I. Yermakov and A. I. Min'kov: *USSR Patent* 468,503 (1978); *Chem. Abs.* 88, 90283 (1978).
224. R. J. Uriarte and D. W. Meek: *Inorg. Chim. Acta* 44, L283 (1980).
225. S. Lamalle, H. Mortreux, M. Evrard, F. Petit, J. Grimblot and J. P. Bonnelle: *J. Mol. Catal.* 6, 11 (1979).
226. O.-E. Brust, I. Sebastian and I. Halasz: *J. Chromatog.* 83, 15 (1973).
227. C. R. Hastings, W. A. Aue and F. N. Larsen: *J. Chromatog.* 60, 329 (1971).
228. J. J. Kirkland and J. J. Destefano: *J. Chromatog. Sci.* 8, 309 (1970).
229. D. C. Locke, J. T. Schmermund and B. Banner: *Anal. Chem.* 44, 90 (1972).
230. T. G. Waddell, D. E. Leyden and M. T. De Bello: *J. Amer. Chem. Soc.* 103, 5303 (1981).
231. V. Z. Sharf, A. S. Gurovets, L. P. Finn, I. B. Slinyakova, V. N. Krutii and L. Kh. Freidlin: *Izv. Akad. Nauk SSSR* 104 (1979).
232. K. Tanaka, S. Shinoda and Y. Saito: *Chem. Lett.* 179 (1979).
233. S. Shinoda and Y. Saito: *Inorg. Chim. Acta* 63, 23 (1982).
234. T. Catrillo, H. Knözinger and M. Wolf: *Inorg. Chim. Acta* 45, L235 (1980).
235. F. R. W. P. Wild, G. Gubitosa and H. H. Brintzinger: *J. Organometal. Chem.* 148, 73 (1978).
236. W. Parr and M. Novotny: *Bonded Stationary Phases in Chromatography* (Ed. E. Grushka), Ann Arbor Science, Michigan, 173 (1974).
237. B. Marciniec, Z. W. Kornetka and W. Urbaniak: *J. Mol. Catal.* 12, 221 (1981).
238. G. D. Shields and L. J. Boucher: *J. Inorg. Nucl. Chem.* 40, 1341 (1978).
239. E. D. Nyberg and R. S. Drago: *J. Amer. Chem. Soc.* 103, 4966 (1981).
240. V. A. Semikolenov, V. A. Likholobov and Yu. I. Yermakov: *Kinet. Katal.* 20, 269 (1979).
241. V. A. Semikolenov, V. A. Likholobov and Yu. I. Yermakov: *Kinet. Katal.* 18, 1294 (1977).
242. V. V. Skopenko, T. P. Lishko, T. A. Sukhan and A. K. Trofimchuk: *Ukr. Khim. Zh.* 46, 1028 (1980); *Chem; Abs.* 94, 10560 (1981).
243. J. Conan, M. Bartholin and A. Guyot: *J. Mol. Catal.* 1, 375 (1976).
244. M. Bartholin, J. Conan and A. Guyot: *J. Mol. Catal.* 2, 307 (1977).
245. M. Capka and J. Hetflejs: *Coll. Czech. Chem. Commun.* 39, 154 (1974).
246. H. B. Kagan, T. Yamagishi, J. C. Motte and R. Setton: *Isr. J. Chem.* 17, 274 (1979).

CHAPTER 3

INTRODUCTION OF METALS ONTO SUPPORTS

In this chapter we shall be concerned with the introduction of the catalytically active metal site on to the support. Whilst the first supported catalysts involved ion-exchange resins in which a cationic metal complex ion was linked electrostatically to the support, the majority of supported metal complex catalysts involve covalent bonds between the metal and the support. Accordingly, after a brief look at ion-exchange based catalysts, this chapter will consider the coordination of metal complexes by functionalised supports, then the polymerisation of functionalised monomers, before examining the reactions between organometallic and metal-carbonyl complexes with metal oxide supports. The chapter concludes with a brief consideration of supported Ziegler–Natta catalysts, surface supported metal salts and finally complexes formed by transition metal oxides with main Group metal oxides.

3.1. Ion-Exchange-Based Catalysts

The subject of transition metal complex catalysts really began when Haag and Whitehurst [1–6] described the use of sulphonated cross-linked polystyrenes into which cationic $[Pd(NH_3)_4]^{2+}$ had been introduced by ion-exchange [7]. Similarly quaternary ammonium functionalised anion-exchange resins have been used to support $[PdCl_4]^{2-}$, $[PtCl_4]^{2-}$ and $[RhCl_6]^{3-}$ [8–10]. The fraction of hydroxide ions on the resin displaced by $[PdCl_4]^{2-}$ from aqueous $K_2[PdCl_4]$ is generally very small (<5%); it increases with resin reticulation and palladium (II) concentration. The exchange is largely limited to the surface layers of the resin beads and quickly leads to a false equilibrium [10]. $(NHEt_3)^+[Fe_3H(CO)_{11}]^-$ has been supported on polystyrene-divinylbenzene resins functionalised with $-CH_2NMe_3^+Cl^-$ and on silica functionalised with $-(CH_2)_3NEt_3^+Cl^-$ [11]. Quaternary ammonium tungstate on Amberlite IRA 400, prepared by ion-exchange (reaction (1)), is effective in the epoxidation of maleic acid by hydrogen peroxide [12].

$$\text{ⓅNR}_3^+\text{Cl}^- + \text{NaHWO}_4 \longrightarrow \text{ⓅNR}_3^+[\text{HWO}_4]^- + \text{NaCl} \qquad (1)$$

Macroporous strong acid ion-exchange resins have been treated with functionalised tertiary phosphines to produce supports which react as nitrogen
donors to a range of cobalt, rhodium and platinum complexes (reaction (2))
[13]. The resulting supported complexes in which ^{31}P NMR and ESCA showed
nitrogen-metal bonding only were active hydroformylation catalysts although
some rhodium leaching occurred.

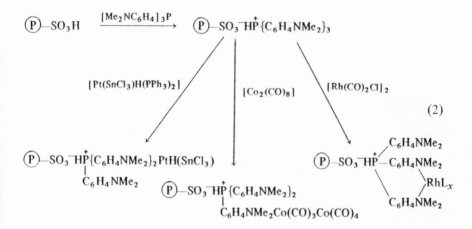

$$(2)$$

3.2. Functionalised Supports

By far the commonest functionalised supports are phosphinated supports and
so this section inevitably draws many of its examples from them. However
other functional groups including nitrogen, oxygen and sulphur donors are also
included. There are broadly three ways of introducing metal complexes on to
functionalised supports: direct reaction of a metal salt with the support, ligand
displacement from the metal complex and bridge splitting of a binuclear metal
complex.

3.2.1. DIRECT REACTION OF A METAL SALT WITH THE SUPPORT

The direct reation of a metal halide with a functionalised support has been used
to prepare supported metal complexes on a number of occasions (reactions (3–
6)). The conditions used are essentially those that would have been used had the
ligands been simple rather than polymeric. However in some cases such as the
interaction of rhodium(III) with polymeric amine supports [15] the products
have not been fully characterised; in other cases such as supported anthranilic
acid the RhCl$_3$ reacts with the support to yield a supported rhodium(I) complex

[16]. Metal complexes of supported porphyrins have been prepared by direct reaction of a metal salt such as cobalt(II) acetate with the supported porphyrin [19, 20].

$$MCl_2 \quad + 2 \,(P)\!\!-\!\!PPh_2 \xrightarrow[\text{(M = Co, Ni) [14]}]{\text{reflux, thf}} \; [(\,(P)\!\!-\!\!PPh_2)_2 MCl_2] \qquad (3)$$

(anhydrous)

$$2 \,(P)\!\!-\!\!CH_2 CN + PdCl_2 \xrightarrow[\substack{\text{PdCl}_2 \text{ down column} \\ \text{of polymer [17]}}]{\text{pass aq. soln. of}} \; [(\,(P)\!\!-\!\!CH_2CN)_2PdCl_2] \qquad (4)$$

$$2 \,(P)\!\!-\!\!PPh_2 + RuCl_3 \xrightarrow{\text{dma}} \; [(\,(P)\!\!-\!\!PPh_2)_2 RuCl_3] \qquad (5)$$

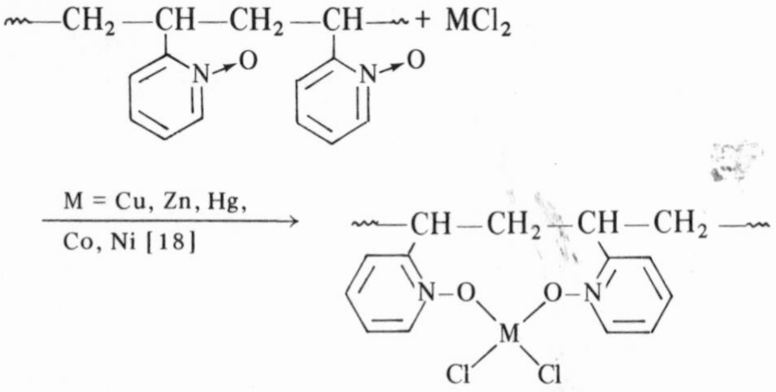

(6)

3.2.2. DISPLACEMENT OF A LIGAND FROM THE METAL COMPLEX

By far the commonest route for preparing supported complex catalysts is to displace a ligand already coordinated to the metal complex by a ligand from the support. Some examples of this are given in reactions (7–12), where it should be appreciated that some of the products are idealised in that the reactivities of donor groups vary according to their position within the polymer.

$$[Rh(acac)(CO)_2] + (P)\!\!-\!\!PPh_2$$

$$\xrightarrow[\text{8 hrs. [14, 21–23]}]{\text{heptane, 20 °C,}} \; [(\,(P)\!\!-\!\!PPh_2)Rh(acac)(CO)] + CO \qquad (7)$$

$$[Rh(acac)nbd] + 2 \,(P)\!\!-\!\!PRR'$$

$$\xrightarrow[\text{70\% HClO}_4 \text{ in thf [24]}]{\text{thf, 20 °C then}} \; [(\,(P)\!\!-\!\!PRR')_2 Rh(nbd)] + acacH \qquad (8)$$

$[(PPh_3)_3RhCl] + (P)-PPh_2$

$$\xrightarrow[\text{2--4 weeks } [25-27]]{\text{benzene, 20 °C}} [((P)-PPh_2)Rh(PPh_3)_2Cl] + PPh_3 \qquad (9)$$

$2 (P)-PPh_2 + [(PhCN)_2MCl_2]$

$$\xrightarrow[\text{60 hrs. (M = Pd, Pt) } [28-30]]{\text{reflux in acetone}} \textit{trans-}[((P)-PPh_2)_2MCl_2] + 2PhCN \qquad (10)$$

$(P)-PPh_2 + [Pd(PPh_3)_4] \xrightarrow[\text{17 hrs. } [29]]{\text{benzene, 25 °C,}} [((P)-PPh_2)Pd(PPh_3)_3] + PPh_3 \quad (11)$

Thus, for example, where chelation is shown not every metal atom may be chelated. It should however be noted that on occasions it has proved possible to influence the presence or absence of chelation by altering the metal complex. Thus treatment of phosphinated polystyrene with $[M(PPh_3)_2(CO)Cl]$, where M = Rh and Ir, always led to two equivalents of triphenylphosphine being displaced; this was shown by recovering the triphenylphosphine from the effluent [33]. Monodentate bonding could be obtained using $[Rh(cod)Cl]_2$ [33]. A somewhat similar situation applies when nickel(0) carbonyl is supported on phosphinated silicas [34]. Thus phosphinated silica reacts with $[Ni(CO)_4]$ to from $[(\{Si\}-(CH_2)_2PPh_2)Ni(CO)_3]$ only, even at the highest ratios of P : Ni. $[\{\{Si\}-(CH_2)_2PPh_2\}_2Ni(CO)_2]$ could only be prepared by preforming $[\{(EtO)_3Si(CH_2)_2PPh_2\}_2Ni(CO)_2]$ and reacting this with silica [34].

In ligand displacement reactions the ligand that is displaced may either have the same donor atom as that on the support, as in reaction (9), or may be different. Although many workers have used the equilibration of a phosphine complex with a phosphinated support, as in reaction (9), there are serious drawbacks to this technique. The original phosphine ligand is not normally volatile and therefore it is difficult to completely remove all the liberated phosphine ligands from the support. Within the support they inevitably act as competitors for the anchored phosphine ligands and as such facilitate leaching of the metal from the support. For this reason a number of groups, including our own, have now turned to displacing volatile ligands which are readily removed from the supports as in reaction (7) [14, 21–23a]. Such ligands can be displaced both thermally and also photochemically as in reaction (12) [31].

A study of the kinetics of the interaction of phosphinated polystyrene with a series of rhodium(I) and iridium(I) complexes [35] showed that the reaction mechanism divides into two distinct parts: (i) mass transfer of the reagents into the polymer, followed by (ii) reaction with the polymer. This approach enables what were otherwise apparently random reactivities to be understood, since these two parts of the reaction will vary in rate as the nature of the metal complex is varied. Thus for a series of rhodium and iridium complexes reacted with the same phosphinated polystyrene the percentage of phosphine sites on the resin that complexed the metal depended on the complex used in the order $[Rh(cod)Cl]_2$ (2.4%) < $[Ir(PPh_3)_2(CO)Cl]$ (10.7%) < $[Rh(PPh_3)_2(CO)Cl]$ (13.3%) < $[Rh(PPh_3)_3Cl]$ (16.6%) < $[Rh(CO)_2Cl]_2$ (99%) [35]. The reaction of $[Rh(PPh_3)_3Cl]$ with phosphinated polystyrene showed no retardation effect arising from the displaced triphenylphosphine [36]. This indicates that the mechanism is associative and that $[Rh(PPh_3)_2Cl]$, which might have been expected to be the active intermediate, is not involved.

The distribution of the metal complex on the support can be varied by altering the reaction conditions. Thus if very short reaction times are used with an excess of the supported ligand present, then it is possible to achieve a high concentration of the metal complex on the outer regions of the support and very low concentrations inside [37]. This is good for promoting high activity, but prevents diffusion within the support being used to enhance specificity. Photochemical reactions can be used to favour surface distribution of the metal complex. However it must be borne in mind that under many catalytic conditions there is a redistribution of the metal complex within the support during the catalytic run; this is particularly true of phosphine supported complexes.

Although a phosphinated support can always act as a multidentate ligand providing there are sufficient phosphine sites in the correct positions relative to one another, a number of supported complexes have been prepared using bi- or multi-dentate phosphines. Thus diphos-substituted polystyrene resins have been used to support both rhodium and iron complexes (reaction (13)) [38–40], as well as both mono- and poly-nuclear palladium(II) complexes [41]. A number of multidentate nitrogen ligands have been used as supports. These

$$\text{(P)}-\!\!\!\left\langle\;\right\rangle\!\!\!-\text{PPh}\quad\text{PPh}_2 \xrightarrow[\;[Fe(CO)_5]\;]{[Rh(PPh_3)_3(CO)H]} \begin{bmatrix}\text{(P)}-\!\!\!\left\langle\;\right\rangle\!\!\!-\text{PPh}\quad\text{PPh}_2\quad\quad\text{PPh}_3\\ \quad\quad\quad\quad\quad\quad\;Rh\\ \quad\quad\quad\quad\;H\quad\;\overset{|}{CO}\quad\text{PPh}_3\end{bmatrix} + 2\text{PPh}_3$$

$$\begin{bmatrix}\text{(P)}-\!\!\!\left\langle\;\right\rangle\!\!\!-\text{PPh}\quad\text{PPh}_2\\ \quad\quad\quad\quad Fe(CO)_3\end{bmatrix} + \begin{bmatrix}\text{(P)}-\!\!\!\left\langle\;\right\rangle\!\!\!-\text{PPh}\quad\text{PPh}_2\\ \quad\quad\quad\quad Fe(CO)_4\;\;Fe(CO)_4\end{bmatrix}$$

(13)

include polstyrene supported bipyridine [42–45]. Schiff bases [46], poly-4-vinylpyridine [47–55] and poly-2-vinylpyridine [56, 57]. Poly-4-vinylpyridine resins can be cross-linked with 1,4-dibromobutane (reaction (14)) which leads to

$$\text{(pyridine resin)} + Br(CH_2)_4Br \longrightarrow \text{(cross-linked N}^+Br^- \text{ resin, (CH}_2)_4\text{ bridge)}$$

(14)

polymers that bind copper(II), for example, more tightly that the non-cross-linked polymer; however the metal ion uptake is reduced due to the positive charge introduced on to the resin by this method of cross-linking [48].

In addition to mononuclear metal complexes, polynuclear complexes have also been supported using simple ligand displacement reactions [58–59a]. Thus clusters such as $[Fe_2Pt(CO)_{10}]$ [60], $[H_4Ru_4(CO)_{12}]$ [60, 61], $[RuPt_2(CO)_8]$ [61], $[H_2Os_3(CO)_{10}]$ [59], $[HAuOs_3(CO)_{11}]$ [61], $[Rh_6(CO)_{16}]$ [62–64] and $[Ir_4(CO)_{12}]$ [65, 66] have been supported on phosphinated polymers and phosphinated silica. The number of carbonyl ligands displaced by phosphine groups depends on the metal complex and the support; thus $[Rh_6(CO)_{16}]$ can be linked by up to three phosphine ligands [64], whereas $[Ir_4(CO)_{12}]$ attaches itself to only one phosphine group on phosphinated silica [66]. $[Fe_4(\mu_3\text{-}S_4)\text{-}(SR)_4]^{2-}$, where R = t-Bu and Bz have been supported on polystyrene functionalised with $-CH_2SH$ groups. The structures of the clusters are preserved but it is not known what proportion of the original alkylsulphide ligands are replaced by the supported sulphur donors [67].

Commercial polystyrene resins containing dimethylbenzylamino groups have been used to anchor $[Rh_4(CO)_{12}]$ and $[Rh_6(CO)_{16}]$ [68–69]. The resulting rhodium cluster anions undergo a variety of equilibrium reactions on the resin, depending on the exact reaction conditions [68–70]. Polyacrylic acid, alternate copolymers of maleic acid with vinylic monomers and poly-β-diketones react with ruthenium and rhodium phosphine complexes to give carboxylate or β-diketonate supported complexes by phosphine displacement [70a].

3.2.3 BRIDGE SPLITTING REACTIONS

The third basic method for introducing metal complexes onto functionalised supports is to use a bridge splitting reaction. Thus chloro-bridged dimeric rhodium(I) complexes such as $[Rh(CO)_2Cl]_2$ [70–73] and $[Rh(cod)Cl]_2$ [33] react with polymeric resins to give monomeric polymer-bound complexes (reactions (15) and (16)) with phosphine and amine supports. However with thiol supports the bridge is not split, the thiol group being deprotonated and then binding in the bridge position [72] (reaction (17)).

$$[Rh(CO)_2Cl]_2 + 2\,\text{(P)}\!-\!Y \xrightarrow[\text{(Y = NR}_2 \text{ or PPh}_2)]{\text{reflux, benzene, 6 hrs.}} 2\,cis\text{-}[(\text{(P)}\!-\!Y)Rh(CO)_2Cl] \quad (15)$$

$$(16)$$

$$[Rh(CO)_2Cl]_2 + \text{(P)}\!-\!SH \longrightarrow \quad + HCl \quad (17)$$

Although, at present bridge-splitting reactions have only been used in the preparation of rhodium(I) supported catalysts, they clearly have a greater potential than this and are likely to be used for other d^8 transition metal complexes in the future.

3.3. Metal Complexes Bound to Polymeric Supports Through Metal-Carbon Bonds

A number of metal complexes have been supported on materials containing cyclopentadienyl side-chains. The preparation of the supports has already been described in Section 2.2.1.d. The metal complexes can either be linked by reaction with a supported metal (e.g. lithium) cyclopentadienide, in which case a metal chloride is usually formed (reaction (18)) [74–76] or by an oxidative-addition reaction with the neutral cyclopentadiene linked to the support, in which case supported hydrido-complexes may be formed (reaction (19)) [74, 77, 78]. The lithium cyclopentadienide route of reaction (18) cannot be used with rhodium trichloride because of the water present, so that reaction (19) must be used [78a].

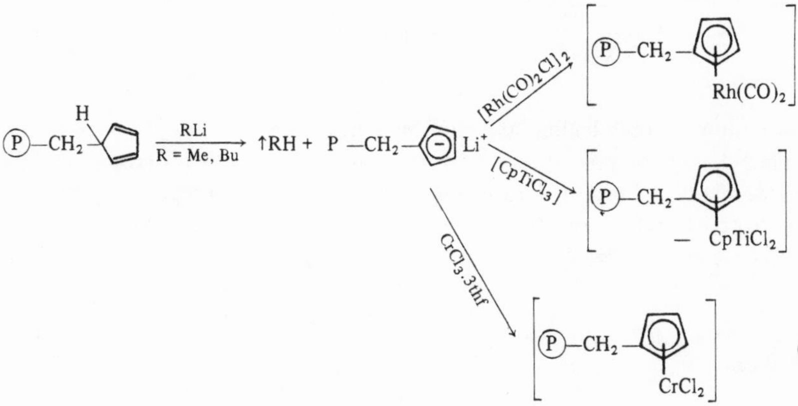

(18)

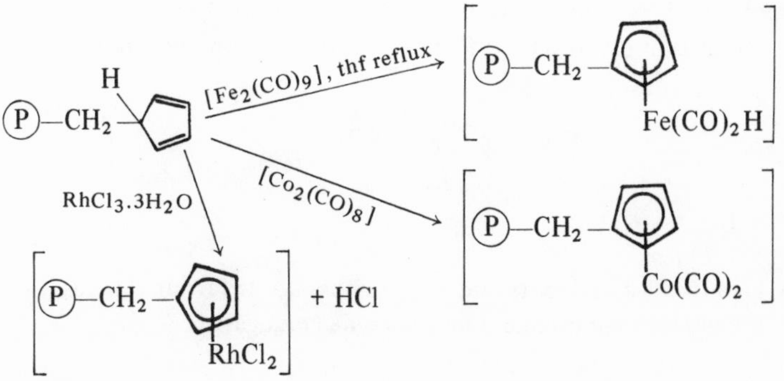

(19)

Oxidative-addition reactions provide a very convenient route for the formation of supported complexes linked directly to the backbone of the polymer. This approach has been used to prepare a number of supported nickel(II), palladium(II) and platinum(II) complexes (reactions (20) and (21)) [79–83], sometimes, as in reaction (21), generating the active metal(0) species by reductive-elimination of two of the original ligands during the reaction [83].

$$(P)-\langle\quad\rangle-X + [M(PPh_3)_4]$$

$$\xrightarrow[M = Ni, Pd, Pt]{X = Br, I} \quad [(P)-\langle\quad\rangle-M(PPh_3)_2 X] + 2PPh_3 \tag{20}$$

$$(P)-\langle\quad\rangle-Cl + [(bipyr)Ni(Et)_2]$$

$$\longrightarrow \quad [(P)-\langle\quad\rangle-Ni(bipyr)Cl] + C_4H_{10} \tag{21}$$

Chromium, molybdenum and tungsten hexacarbonyls react directly with the phenyl rings of polystyrene to form supported complexes (reaction (22)) [84]. Similarly a supported ruthenium catalyst has been prepared by reacting [Ru(cod)(cot)] with polystyrene to give a product in which both 1,5-cyclooctadiene and 1,3,5-cyclooctatriene have been displaced [85]. Another example

$$(P)-\langle\quad\rangle + [M(CO_6] \xrightarrow{M = Cr, Mo, W} \left[(P)-\langle\underset{M(CO)_3}{\bigcirc}\rangle\right] + 3\,CO \tag{22}$$

of direct complexation of a metal by the polymer involves the reaction of 1,4-polybutadiene with iron carbonyls to form (η^4-diene)tricarbonyliron units with

1

the polymer **1** [86]. Basic solvents and high temperatures favour this reaction, in which the initially isolated double bonds become conjugated.

Metal-carbon bond formation has been used to link metal carbonyl anions to polystyrene by treating chloromethylated polystyrene with the sodium salt of a carbonyl anion (reaction (23)) [87–89].

$$(23)$$

3.4. Polymerisation of Functionalised Monomers

A number of workers have prepared transition-metal complexes which contain either a free olefin or a trichlorosilyl or trimethylsiloxy functional group that can be polymerised. The advantages that such an approach offers are:

(i) The concentration of the metal-ligand monomer can be controlled so that resins containing a wide range of ligand concentrations can be synthesised.

(ii) The nature of the polymer matrix can be readily varied by altering the co-monomer. Thus hydrophilic or hydrophobic matrices can be constructed.

(iii) The distribution of the metal complex throughout the resin beads can be predetermined.

The main disadvantage of this approach lies in the fact that the transition metal complex containing monomers may undergo undesirable reactions with radical, basic or acidic initiators, so either preventing their use or precluding optimal polymerisation behaviour.

Transition metal complexes that have been copolymerised [90, 91] include vinylferrocene, 2 [92–94], (η^5-vinylcyclopentadienyl)manganesetricarbonyl, 3 [95–97], (η^5-vinylcyclopentadienyl)nitrosylchromium dicarbonyl, 4 [98], (η^5-vinylcyclopentadienyl)methyltungsten tricarbonyl, 5 [99, 100], (η^6-styrene)chromium tricarbonyl, 6 [84], (η^6-benzylacrylate)chromium tricarbonyl, 7 [101], and its 2-phenylethyl analogue, 8 [102, 103], [M(PBu$_3$)$_2$X(η^1-p-C$_6$-H$_4$CH=CH$_2$)], where M = Pd, X = Cl, Br, CN, Ph and M = Pt, X = Cl [104, 105], and trans-[Pt(PBu$_3$)$_2$(1,4-butadiyne)] [106].

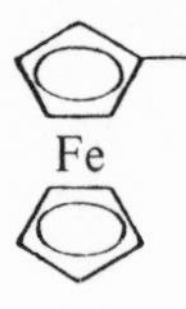

$$
\begin{array}{ccccc}
\underset{\text{Fe}}{\overset{}{\text{(ferrocenyl vinyl)}}} & \text{Mn(CO)}_3 & \text{Cr(CO)}_2\text{NO} & \text{W(CO)}_3\text{Me} & \text{Cr(CO)}_3 \\
\textbf{2} & \textbf{3} & \textbf{4} & \textbf{5} & \textbf{6}
\end{array}
$$

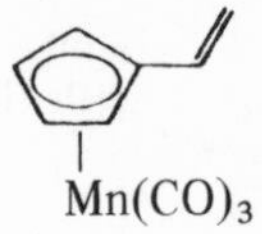

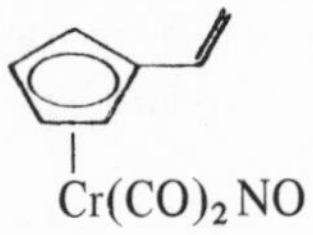

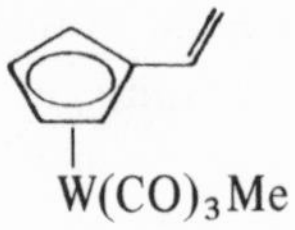

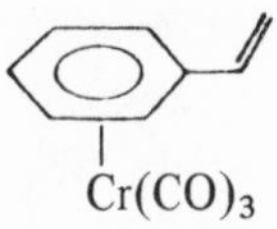

$$
\underset{\text{Cr(CO)}_3}{\bigcirc}-\text{CH}_2\text{OCOCH}=\text{CH}_2 \qquad \underset{\text{Cr(CO)}_3}{\bigcirc}-\text{CH}_2\text{CH}_2\text{OCOCH}=\text{CH}_2
$$

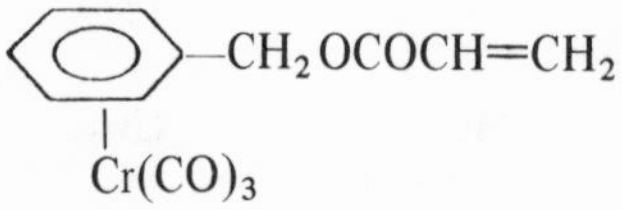

$$
\textbf{7} \qquad\qquad\qquad\qquad\qquad \textbf{8}
$$

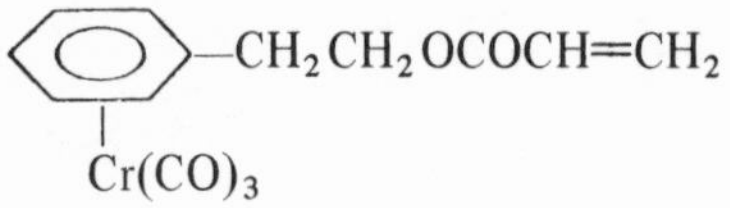

The homopolymerisation kinetics of **2**, **3** and **5** have been examined in detail and found to be unusual [90–92, 100, 107]. **2**–**5** all exhibit very electron-rich vinyl groups in radical initiated copolymerisations.

(η^4-2,4-Hexadienyl acrylate)iron tricarbonyl, **9**, has been prepared as in reaction (24) and polymerised as in reaction (25) to yield polymer-supported (η^3-allyl)iron tricarbonyl species [103, 107].

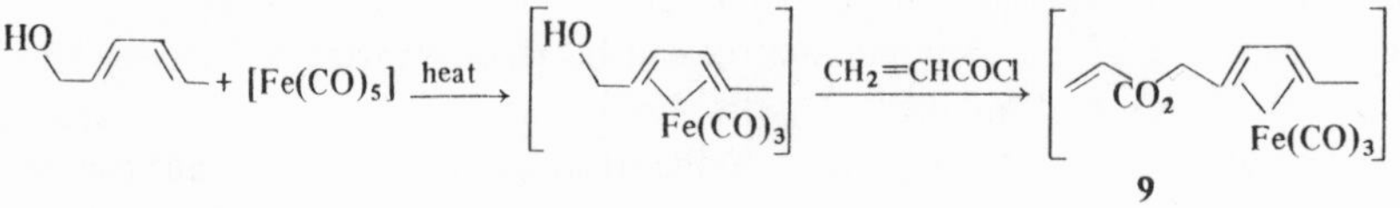

$$
\text{HO} \diagdown\diagup\diagdown + [\text{Fe(CO)}_5] \xrightarrow{\text{heat}} \left[\underset{\text{Fe(CO)}_3}{\text{HO}\diagdown\diagup\diagdown}\right] \xrightarrow{\text{CH}_2=\text{CHCOCl}} \left[\underset{\text{Fe(CO)}_3}{\diagup\text{CO}_2\diagdown\diagup\diagdown}\right] \quad (24)
$$

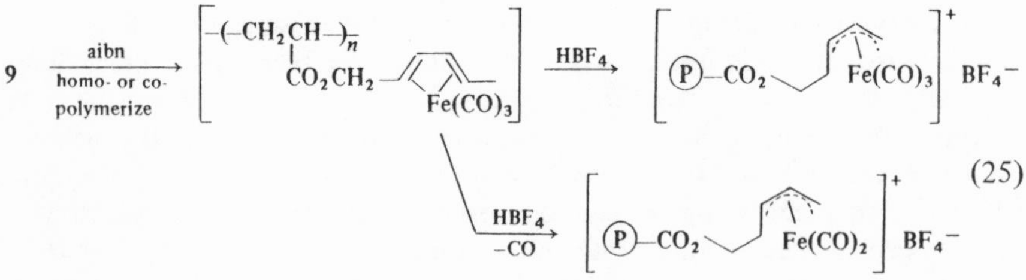

$$
\textbf{9}
$$

$$
\textbf{9} \xrightarrow[\substack{\text{homo- or co-}\\\text{polymerize}}]{\text{aibn}} \left[\underset{\substack{\text{CO}_2\text{CH}_2 \\ \text{Fe(CO)}_3}}{(-\text{CH}_2\text{CH})_n}\right] \xrightarrow{\text{HBF}_4} \left[\underset{\text{Fe(CO)}_3}{(\text{P})-\text{CO}_2}\right]^+ \text{BF}_4^-
$$

$$
\xrightarrow[-\text{CO}]{\text{HBF}_4} \left[\underset{\text{Fe(CO)}_2}{(\text{P})-\text{CO}_2}\right]^+ \text{BF}_4^- \qquad (25)
$$

The tricobalt nonacarbonyl-substituted styrene, **10**, has been prepared and copolymerised with styrene, methyl methacrylate and other monomers [108].

$$(CO)_3Co \overset{\displaystyle \overset{\Large\bigcirc}{\underset{\displaystyle C}{|}}}{\underset{\displaystyle \underset{(CO)_3}{Co}}{\diagup\!\!\!\diagdown}} Co(CO)_3$$

10

Recently metal vapour chemistry has been applied to the formation of polymer supported metals and metal clusters. The first example of the use of this technique is shown in equation (26) [109].

$$Cr(g) + \left[\!\!\!\begin{array}{c}\end{array}\!\!\!\right] + PF_6 \xrightarrow{77\ K} \left[\begin{array}{c} \end{array}\!\!-Cr(PF_3)_3\right] + \left[\begin{array}{c} \end{array}\!\!-Cr(PF_3)_3\right]_n \qquad (26)$$

$$1 \quad : \quad 6$$

However when styrene is used the vinyl group often interacts with the metal to give an η^2-product and so prevent polymerisation. Alternatively metal vapours often promote olefin or acetylene polymerisation without incorporation of the metal. Polystyrene and poly(t-butylstyrene) in the form of a 35% solution in diethyleneglycol dibutyl ether have been reacted with chromium metal vapour at 240 K to form bis(arene)chromium complexes through the phenyl rings [110]. A similar approach has been used to incorporate Ti, V, Cr, Mo and W into polyphenylsiloxanes through η^6-coordination [110]. However this approach is not easy. It requires sophisticated apparatus and it is difficult to find solvents which keep the polymers in solution and yet possess a sufficiently low vapour pressure to enable the metal vapour reaction to be carried out.

A number of research groups have prepared transition metal complexes containing $-Si(OR)_3$, $-SiCl_3$ or occasionally $-Ge(OMe)_3$ terminal groups within their ligands (reaction (27)) [22, 111–114a] and then either reacted these with silica (reaction (28)) [22, 111–113] or hydrolysed them in the presence of methyltrichlorosilane using aqueous dioxane to form insoluble polymeric globules (reactions (29–31)) [115].

$$(EtO)_3Si(CH_2)_nPPh_2 + ML_n \longrightarrow [(EtO)_3Si(CH_2)_nPPh_2ML_{n-1}] + L \qquad (27)$$

$$[(EtO)_3Si(CH_2)_nPPh_2ML_{n-1}] + \{Si\}-OH$$

$$\longrightarrow \quad [\{Si\}-O-\underset{|}{\overset{|}{Si}}-(CH_2)_nPPh_2ML_{n-1}] \qquad (28)$$

$$[\{Cl_3Si(CH_2)_nPPh_2\}_3RhCl] + m\ CH_3SiCl_3$$

$$\xrightarrow[n=2,\,8;\ m=0-200]{H_2O} \quad [\{O_{3/2}Si(CH_2)_nPPh_2\}_3RhCl.(O_{3/2}SiCH_3)_m] \qquad (29)$$

$$[\{Cl_3Si(CH_2)_2PPh_2\}_2RhCl]_2 + m\ CH_3SiCl_3$$

$$\xrightarrow[m=0-200]{H_2O} \quad [\{O_{3/2}Si(CH_2)_2PPh_2\}_4Rh_2Cl_2.(O_{3/2}SiCH_3)_m] \qquad (30)$$

$$[\{Cl_3Si(CH_2)_2PPh_2\}_3RhCl] + m\ Cl_3Si(CH_2)_2PPh_2$$

$$\xrightarrow[m=0.7]{H_2O} \quad [\{O_{3/2}Si(CH_2)_2PPh_2\}_3RhCl.(O_{3/2}Si(CH_2)_2PPh_2)_m] \qquad (31)$$

3.5. Direct Reaction Between Organometallic Compounds and Inorganic Oxide Surfaces

The mechanism of the Ziegler-Natta polymerisation of olefins is traditionally believed to involve insertion of an olefin into a metal-carbon bond [116]. As part of early attempts to provide support for this mechanism solid organometallic compounds of the traditional Ziegler-Natta transition metals, Ti, Zr, Hf, V, Nb and Cr were investigated as catalysts. The number of active centres on these compounds was not high and accordingly their activity was low. However highly active catalysts were prepared by supporting these organometallic compounds on silica, alumina and aluminosilicates [117-140]. The organic ligands present were allyl (Zr, Hf, Cr, Ni), benzyl (Ti, Zr, Hf), neopentyl (Ti, Zr), cyclopentadienyl (Cr) and arene (Cr). The reaction between support and organometallic depends on the moisture sensitivity of the organometallics which enables them to react with hydroxyl groups on the oxide surface (reaction (32)).

$$Support\text{-}OH + MR_n \longrightarrow Support-O-MR_{n-1} + RH \qquad (32)$$

The nature of the support is very important. Thus highly active olefin polymerisation catalysts are obtained by supporting the benzyl complexes of titanium and zirconium on alumina [117, 120, 141] and the tetrakis-η^3-allyl complex of zirconium on both silica and alumina [142] whereas $[Cr(\eta^3\text{-}C_3H_5)_3]$ and $[Cr(\eta^5\text{-}C_5H_5)_2]$ are more active when supported on silica [118, 119, 143]. The conditions under which the oxide has been pretreated before introduction of the organometallic are very important since the concentration and predominant types of surface hydroxyl groups are very dependent on this [120, 121, 144, 145]. In the case of silica heating in vacuo at 200°C for 3 hours removes all the surface water. Further heating then successively removes more and more surface hydroxyl groups, until at 1200°C all such groups are removed. Not all the hydroxyl groups on a metal oxide can react with a transition metal alkyl, but the number of those that can react may readily be determined by reacting the oxide with a solution of methyl magnesium iodide [120] or methyl lithium [146] in toluene and measuring the amount of methane gas evolved (reaction (33)).

$$\text{ʃ—OH + MeM} \xrightarrow[\text{M = MgI or Li}]{} \text{ʃ—OM + CH}_4 \qquad (33)$$

Since only a fraction of the metal atoms generate catalytically active sites, it is the nature rather than the number of surface immobilised transition metal centres that is important. Accordingly there is an optimum dehydration temperature for maximum polymerisation activity that depends on both the nature of the metal and its ligands (Table I).

TABLE I

Dehydration temperatures necessary in the pretreatment of oxide supports to yield maximum activity in the subsequent supported organometallic ethylene polymerisation catalyst (data from [138]).

Support	Metal Complex	Optimum dehydration temperature (°C)
SiO_2	$[Cr(C_3H_5)_3]$	400
SiO_2	$[Cr(\eta^5\text{-}C_5H_5)_2]$	670
SiO_2	$[Zr(C_3H_5)_4]$	25
SiO_2	$[Zr(C_3H_5)_3X]$ (X = Cl, Br, I)	750
Al_2O_3	$[Zr(C_3H_5)_4]$	400
Al_2O_3	$[Ti(CH_2Ph)_4]$	600

Deposition of chromocene from hydrocarbon solution on to amorphous silica of high surface area yields a highly active olefin polymerisation catalyst [119, 122, 147–150] (reaction (34)), in spite of the fact that chromocene itself is inactive.

$$\{Si\}-Si-OH + \quad Cr \quad \longrightarrow \quad \{Si\}-Si-O-Cr- \quad + C_5H_6 \qquad (34)$$

During deposition the silica changes in colour from white to jet black and the hydrocarbon solution from wine-red to colourless. The silica in the product serves two functions: it anchors the chromium and stabilises the chromium in a coordinatively unsaturated state. The act of anchoring prevents mutual interaction and destruction of these coordinatively unsaturated species. The high silica dehydration temperature noted in Table I reflects the importance of isolated chromium species as shown in reaction (33).

When organometallic compounds such as $[Zr(C_3H_5)_4]$ react with surface hydroxyl groups, attachment can be at one or two sites (reactions (35) and (36)) [120, 123, 141].

$$(35)$$

$$(36)$$

Three-site attachment is extremely unlikely because of the low surface density of hydroxyl groups. Silica dehydrated at 200°C gives primarily two site attachment (reaction (35)) as shown by quantitative determination of the volume of propene released. Single site attachment (reaction (34)) becomes successively more common as the temperature at which the metal oxide is pretreated is increased [120, 151].

In addition to using supported organometallic complexes directly as catalysts, they may also be both reduced and oxidised to prepare further active catalysts. Reduction is usually effected by hydrogen (reaction (37)) to yield how valent supported metal ions or occasionally hydride complexes [152], whilst oxidation is usually achieved using oxygen (reaction (38)).

$$\mathrm{)-OMR}_n + n/2\ H_2 \longrightarrow \mathrm{)-OM} + n\ RH \qquad (37)$$

$$\mathrm{)-OMR}_n + O_2 \longrightarrow \mathrm{)-OMO}_n + \text{oxidation products of R} \qquad (38)$$

The reduced complexes are of value as hydrogenolysis catalysts [153, 154], including ammonia synthesis [155], whilst the oxidised molybdenum catalyst promotes the dehydrogenation of cyclohexene to yield benzene and the oxidation of carbon monoxide to carbon dioxide [156].

Metal oxide-supported rhodium complexes have been prepared from both rhodium(I)-hydride complexes [157, 158] (reaction (39)) and rhodium(III)-

$$Al_2O_3 + [RhH(CO)(PPh_3)_3] \longrightarrow [)-O-Rh(CO)(PPh_3)_3] + H_2 \qquad (39)$$

allyl complexes (Scheme 1). When 'Rh(CO)(PPh$_3$)$_3$' is bound to alumina there is a substantial interaction between the triphenylphosphine ligands and the surface oxide ions which promotes dissociation of triphenylphosphine and hence hydroformylation activity [158]. All the rhodium complexes in Scheme 1, except **18**, are active olefin hydrogenation catalysts.

It is not necessary to preform the organometallic complex in order to deposit it on a metal oxide. Thus nickel has been supported on alumina by cocondensing nickel atoms and an arene on alumina at $-196°C$ to produce a hydrogenation catalyst [162], as well as by simultaneously treating γ-alumina with [Ni(CO)$_4$] and allylchloride [163].

3.6. Surface Bonding of Metal Carbonyls on Inorganic Oxides

Metal carbonyls can be immobilised by impregnating inorganic oxides [164–168]. The method typically involves evacuation of the support to ensure penetration of the solution into the pores. After stirring a suspension of the oxide in the impregnating solution excess solution is removed before drying under an inert atmosphere. Frequently the product so formed is further treated, sometimes just by heating, to activate it. The critical variables in impregnation are pretreatment of the inorganic oxide, the contact time and the temperature and pH of the impregnating solution. An alternative procedure that can be used with volatile carbonyls involved sublimation of the carbonyl in a flow of an inert

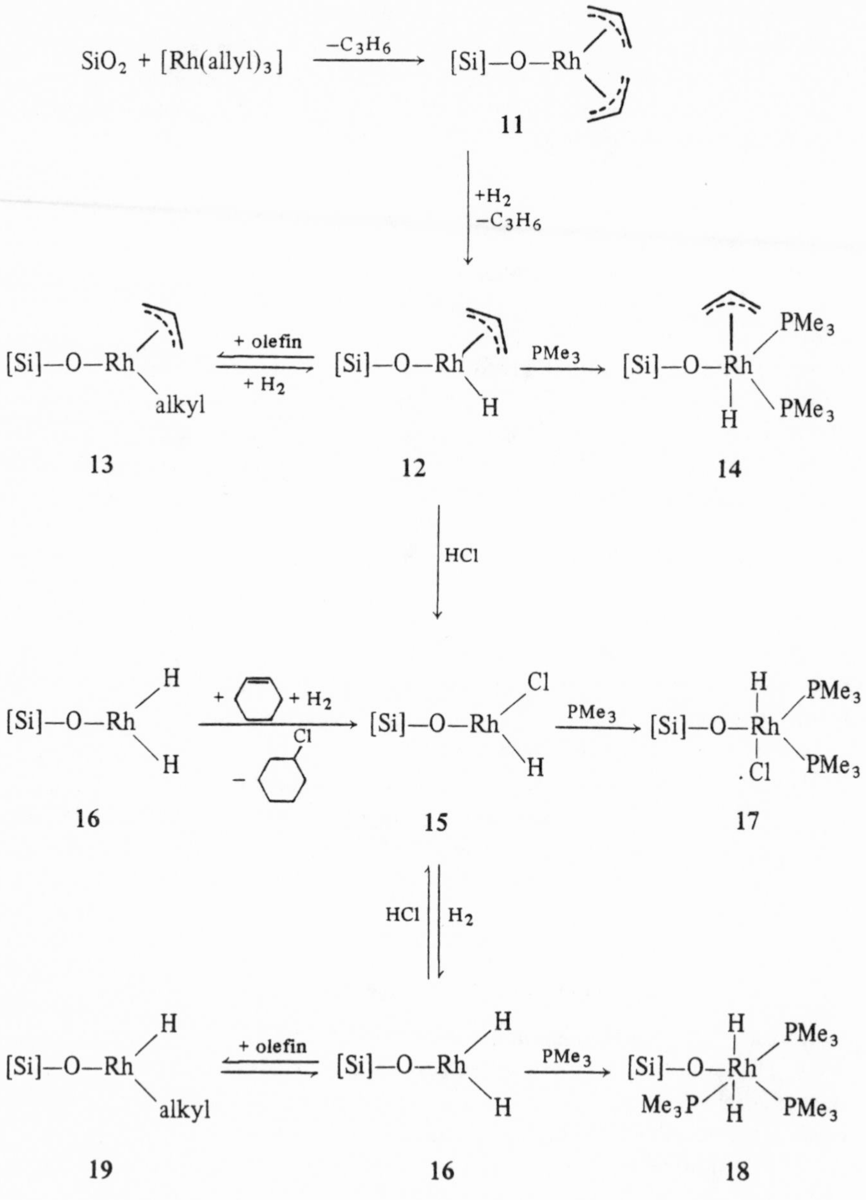
SiO2 + [Rh(allyl)3]
−C3H6
[Si]—O—Rh
11
+H2
−C3H6
[Si]—O—Rh
alkyl
+ olefin
+ H2
[Si]—O—Rh
H
PMe3
[Si]—O—Rh
H
PMe3
PMe3
13
12
14
HCl
[Si]—O—Rh
H
H
+ H2
Cl
[Si]—O—Rh
Cl
H
PMe3
H
PMe3
[Si]—O—Rh
. Cl
PMe3
16
15
17
HCl
H2
H
[Si]—O—Rh
alkyl
+ olefin
[Si]—O—Rh
H
H
PMe3
H
PMe3
[Si]—O—Rh
Me3P
H
PMe3
19
16
18

Scheme 1

gas onto the support [169]. Careful grinding together of the metal oxide and carbonyl under an inert atmosphere has been used to support metal carbonyls, although the heat generated during the grinding can lead to some decomposition of the carbonyl [170]. If the grinding is done in air [171, 172] oxidation of the carbonyl may occur [169].

After the the initial physical adsorption many metal carbonyls subsequently react with surface sites as in reaction (40).

$$[Mo(CO)_6] + \text{$\}$—O} \longrightarrow [Mo(O—\text{$\}$})(CO)_5] + CO \qquad (40)$$

This can be promoted by heating in vacuum, or in the presence of an inert gas or even in oxygen, so forming an oxidised subcarbonyl species. Such species may themselves react further with hydroxyl surface groups with evolution of both carbon monoxide and hydrogen as in reaction (41) [169].

$$\text{$\}$—Mo(CO)}_3 + 2 \text{ $\}$—OH} \longrightarrow [(\text{$\}$—O})_2 Mo^{II}] + H_2 + 3 CO \qquad (41)$$

The net product is not a supported metal carbonyl; it is however often a highly active, well dispersed supported metal catalyst, so that impregnation of inorganic oxides by metal carbonyls has often been used as the route to such catalysts. Temperature programmed decomposition chromatography provides a very valuable technique for monitoring the progress of chemical interaction between physically adsorbed metal carbonyls and the oxide surface [173–181]. In this a gas, either inert such as helium or active such as hydrogen, is swept over the metal carbonyl impregnated oxide, which is heated according to a predetermined temperature programme. The effluent gas is then sampled, through a gas sampling valve, and analysed using a gas chromatograph. Samples can be taken as regularly as desired. Thus a temperature programmed decomposition chromatogram contains information on the temperature regions at which a carbonyl complex first begins to decompose and bind to the support, the existence of zero-valent subcarbonyl species where carbon monoxide evolution alone takes place, and the temperature and extent of catalyst oxidation during which hydrogen and sometimes hydrocarbons are evolved. This technique has demonstrated that literature claims [182–187], based on infrared studies, that metal carbonyls adsorbed on inorganic oxides decompose on heating to yield the free metal, are erroneous [180].

The nature of the product of reaction between a metal carbonyl and an oxide support depends on the nature of the oxide. Thus $[Cr(CO)_6]$ on alumina undergoes initial loss of carbon monoxide followed at a higher temperature by loss of further carbon monoxide accompanied by hydrogen evolution [188]. In contrast, when $[Cr(CO)_6]$ on silica is heated, all six carbonyl ligands are lost in rapid succession over a narrow temperature range [176, 177]; this loss is accompanied by chromium oxidation and consequent hydrogen evolution. Both alumina and silica supported $[Cr(CO)_6]$ complexes catalyse olefin hydrogenation [176, 189, 190].

The extent of carbonyl decomposition on heating oxide-supported $[Mo(CO)_6]$ depends on the basicity of the hydroxyl groups on the support [191]. Decomposition occurs most readily on silica which does not stabilise subcarbonyl species; by contrast heating γ-alumina adsorbed $[Mo(CO)_6]$ results in most of the $[Mo(CO)_6]$ being desorbed, although the remainder forms a variety of subcarbonyl species including $[(Al\}-O)(CO_2Mo(\mu\text{-}CO\rightarrow Al\text{-})_2Mo(CO)_2(O\text{-}\{Al)]$, in which the bridging carbonyl ligands are coordinated to Lewis acid sites on the alumina [187, 192, 193].

The interaction of $[W(CO)_6]$ with metal oxides is very similar to that of $[Mo(CO)_6]$ [194]. Heating $[W(CO)_6]$ on silica below 200°C yields the highly dispersed subcarbonyl $W(CO)_3$ [174]. $[W(CO)_5L]$, where $L = P(OPh)_3$, PPh_3 and $PBu_3\text{-}n$ reacts with γ-alumina to form a highly active olefin metathesis catalyst in which $W-C{=}O{\rightarrow}Al \leqslant$ coordination reduces the electron density at tungsten such that the ligand L is released to yield the coordinatively unsaturated $W(CO)_5$ species on the surface [182]. An alternative approach to supporting tungsten carbonyls on silica using an intermediate tin or germanium atom (reaction (42)) yields olefin isomerisation catalysts rather than the olefin metathesis catalyst obtained by direct linkage of tungsten to silica or alumina [195].

$$(42)$$

Silica and alumina have been functionalised with $-Co(CO)_4$ by first treating the oxides with $(EtO)_3SiH$, Me_2ClSiH or Cl_3SiH to introduce the $-SiH$ functionality, followed by reaction of this with $[Co_2(CO)_8]$ [196]. The photochemistry, in particular the carbonyl loss of the products, has been studied. $[Co_3(CO)_9CCH_3]$ on deposition on alumina, silica and NaY zeolites yields formic acid together with supported cobalt carbonyls whose exact nature depended on both the support and its pretreatment [197].

Transition metal cluster carbonyls are currently being widely studied as potential catalysts that are intermediate in character between metal surfaces and homogeneous catalysts [198–203]. They do, however, often have unique rather than intermediate properties. Because of their intrinsic interest, a considerable amount of effort has been devoted to supporting them on inorganic oxides. Temperature programmed decomposition studies of $[Fe(CO)_5]$, $[Fe_2(CO)_9]$ and $[Fe_3(CO)_{12}]$ on alumina suggest that the nuclearity of the precursor carbonyl affects the chemistry of the surface-bonded product [178, 204]. However infrared studies suggest that both $[Fe(CO)_5]$ and $[Fe_3(CO)_{12}]$ form the anionic hydride cluster $[HFe_3(CO)_{11}]^-$ on magnesium oxide and alumina, with $[Fe(CO)_5]$ being the slower [205]. $[Fe_2(CO)_9]$ is known to undergo some disproportionation to the mono- and tri-nuclear species [178]. Addition of HCl to supported $[HFe(CO)_{11}]^-$ regenerates $[Fe_3(CO)_{12}]$ [205]. When $[Fe(CO)_5]$, $[Fe_2(CO)_9]$ and $[Fe_3(CO)_{12}]$ are introduced into NaY and HY zeolites, warming to $60°C$ results in the formation of $[H_2Fe_3(CO)_{11}]$ which is held by hydrogen bonding within the zeolite supercage [185, 206]. The HY zeolite supported product decomposes at $200°C$ with evolution of carbon monoxide to yield supported iron(II) whereas the NaY supported product loses carbon monoxide but does not suffer iron oxidation. Irradiation of silica-adsorbed $[Fe(CO)_5]$ yields only $[Fe_3(CO)_{12}]$ in significant amount in contrast to irradiation in the gas, liquid or solution phases when $[Fe_2(CO)_9]$ is the major product [207].

The carbonyl clusters of ruthenium and osmium have been reviewed thoroughly [208]. $[Ru_3(CO)_{12}]$ supported on silica is an active olefin isomerisation and hydrogenation catalyst [209] as well as a hydrogenolysis catalyst for converting ethylbenzene into toluene and methane under conditions in which the xylenes scarcely react, allowing ethylbenzene to be separated from a mixture of xylenes [210]. The initial products are subcarbonyl species which subsequently form $[(Si\text{-}O)_x Ru_3 H_y(CO)_5]$ [209, 211]. When $[Ru_3(CO)_{12}]$ adsorbed on HY zeolite is heated to about $200°C$ the stable subcarbonyl $Ru_3(CO)_9$ is formed within the zeolite supercage; this does not completely lose carbon monoxide until heated to almost $500°C$ [186]. The retention or loss of cluster geometry when $[Ru_3(CO)_{12}]$ adsorbed on to metal oxides is decomposed depends very much on the support [212].

$[Os_3(CO)_{12}]$, $[Os_6(CO)_{18}]$ and $[H_2Os_3(CO)_{10}]$ are weakly physisorbed on to silica, titanium dioxide, magnesium oxide, alumina, zince oxide and HX zeolite at room temperature [175, 213–220]. On heating, thermal decomposition takes place, the exact mode of decomposition depending critically on the pretreatment given to the support as well as the degree of loading. If the supports are not extensively dried then hydrocarbons are formed together with osmium clusters linked to the metal oxide as in **20** [181, 214–218].

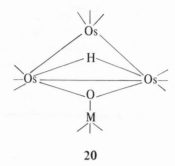

20

Although the majority of the hydrocarbons formed is methane, some ethylene and ethane are formed together with traces of higher hydrocarbons none of which are C_3 compounds [181]. The osmium complexes are active olefin hydrogenation catalysts [221] as well as active Fischer–Tropsch catalysts with a high selectivity for methane [215, 222]. Compound **20** catalyses the water gas shift reaction at 140°C under 1 atmosphere of carbon monoxide in the presence of excess water, at the end of the reaction $[(\geqslant M\text{-}O)_2 Os(CO)_3]_n$ appears to remain [217]. On heating **20** in vacuo, the cluster decomposes and $[(\geqslant M\text{-}O)_2 Os(CO)_3]_2$ and $[(\geqslant M\text{-}O)_2 Os(CO)_2]_n$, which can be interconverted by carbonylation/decarbonylation, are formed [214, 216, 218, 219]. When $[Os_3(CO)_{12}]$ and $[Os_6(CO)_{18}]$ were impregnated at high loading on extensively dehydroxylated supports the main products of temperature programmed decomposition were carbon dioxide and $[Os_n(CO)_{xn}C_{yn}]$ where the most likely value of n is 12, $2.0 \leq x \leq 3.0$ and $0 \leq y \leq 0.4$; some C_2 hydrocarbons were also formed [220].

$[\{Co(CO)_3\}\mu_3\text{-}CCH_3]$ is reversibly adsorbed on to metal oxides with loss of a carbonyl group. On heating it decomposes to Co^{2+} and $[Co(CO)_4]^-$; the ease of this decomposition depends on the surface in the order $\gamma\text{-}Al_2O_3 > NaY$ zeolite $\gg SiO_2$ [223]. $[Co_4(CO)_{12}]$ has been supported on zinc oxide to yield an olefin hydroformylation catalyst which, although less active than those derived from $[RhCo_3(CO)_{12}]$, $[Rh_2Co_2(CO)_{12}]$, $[Rh_4(CO)_{12}]$ and $[Rh_6\text{-}(CO)_{16}]$, gives higher n-aldehyde yields than the rhodium containing compounds [224, 225]. Infrared studies indicate that $[Rh_2Co_2(CO)_{12}]$ is initially adsorbed on to γ-alumina with loss of bridging carbonyls, followed by loss of the remaining carbonyl ligands at temperatures above 27°C or on standing in air at room temperature [226].

At room temperature $[Rh_6(CO)_{16}]$ retains its character when adsorbed on to silica, γ-alumina or zeolites [227–229]. On heating decarbonylation occurs but the Rh_6 cluster is retained. On exposure to air or oxygen, carbon dioxide is evolved, but the Rh_6 cluster is largely retained [227, 229–231]; on γ-alumina

$[Rh_6(CO)_{16-n}(O_2)_n]/\gamma\text{-}Al_2O_3$ is the product formed [227], whereas $[Rh_6\text{-}(CO)_{16-n}H_{2n}]$ has been suggested as being formed on silica [183], although this may be in doubt [180]. On alumina complete decarbonylation can be achieved by oxygen at room temperature and, provided sufficient adsorbed water is present, complete recarbonylation can be accomplished on re-exposure to carbon monoxide [230]. If sufficient water is present the surface hydroxyl groups readily oxidise the rhodium(0) to a rhodium(I) carbonyl species, **21** (reaction (43)). This reacts with carbon monoxide in the presence of excess water to regenerate $[Rh_6(CO)_{16}]$ [231]. A facile reduction of **21** under hydrogen or

$[Rh_6(CO)_{16}]/\gamma\text{-}Al_2O_3 + 6\ Al\text{-}OH$

$$\longrightarrow 3 \quad [\ldots] \quad + 4\ CO + 3\ H_2 \tag{43}$$

excess water in the absence of carbon monoxide gives metallic rhodium together with an intramolecular rearrangement of linear to doubly bridged carbonyl ligands (reaction (44)) [231].

$$n \quad [\ldots] \quad + n/2\ H_2$$

$$\longrightarrow (Rh\text{---}Rh)_n + n \quad [\ldots] \tag{44}$$

This subsequently forms $[Rh_4(CO)_{12}]$, $[Rh_6(CO)_{16}]$ in the presence of high concentrations of carbon monoxide and water, or larger polymeric species in the presence of low concentrations of water [232]. When $[Rh_6(CO)_{16}]$ adsorbed on to γ-alumina is heated in vacuo above $250°C$ it loses its ability to reversibly decarbonylate and recarbonylate [230]. Not surprisingly, in view of the chemistry just described, $[Rh_6(CO)_{16}]$ adsorbed on to metal oxides catalyses the water gas shift reaction.

The high surface mobility of rhodium carbonyl fragments is indicated by the fact that when $[Rh_4(CO)_{12}]$ or $[Rh(CO)_2Cl]_2$ are supported on alumina, decarbonylated and exposed to carbon monoxide the resulting adsorbed species is $[Rh_6(CO)_{16}]$ [227]. A similar reaction (45) occurs at room temperature between $[Rh_4(CO)_{12}]$ on silica or η-alumina [232]. This reaction is severely

$$3 \, [Rh_4(CO)_{12}] \quad \longrightarrow \quad 2 \, [Rh_6(CO)_{16}] + 4 \, CO \qquad (45)$$

retarded by added carbon monoxide. Both $Rh_4(CO)_{12}/SiO_2$ and $Rh_6(CO)_{16}/SiO_2$ can be oxidised at $80°C$ to $Rh^I(CO)_2$ species, such as those above, in the absence of water [231–233].

Rhodium carbonyl cluster anions with between 4 and 13 rhodium atoms such as $[Rh_6(CO)_{15}]^{2-}$, $[Rh_7(CO)_{16}]^{3-}$, $[Rh_{13}(CO)_{23}H_3]^{2-}$ and $[Rh_{13}(CO)_{23}H_2]^{3-}$, impregnated on to zirconium dioxide and titanium dioxide-containing silica provide highly dispersed rhodium catalysts that are very active for the Fischer-Tropsch conversion of carbon monoxide and hydrogen to ethanol [234]. There is a strong interaction between the carbonyl cluster anion and surface hydroxyl groups to form supported hydride carbonyl clusters such as $[Rh_6(CO)_{15}H]^-$. The function of the zirconium and titanium dioxides is to prevent rhodium aggregation.

Although most supported carbonyl catalysts have been prepared from pre-formed metal carbonyls this is not necessary. Rhodium trichloride absorbed on silica reacts with carbon monoxide at $100°C$ to form surface rhodium(I) dicarbonyls. On magnesia both rhodium(I) dicarbonyl and rhodium(I) mono-carbonyl species are formed [235].

When $[Ir_4(CO)_{12}]$ adsorbed on to alumina is heated to $350°C$ for 5 hours complete decarbonylation occurs to give < 1 nm iridium crystallites [184, 226, 236]. A similar decarbonylation occurs on silica although hydrocarbons are evolved in this case [175, 213]. The difference is in part due to the formation of a fraction of oxidised iridium species on alumina as opposed to solely metallic iridium on silica [237]. Catalysts derived from surface-bonded $[Ir_4(CO)_{12}]$ have been used for the hydrogenolysis of paraffins [238] and as Fischer-Tropsch catalysts have also been prepared by reducing supported $K[Ir(CO)_4]$ [240, 241].

$[Ni(CO)_4]$ physically adsorbed on hydroxylated alumina decomposes upon evacuation to release carbon monoxide and form small carbonylated metal clusters [241]. These appear to be bound through direct nickel-surface bonds [242, 243]. When $[Ni(CO)_4]$ adsorbed on dehydrated alumina is heated no hydrogen evolution and consequently no nickel oxidation occurs [183,

243]. [Ni(CO)$_4$] is not readily adsorbed by silica nor does it readily react with silica [241], whereas it does react with type X [244–247] and type Y zeolites being adsorbed at acidic electron acceptor sites on HY zeolites but at electron donor oxygen anion sites on NaY zeolites [242]. [Ni$_2$(Cp)$_2$(CO)$_2$] and [Ni$_3$(Cp)$_3$(CO)$_2$] adsorbed on silica gel and Vycor glass are active olefin hydrogenation and hydroformylation catalysts [248]. On heating to 120°C under vacuum carbon monoxide is evolved from silica supported [Ni$_3$(Cp)$_3$-(CO)$_2$], but on cooling to 25°C under carbon monoxide the original complex is reformed, suggesting that the decarbonylated species retains its 3-centred structure throughout.

Clusters anions such as [{Pt$_3$(CO)$_6$}$_n$]$^{2-}$ (n = 1–6) have been deposited on silica [249–251], γ-alumina [249–251], zinc oxide [239] and magnesium oxide [239] and decomposed by pyrolysis in vacuo or under hydrogen to produce platinum metal catalysts active for the dehydrocyclisation of n-hexane [249] and Fischer-Tropsch reactions [239, 250]. It is sometimes assumed when cluster compounds are impregnated onto metal oxides that the structure of the cluster is essentially unchanged. However, this is not always true. When [Pt$_3$(μ-CO)$_3$(PPh$_3$)$_4$] is impregnated onto metal oxides it may be transformed into [Pt$_5$(μ-CO)$_5$(CO)(PPh$_3$)$_4$]; transformation depends on the nature of the support and its pretreatment, acidic supports promoting the transformation and basic supports suppressing it, as well as the solvent used, a hydrophilic solvent such as ether or tetrahydrofuran decreasing transformation by removing water and decreasing the number of Lewis acid sites [252].

3.7. Supported Ziegler-Natta Catalysts

Although they lie largely outside the scope of the present book we briefly mention supported Ziegler-Natta catalysts because of their close relation to other supported metal complex catalysts. When titanium complexes are supported on magnesium compounds the resulting Ziegler-Natta catalysts have remarkably high activities with yields of up to 200 000 g of polyethylene per gram of titanium being claimed [116]. When TiCl$_4$ vapour interacts with partially hydroxylated magnesium oxide the titanium bonds to one or two oxygen atoms (reactions (46) and (47)) [253]. In the presence of triethylaluminium the singly bound titanium species is alkylated and reduced (reaction (48)) to form a

$$Mg\!\!\!\!\diagdown\!\!-OH + TiCl_4 \longrightarrow Mg\!\!\!\!\diagdown\!\!-O-TiCl_3 + HCl \qquad (46)$$

$$Mg\!\!\!\!\diagdown\begin{array}{c} -OH \\ -OH \end{array} + TiCl_4 \longrightarrow Mg\!\!\!\!\diagdown\begin{array}{c} -O \\ -O \end{array}\!\!\!Ti\diagdown\begin{array}{c} Cl \\ Cl \end{array} + 2\,HCl \qquad (47)$$

$$Mg \substack{\\\\} -O-TiCl_3 + AlEt_3 \xrightarrow{-AlEt_2Cl} Mg \substack{\\\\} -O-TiEtCl_2 \xrightarrow{-Et} Mg \substack{\\\\} -O-TiCl_2$$

$$\begin{array}{c} + AlEt_3 \\ - AlEt_2Cl \end{array} \Bigg\downarrow \qquad\qquad (48)$$

$$Mg \substack{\\\\} -O-Ti \underset{Et}{\overset{Cl}{<}}$$

titanium (III)-alkyl compound on the surface of the magnesium oxide [253]. It is interesting that the doubly oxygen-bonded species both on magnesium oxide and on γ-alumina and silica is not reduced by aluminium trialkyls [254, 255]. Tetrabenzyltitanium and tribenzyltitaniumchloride on Mg(OH)Cl also form highly active olefin polymerisation catalysts (reactions (49) and (50)) that in the presence of ether give 100% yields of stereoregular polymers [256].

$$[Ti(CH_2Ph)_4] + Mg(OH)Cl \longrightarrow Mg \substack{\\\\} \begin{array}{c} -O \\ -O \end{array} \overset{CH_2Ph}{\underset{CH_2Ph}{>}} Ti \qquad\qquad (49)$$

$$[TiCl(CH_3Ph)_3] + Mg(OH)Cl \longrightarrow Mg \substack{\\\\} \begin{array}{c} -O \\ -O \end{array} \overset{CH_2Ph}{\underset{Cl}{>}} Ti \qquad\qquad (50)$$

TiCl$_4$ [257, 258], TiCpCl$_3$ [259, 260] and TiCp$_2$Cl$_2$ [261] all react with silica and alumina in a manner analogous to that shown in reaction (46). On reduction with aluminium alkyls olefin polymerisation catalysts are formed, although these are less active than their magnesium based analogues. A similar comment applies to [Ti(CH$_2$Ph)$_4$] on silica and alumina [262], as well as to [Ti(OEt)$_4$] and [Ti(O^{i}Pr)$_4$] on silica [263, 264]. VCl$_4$ has been supported on silica and alumina in exactly the same way as TiCl$_4$ [265–267]; on silica a mixture of monodentate and bidentate attachment occurs (cf. reactions (46) and (47)) whereas, on alumina, bidentate attachment predominates. When VOCl$_3$ is supported on silica cleavage of Si$-$O$-$V bonds occurs during hydrolysis of the surface [268]. Vanadium can also be introduced by impregnation of surfaces with aqueous vanadate solutions, solutions of bis(bidentate)oxovanadium(IV) salts or [V$_2$O$_3$(OH)$_4$] vapour at 600 C [269, 270].

When bis(triphenylsilyl)chromate is supported on silica or aluminosilicate supports, chromium is bound through a hydroxylic oxygen (reaction (51)) [271].

$$2 \ Ph_3SiOH + CrO_3 \longrightarrow (Ph_3SiO)_2CrO_2$$

$$\xrightarrow{SiO_2} \ Si\!\!\!\big\rbrace\!\!-O-\overset{O\ \ O}{\underset{}{Cr}}-OSiPh_3 + Ph_3SiOH \tag{51}$$

$[Mo(OEt)_5]_2$ reacts with surface hydroxyl groups of silica and alumina. On silica the product contains mainly surface anchored binuclear molybdenum complexes (reaction (52)) [140b, 272], with very few mononuclear molybdenum species, whereas the reverse is true on alumina (reaction (53)) [140b].

$$4 \ Si\!\!\!\big\rbrace\!\!-OH + [Mo(OEt)_5]_2 \longrightarrow [(Si\!\!\!\big\rbrace\!\!-O)_4Mo_2(OEt)_6] + 4 \ EtOH \tag{52}$$

These compounds can be subsequently reduced or oxidised to form a range of catalysts with molybdenum in a variety of oxidation states including II, IV, V and VI. When silica, dehydrated at 180°C, is treated with a solution of $MoCl_5$ in carbon tetrachloride tridentate attachment of molybdenum occurs (reaction (54)) [273].

$$4 \ Al\!\!\!\big\rbrace\!\!-OH + [Mo(OEt)_5]_2 \longrightarrow 2 \ [(Al\!\!\!\big\rbrace\!\!-O)_2Mo(OEt)_3] + 4 \ EtOH \tag{53}$$

$$3 \ Si\!\!\!\big\rbrace\!\!-OH + MoCl_5 \longrightarrow [(Si\!\!\!\big\rbrace\!\!-O)_3MoCl_2] + 3 \ HCl \tag{54}$$

The product may be converted by water vapour and subsequent heating to an oxy-molybdenum (V) species (reaction (55)).

$$[(Si\!\!\!\big\rbrace\!\!-O)_3MoCl_2] + 2H_2O \xrightarrow{-2HCl} [(Si\!\!\!\big\rbrace\!\!-O)_3Mo(OH)_2]$$

$$\xrightarrow[-H_2O]{180°C} [(Si\!\!\!\big\rbrace\!\!-O)_3Mo\!=\!O] \tag{55}$$

3.8. Surface Supported Metal Salts

Whilst Ziegler-Natta activity of surface supported metal salts largely ceases at Group VI, salts of metals from Groups VII and VIII have also been supported on metal oxides. $[Re(OEt)_3]$ reacts with silica to form direct Si—O—Re bonds (reaction (56)) [140b].

$$n \; Si\!\!\not\!\!\!\mid\!\!-OH + [Re(OEt)_3]_3 \longrightarrow$$

$$(56)$$

The immobilised rhenium atoms can be subsequently oxidised or reduced [140b, 274]. $ReCl_5$ and $ReOCl_4$ have also been immobilised on γ-alumina by reaction with surface hydroxyl groups [275].

$[Ru(NH_3)_6]^{3+}$ adsorbed on X- and Y-type zeolites decomposes on heating at 450°C in flowing oxygen to form a mixture of products of which the predominant one is $[Ru(O_{zeol})_3(NH_3)_x(NO)]$, where $x = 1$ or 2 [276]. At 530°C this decomposes to form either RuO_2 in the presence of oxygen or metallic ruthenium in its absence. Cobalt(II) phthalocyanate and cobalt(II) tetraphenyl-porphyrin have been adsorbed on to γ-alumina and titanium dioxide respectively [277, 278]. Electron transfer occurs from cobalt to the support. When a series of $[MCl_6]^{2-}$ and $[MCl_4]^{2-}$ salts were adsorbed on to alumina to form $[MCl_m-(OAl\leqslant)]^-$ (m = 5 and 3 respectively) the relative rates of reation, $M = Pd^{IV} > Pd^{II} > Rh^{IV} > Ru^{II} \gg Pt^{II} \sim Pt^{IV} \sim Ir^{IV}$, were determined partly by the rates of diffusion in to the pores and partly by the rates of the substitution reaction [279]. $[Pd(hfacac)_2]$ (hfacac = $CF_3COCHCOCF_3$) reacts with alumina in benzene to form 22, which on heating with hydrocarbons such as benzene and propene gives hydrocarbon cracking and reduction of palladium(II) to palladium metal [280].

3.9. Surface Complexes of Transition Metal Oxides on Oxide Supports

The surface complexes formed by interaction of transition metal oxides with metal oxides such as silica, alumina, magnesia and zinc oxide really lie outside the scope of the present book. The interested reader is referred to recent reviews [281–284] and papers [285–293]. Although the study of these catalysts was originally initiated in order to gain a fundamental understanding of the

22

mechanisms of catalytic action, a number of companies are now studying them for commercial reasons, particularly olefin metathesis [283, 284, 291–295]. They have also been used to study the decomposition of nitrous oxide, the oxidation of carbon monoxide, hydrogen, methanol and olefins, the oxychlorination of ethylene and the dehydrogenation of isobutane and ethylbenzene.

A typical olefin metathesis catalyst, such as the Phillips catalyst [296, 297], is prepared by impregnating alumina or silica with an aqueous solution of ammonium paramolybdate, ammonium tungstate or ammonium perrhenate, drying, and calcining the product at 500–600°C for 5 hours in the presence of a stream of dry air or nitrogen. However in addition to molybdenum, tungsten and rhenium oxides, a wide range of other oxides have been used, including those of V, Nb, Ta, Ru, Os, Rh, Ir, La, Sr, Ba, Se and Te [281]. In general, the promoter oxide is present at between 1 and 15% by weight. More than one oxide is frequently added, for example highly efficient propene disproportionation has been obtained with a catalyst containing 3.4% Co_2O_3, 11.0% MoO_3 and 85.6% Al_2O_3 (by weight) in which cobalt oxide reduces coke formation [296].

References

1. W. O. Haag and D. D. Whitehurst: *German Patent* 1,800,371 (1969); *Chem. Abs.* 71, 114951 (1969).
2. W. O. Haag and D. D. Whitehurst: *German Patent* 1,800,379 (1969); *Chem. Abs.* 72, 31192 (1970).
3. W. O. Haag and D. D. Whitehurst: *German Patent* 1,800,380 (1969); *Chem. Abs.* 71, 33823 (1969).
4. W. O. Haag and D. D. Whitehurst: *Belgian Patent* 721,686 (1969).
5. W. O. Haag and D. D. Whitehurst: *Proc. 2nd North Amer. Meeting Catal. Soc.* 16 (1971).

6.   W. O. Haag and D. D. Whitehurst: *Proc. 5th Int. Congr. Catal.* 465 (1972); published as *Catalysis vol. 1* (Ed. by J. H. Hightower) North-Holland Publ., Amsterdam (1973).
7.   D. L. Hanson, J. R. Kratzer, B. C. Gates and G. C. A. Schuit: *J. Catal.* **32**, 204 (1974).
8.   R. Linarte-Lazcano, M. P. Pedrosa, J. Sabadie and J. E. Germain: *Bull. Soc. Chim. France*, 1129 (1974).
9.   J. F. McQuillin, W. O. Ord and P. L. Simpson: *J. Chem. Soc.* 5996 (1963).
10. H. Dupin, J. Sabadie, D. Barthomeuf and J. E. Germain: *Bull. Soc. Chim. France*, I-86 (1979).
11. J.-B. N'G. Effa, J. Lieto and J.-P. Aune: *Inorg. Chim. Acta* **65**, L105 (1982).
12. G. G. Allan and A. N. Neogi: *J. Catal.* **19**, 256 (1970).
13. S. C. Tang, T. E. Paxson and L. Kim: *J. Mol. Catal.* **9**, 313 (1980).
14. K. G. Allum, R. D. Hancock, I. V. Howell, R. C. Pitkethly and P. J. Robinson: *J. Organometal. Chem.* **87**, 189 (1975).
15. M. Capka, P. Svoboda and J. Hetflejs: *Chem. Ind. (London)* 650 (1972).
16. N. L. Holy: *J. Org. Chem.* **44**, 239 (1979).
17. M. Kraus and D. Tomanova: *J. Poly. Sci., Poly. Chem. Ed.* **12**, 1781 (1974).
18. K. Alshaikh-Kadir and P. Holt: *Makromol. Chem.* **177**, 311 (1976).
19. L. D. Rollmann: *J. Amer. Chem. Soc.* **97**, 2132 (1975).
20. R. B. King and E. M. Sweet: *J. Org. Chem.* **44**, 385 (1979).
21. K. G. Allum and R. D. Hancock: *Brit. Patent* 1,295,673 (1972).
22. K. G. Allum, R. D. Hancock, I. V. Howell, S. McKenzie, R. C. Pitkethly and P. J. Robinson: *J. Organometal. Chem.* **87**, 203 (1975).
23. F. R. Hartley, L. S. G. Murray and P. N. Nicholson: *J. Mol. Catal.* **16**, 363 (1982).
23a. A. Luchetti, L. F. Wieserman and D. M. Hercules: *J. Phys. Chem.* **85**, 549 (1981).
24. G. Strukul, M. Bonivento, M. Graziani, E. Cernia and N. Palladino: *Inorg. Chim. Acta* **12**, 15 (1975).
25. R. H. Grubbs and L. C. Kroll: *J. Amer. Chem. Soc.* **93**, 3062 (1971).
26. R. H. Grubbs, L. C. Kroll and E. M. Sweet: *J. Macromol. Sci., Chem* **7**, 1047 (1973).
27. C. U. Pittman, L. R. Smith and R. M. Hanes: *J. Amer. Chem. Soc.* **97**, 1742 (1975).
28. H. S. Bruner and J. C. Bailar: *Inorg. Chem.* **12**, 465 (1973).
29. M. Capka, P. Svoboda and J. Hetflejs: *Coll. Czech. Chem. Comm.* **38**, 1242 (1973).
30. A. R. Sanger, L. R. Schallig and K. G. Tan: *Inorg. Chim. Acta* **35**, L325 (1979).
31. G. O. Evans, C. U. Pittman, R. McMillan, R. T. Beach and R. Jones: *J. Organometal. Chem.* **67**, 295 (1974).
31a. R. D. Sanner, R. G. Austin, M. S. Wrighton, W. D. Honnick and C. U. Pittman: *Inorg. Chem.* **18**, 928 (1979).
32. W. Beck, R. Höfer, J. Erbe, H. Menzel, U. Nagel and G. Platzen: *Z. Naturforsch.* **B29**, 567 (1974).
32a. C. U. Pittman, B. T. Kim and W. M. Douglas: *J. Org. Chem.* **40**, 590 (1975).
32b. S. Warwel and P. Buschmeyer: *Angew. Chem. Int. Ed.* **17**, 131 (1978).
32c. R. A. Awl, E. N. Frankel, J. P. Friedrich and E. H. Pryde: *J. Amer. Oil Chem. Soc.* **55**, 577 (1978).
32d. H. B. Gray and C. C. Frazier: *US Patent* 4,228,035 (1980); *Chem. Abs.* **94**, 37128 (1981).
33. J. P. Collman, L. S. Hegedus, M. P. Cooke, J. R. Norton, G. Dolcetti and D. N. Marquardt: *J. Amer. Chem. Soc.* **94**, 1789 (1972).
34. A. K. Smith, J. M. Basset and P. M. Maitlis: *J. Mol. Catal.* **2**, 223 (1977).
35. H. Hershcovitz and G. Schmuckler: *J. Inorg. Nucl. Chem.* **41**, 687 (1979).

36. A. Modelli, F. Scagnolari, G. Innorta, S. Torroni and A. Foffani: *Inorg. Chim. Acta* 76, L149 (1983).
37. R. H. Grubbs and E. M. Sweet: *Macromolecules* 8, 241 (1975).
38. C. U. Pittman and C. C. Lin: *J. Org. Chem.* 43, 4928 (1978).
39. C. U. Pittman and A. Hirao: *J. Org. Chem.* 43, 640 (1978).
40. R. D. Sanner, R. G. Austin, M. S. Wrighton, W. D. Honnick and C. U. Pittman: *Adv. Chem. Ser.* 184, 13 (1980).
41. V. A. Semikolenov, V. A. Likholobov and Yu. I. Ermakov: *Kinet. Katal.* 22, 1026 (1981).
42. R. J. Card and D. C. Neckers: *Inorg. Chem.* 17, 2345 (1978).
43. R. J. Card, C. E. Liesner and D. C. Neckers: *J. Org. Chem.* 44, 1095 (1979).
44. R. J. Card and D. C. Neckers: *J. Org. Chem.* 43, 2958 (1978).
45. R. J. Card and D. C. Neckers: *J. Amer. Chem. Soc.* 99, 7733 (1977).
46. R. Linarte-Lazcano and J. E. Germain: *Bull. Soc. Chim. France* 1869 (1971).
47. E. Cernia and F. Gasparini: *J. Appl. Polymer Sci.* 19, 915 (1975).
48. H. Nishide and E. Tsuchida: *Makromolek. Chem.* 177, 2295 (1976).
49. J. M. Calvert and T. J. Meyer: *Inorg. Chem.* 20, 27 (1981).
50. G. Braca, C. Carlini, F. Ciardelli, G. Sbrana and R. Arbuatti: *Chim. Ind. (Milan)* 55, 373 (1973).
51. G. Carlini, G. Sbrana and L. Casucci: *Chim. Ind. (Milan)* 57, 130 (1975).
52. T. Sasaki and F. Matsunaga: *Bull. Chem. Soc. Japan* 41, 2440 (1968).
53. E. Tsuchida and K. Honda: *German Patent* 2,511,088 (1975); *Chem., Abs.* 84, 18101 (1976).
54. Imperial Chemical Industries Ltd., *French Patent* 2,013,481.
55. E. Tsuchida, H. Nishide and T. Nishiyama: *J. Poly. Sci., Poly. Symp. Ed.* 47, 35 (1974).
56. A. J. Moffat: *J. Catal.* 19, 322 (1970).
57. A. J. Moffat: *J. Catal.* 18, 193 (1970).
58. B. C. Gates and J. Lieto: *ChemTech* 195 (1980).
59. B. C. Gates and J. Lieto: *ChemTech* 248 (1980).
59a. D. F. Foster, J. Harrison, B. S. Nicholls and A. K. Smith: *J. Organometal. Chem.* 248, C29 (1983).
60. J. Kohnle: *German Patent* 1,938,613 (1971); *Chem. Abs.* 74, 142612 (1971).
61. R. Pierantozzi, K. J. McQuade, B. C. Gates, M. Wolf, H. Knözinger and W. Ruhmann: *J. Amer. Chem. Soc.* 101, 5436 (1979).
62. M. S. Jarrell, B. C. Gates and E. D. Nicholson: *J. Amer. Chem. Soc.* 100, 5727 (1978).
63. E. W. Thornton, H. Knözinger, B. Tesche, J. J. Rafalko and B. C. Gates: *J. Catal.* 62, 117 (1980).
64. K. Iwatate, S. R. Dasgupta, R. L. Schneider, G. C. Smith and K. L. Watters: *Inorg. Chim. Acta* 15, 191 (1975).
65. J. J. Rafalko, J. Lieto, B. C. Gates and G. L. Schrader: *J. C. S., Chem. Commun.* 540 (1978).
66. T. Castrillo, H. Knözinger, J. Lieto and M. Wolf: *Inorg. Chim. Acta* 44, L239 (1980).
67. M. D. Monteil, J. B. N. Effa, J. Lieto, P. Verlaque and D. Benlian: *Inorg. Chim. Acta* 76, L309 (1983).
68. T. Kitamura, T. Joh and N. Hagihara: *Chem. Lett.* 203 (1975).
68a. A. K. Smith, W. Abboud, J. M. Basset, W. Reimann, G. L. Rempel, J. L. Bilhou, V. Bilhou-Bougnol, W. F. Graydon, J. Dunogues, N. Ardoin and N. Duffant In:

Fundamental Research in Homogeneous Catalysis (Ed. by M. Tsutsui) Plenum Press, New York **3**, 621 (1979).

69. R. C. Ryan, G. M. Wilemon, M. P. DalSanto and C. U. Pittman: *J. Mol. Catal.* **5**, 319 (1979).
70. A. T. Jurewicz, L. D. Rollmann and D. D. Whitehurst: *Adv. Chem. Ser.* **132**, 240 (1974).
70a. G. Sbrana, G. Braca, G. Valentine, G. Pazienza and A. Altomare: *J. Mol. Catal.* **3**, 111 (1978).
71. I. Dietzmann, D. Tomanova and J. Hetflejs: *Coll. Czech. Chem. Commun.* **39**, 123 (1974).
72. L. D. Rollmann: *Inorg. Chim. Acta* **6**, 137 (1972).
73. W. R. Cullen, D. J. Patmore, A. J. Chapman and A. D. Jenkins: *J. Organometal. Chem.* **102**, C12 (1975).
74. B. H. Chang, R. H. Crubbs and C. H. Brubaker: *J. Organometal. Chem.* **172**, 81 (1979).
75. R. H. Crubbs, C. Gibbons, L. C. Kroll, W. D. Bonds and C. H. Brubaker: *J. Amer. Chem. Soc.* **95**, 2373 (1973).
75a. C. H. Brubaker in: *Catalysis in Organic Synthesis* (Ed. by G. V. Smith) Academic Press, New York, p. 25 (1977).
76. G. Gubitosa and H. H. Brintzinger: *J. Organometal. Chem.* **140**, 187 (1977).
77. G. Gubitosa, M. Boldt and H. H. Brintzinger: *J. Amer. Chem. Soc.* **99**, 5174 (1977).
78. P. Perkins and K. P. C. Vollhardt: *J. Amer. Chem. Soc.* **101**, 3985 (1979).
78a. H.-S. Tung and C. H. Brubaker: *J. Organometal. Chem.* **216**, 129 (1981).
79. N. Kawata, T. Mizoroki, A. Ozaki and M. Ohkawara: *Chem. Lett.* 1165 (1973).
79a. N. Kawata, T. Mizoroki and A. Ozaki: *J. Mol. Catal.* **1**, 275 (1976).
79b. T. Mizoroki, N. Kawata, S. Hinata, K. Maruya and A. Ozaki in: *Catalysis: Heterogeneous and Homogeneous* (Ed. by B. Delmon and G. Jannes) Elsevier, Amsterdam, 319 (1975).
80. N. Kawata, T. Mizoroki and A. Ozaki: *Bull. Chem. Soc. Japan* **47**, 1807 (1974).
81. D. C. Locke, J. T. Schmermund and B. Banner: *Anal. Chem.* **44**, 90 (1972).
82. J. Manassen: *Isr. J. Chem.* **8**, 5 (1970).
83. S. Ikeda and T. Harimoto: *J. Organometal. Chem.* **60**, C67 (1973).
84. C. U. Pittman, P. L. Grube, O. E. Ayers, S. P. McManus, M. D. Rausch and G. A. Moser: *J. Poly. Sci., Poly. Chem. Ed.* **10**, 379 (1972).
85. P. Pertici, G. Vitulli, C. Carlini and F. Ciardelli: *J. Mol. Catal.* **11**, 353 (1981).
86. M. Berger and T. A. Manuel: *J. Poly. Sci., Poly. Chem. Ed.* **4**, 1509 (1966).
87. W. Beck, R. Hoefer, J. Erbe, H. Menzel, U. Nagel and G. Platzen: *Z. Naturforsch.* **B29**, 567 (1974).
88. C. U. Pittman and R. F. Felis: *J. Organometal. Chem.* **72**, 389 (1974).
89. C. U. Pittman, O. E. Ayers and S. P. McManus: *J. Macromol. Sci. Chem.* **A7**, 1563 (1973).
90. C. U. Pittman in: *Organometallic Reactions and Synthesis* (Ed. by E. I. Becker and M. Tsutsui) Plenum Press, New York **6**, 1 (1977).
91. C. U. Pittman in: *Organometallic Polymers* (Ed. by C. E. Carraher, J. E. Sheats and C. U. Pittman) Academic Press, New York, p. 1 (1978).
92. M. H. George and G. F. Hayes, *J. Poly. Sci., Poly. Chem. Ed.* **13**, 1049 (1975).
93. J. C. Lai, T. D. Rounsefell and C. U. Pittman: *J. Poly. Sci., Part A-1* **9**, 651 (1971).
94. C. U. Pittman and P. L. Grube: *J. Appl. Poly. Sci.* **18**, 2269 (1974).

95. G. V. Marlini, T. D. Rounsefell and C. U. Pittman: *Abs. 24th South Eastern Regional Meeting Amer. Chem. Soc., Birmingham, Ala.* 209 (1972).

96. C. U. Pittman, G. V. Marlin and T. D. Rounsefell: *Marcromolecules* 6, 1 (1973).

97. C. U. Pittman, C. C. Lin and T. D. Rounsefell: *Macromolecules* 11, 1022 (1978).

98. C. U. Pittman, T. D. Rounsefell, E. A. Lewis, J. E. Sheats, B. H. Edwards, M. D. Rausch and E. A. Mintz: *Macromolecules* 11, 560 (1978).

99. D. W. Macomber, M. D. Rausch, T. V. Jayaraman, R. D. Priester and C. U. Pittman: *J. Organometal. Chem.* 205, 353 (1981).

100. C. U. Pittman, T. V. Jayaraman, R. D. Priester, S. Spencer, M. D. Rausch and D. Macomber: *Macromolecules* 14, 237 (1981).

101. C. U. Pittman, R. L. Voges and J. Elder: *Macromolecules* 4, 302 (1971).

102. C. U. Pittman and G. V. Marlin: *J. Poly. Sci., Poly. Chem. Ed.* 11, 2753 (1973).

103. C. U. Pittman, O. E. Ayers and S. P. McManus: *Macromolecules* 7, 737 (1974).

104. N. Fujita and K. Sonogashira: *J. Poly. Sci., Poly. Chem. Ed.* 10, 379 (1972).

105. N. Fujita and K. Sonogashira: *J. Poly. Sci., Poly. Chem. Ed.* 12, 2845 (1974).

106. H. Fujita, M. Motowoka, T. Norisuye and A. Teramoto: *Polymer Preprint, Amer. Chem. Soc. Div. Poly. Chem.* 20, 38 (1979).

107. C. U. Pittman: *Preprints Amer. Chem. Soc. Org. Coatings and Plastics Div.* 41, 27 (1979).

108. C. U. Pittman and S. Massad: unpubl. results, quoted in C. U. Pittman: *Comprehensive Organometallic Chemistry* (Ed. by G. Wilkinson, F. G. A. Stone and E. W. Abel) Pergamon, Oxford, Vol. 8, Ch. 55 (1982).

109. R. Middleton: Ph. D. Thesis, University of Bristol, 1974.

110. C. G. Francis and G. A. Ozin in: *Advances in Organometallic and Inorganic Polymer Science* (Ed. by C. E. Carraher, J. E. Sheats and C. U. Pittman) Marcel Dekker, New York p. 167 (1982).

111. K. G. Allum, R. D. Hancock, S. McKenzie and R. C. Pitkethly: *Proc. 5th Int. Congr. Catal.* 477 (1972) (*Catalysis Vol. 1* (Ed. by J. W. Hightower) North-Holland Publ., Amsterdam, 1973).

111a. K. G. Allum, R. D. Hancock, I. V. Howell, T. E. Lester, S. McKenzie, R. C. Pitkethly and P. J. Robinson: *J. Organometal. Chem.* 107, 393 (1976).

112. K. G. Allum, S. McKenzie and R. C. Pitkethly: *US Patent* 3,726,809 (1973).

112a. R. Jackson, J. Ruddlesden, D. J. Thompson and R. Whelan: *J. Organometal. Chem.* 125, 57 (1977).

112b. P. Panster, W. Bunder and P. Kleinschmit: *German Patent* 2,834,691 (1980); *Chem. Abs.* 93, 32413 (1980).

113. A. A. Oswald, L. L. Murrell and L. J. Boucher: *Abstr. 167th Amer. Chem. Soc. Meeting* PETR 34 and PETR 35 (1974).

114. W. R. Cullen and Z. C. Brzezinska: *Inorg. Chem.* 18, 3132 (1979).

114a. G. V. Lisichkin, A. Ya. Yuffa, A. V. Gur'ev and F. S. Denisov: *Vestn. Mosk. Univ., Khim.* 17, 467 (1976); *Chem. Abs.* 86, 55029 (1977).

115. Z. C. Brzezinska and W. R. Cullen: *Can. J. Chem.* 58, 744 (1980).

116. H. Sinn and W. Kaminsky: *Adv. Organometal. Chem.* 18, 99 (1980).

117. D. G. H. Ballard: 23rd Int. Cong. Pure Appl. Chem., *Spec. Lett.* 6, 219 (1971).

118. Yu. I. Yermakov, A. M. Lazutkin, E. A. Demin, V. A. Zakharov and Yu. P. Grabovskii: *Kinet. Katal.* 13, 1422 (1972).

119. F. J. Karol, G. L. Karapinka, C. Wu, A. W. Dow, R. N. Johnson and W. L. Carrick: *J. Polym. Sci., Polym. Chem. Ed.* 10, 2621 (1972).

120. D. G. H. Ballard: *Adv. Catal.* **23**, 263 (1973).
121. J. P. Candlin and H. Thomas: *Adv. Chem. Ser.* **132**, 212 (1974).
122. G. L. Karapinka: *German Patent* 1,808,388 (1970); *Chem. Abs.* **72**, 80105 (1970).
123. D. G. H. Ballard, E. Jones, A. J. P. Pioli, P. A. Robinson and R. J. Wyatt: *German Patent* 2,040,353 (1971); *Chem. Abs.* **74**, 126364 (1971).
124. Union Carbide: *British Patent* 1,264,393 (1972); *Chem. Abs.* **76**, 154465 (1972).
125. Yu. I. Yermakov, A. M. Lazutkin, E. A. Demin, V. A. Zakharov, E. G. Kushareva and Yu. P. Grabovskii: *USSR Patent* 334,738 (1972); *Chem. Abs.* **78**, 44261 (1973).
126. R. N. Johnson and F. J. Karol: *German Patent* 1,963,256 (1970); *Chem. Abs.* **73**, 77820 (1970).
127. Yu. I. Yermakov and V. Zakharov: *Adv. Catal.* **24**, 173 (1975).
128. Yu. I. Yermakov: *Catal. Rev., Sci. Eng.* **13**, 77 (1976).
129. F. J. Karol in: *Encyclopaedia Polymer Sci. Technol.* Wiley, New York, Vol. Supp. 1, p. 120 (1976).
130. Yu. I. Yermakov, Yu. P. Grabovskii, A. M. Lazutkin and V. A. Zakharov: *Kinet. Katal.* **16**, 787 (1975).
131. Yu. I. Yermakov, Yu. P. Grabovskii, A. M. Lazutkin and V. A. Zakharov: *Kinet. Katal.* **16**, 911 (1975).
132. Yu. I. Yermakov and B. N. Kuznetsov: *Kinet. Katal.* **18**, 1167 (1977).
133. Yu. I. Yermakov, B. N. Kuznetsov, Yu. P. Grabovskii, A. N. Startsev, A. M. Lazutkin, V. A. Zakharov and A. I. Lazutkina: *J. Mol. Catal.* **1**, 93 (1976).
134. W. Skupinski and S. Malinowski: *J. Organometal. Chem.* **117**, 183 (1976).
135. W. Skupinski and S. Malinowski: *J. Mol. Catal.* **4**, 95 (1978).
136. Yu. I. Yermakov and V. A. Likholobov: *Kinet. Katal.* **21**, 1208 (1980).
137. A. P. Shelepin, A. P. Chernyshev, V. I. Koval'chuk, P. A. Zhdan, E. N. Yurchenko, B. N. Kuznetsov and Yu. I. Yermakov: *Kinet. Katal.* **22**, 716 (1981).
138. V. A. Zakharov and Yu. I. Yermakov: *Catal. Rev., Sci. Eng.* **19**, 67 (1979).
139. Yu. I. Yermakov: *Stud. Surf. Sci. Catal.* **7A**, 57 (1981).
140. Yu. I. Yermakov, B. N. Kuznetsov and V. A. Zakharov: *Catalysis by Supported Complexes*, Elsevier, Amsterdam 1981. *a*. Chapter 3; *b*. Chapter 2.
141. D. G. H. Ballard: *J. Polym. Sci., Polym. Chem. Ed.* **13**, 2191 (1975).
142. V. A. Zakharov, V. K. Dudchenko, V. I. Babenko and Yu. I. Yermakov: *Kinet Katal.* **17**, 738 (1976).
143. B. Rebenstorf, B. Jonson and R. Larsson: *Acta Chem. Scand.* **A36**, 695 (1982).
144. H. P. Boehm: *Adv. Catal.* **16**, 179 (1966).
145. L. H. Little: *Infrared Spectra of Adsorbed Species*, Academic Press, London, 1966.
146. J. Schwartz and M. D. Ward: *J. Mol. Catal.* **8**, 465 (1980).
147. G. L. Karapinka: *US Patent* 3,709,853 (1973); *Chem. Abs.* **78**, 85087 (1973).
148. F. J. Karol and R. N. Johnson: *J. Polym. Sci., Polym. Chem. Ed.* **13**, 1607 (1975).
149. F. J. Karol, C. Wu, W. T. Reichle and N. J. Maraschin: *J. Catal.* **60**, 68 (1979).
150. F. J. Karol, W. L. Munn, G. L. Goeke, B. E. Wagner and N. J. Maraschin: *J. Polym. Sci., Chem. Ed.* **16**, 771 (1978).
151. V. A. Zakkarov, V. K. Dudchenko, A. I. Min'kov, O. A. Efimov, L. G. Khomyakova, V. P. Babenko and Yu. I. Yermakov: *Kinet. Katal.* **17**, 643 (1976).
152. V. A. Zakharov, V. K. Dudchenko, E. A. Paukshits, L. G. Karakchiev and Yu. I. Yermakov: *J. Mol. Catal.* **2**, 421 (1977).
153. Yu. I. Yermakov, B. N. Kuznetsov and Yu. A. Ryndin: *Kinet Katal.* **19**, 169 (1978).
154. Yu. I. Yermakov, A. N. Startsev, V. A. Burmistrov and B. N. Kuznetsov: *React. Kinet. Catal. Lett.* **14**, 155 (1980).

155. B. N. Kuznetsov, V. L. Kuznetsov and Yu. I. Yermakov: *Kinet. Katal.* **16**, 790 (1975).
156. Y. Iwasawa, M. Yamagashi and S. Ogasawara: *JCS Chem. Commun.* 246 (1982).
157. S. A. Panichev, G. V. Kudryavtsev and G. V. Lisichkin: *Koord. Khim.* **5**, 1141 (1979).
158. J. Hjortkjaer, M. S. Scurrell and P. Simonsen: *J. Mol. Catal.* **10**, 127 (1980).
159. M. D. Ward and J. Schwartz: *J. Mol. Catal.* **11**, 397 (1981).
160. S. J. DeCanio, H. C. Foley, C. Dybowski and B. C. Gates: *J. C. S. Chem. Commun.* 1372 (1982).
161. H. C. Foley, S. J. DeCanio, K. D. Tau, K. J. Chao, J. H. Onuferko, C. Dybowski and B. C. Gates: *J. Amer. Chem. Soc.* **105**, 3074 (1983).
162. D. H. Ralston: *Diss. Abs.* **B41**, 4128 (1981).
163. L. K. Przheval'skaya, D. B. Furman, V. A. Shvets, S. S. Zhokovskii, T. M. Kharitonova, G. N. Bondarenko, A. M. Taber, V. B. Kazarskii, O. V. Bragin, I. V. Kalechits and V. E. Basserberg: *Kinet. Katal.* **19**, 1036, 1283 (1978).
164. D. C. Bailey and S. H. Langer: *Chem. Rev.* **81**, 109 (1981).
165. J. Evans: *Chem. Soc. Rev.* **10**, 159 (1981).
166. F. G. Ciapetta and C. J. Plank in: *Catalysis* (Ed. by P. H. Emmett) Reinhold, New York, Vol. 1, p. 315 (1954).
167. G. W. Higginson: *Chem. Eng.* **81** (20), 98 (1974).
168. W. B. Innes in: *Catalysis* (Ed. by P. H. Emmett) Reinhold, New York, Vol. 1, p. 245 (1954).
169. A. Brenner and R. L. Burwell: *J. Catal.* **52**, 353 (1978).
170. P. S. Braterman: *Metal Carbonyl Spectra*, Academic Press, New York, 1975.
171. R. F. Howe and C. Kemball: *J. Chem. Soc., Faraday Trans. I* **70**, 1153 (1974).
172. R. F. Howe and I. R. Leith: *J. Chem. Soc., Faraday Trans. I* **69**, 1967 (1973).
173. D. A. Hucul: *Diss. Abs.* **B41**, 1362 (1980).
174. A. Brenner and D. A. Hucul: *J. Catal.* **61**, 216 (1980).
175. A. Brenner and D. A. Hucul: *J. Amer. Chem. Soc.* **102**, 2484 (1980).
176. A. Brenner, D. A. Hucul and S. J. Hardwick: *Inorg. Chem.* **18**, 1478 (1979).
177. E. Guglielminotti: *J. Mol. Catal.* **13**, 207 (1981).
178. A. Brenner and D. A. Hucul: *Inorg. Chem.* **18**, 2836 (1979).
179. A. Brenner: *J. Mol. Catal.* **5**, 157 (1979).
180. D. A. Hucul and A. Brenner: *J. Phys. Chem.* **85**, 496 (1981).
181. D. A. Hucul and A. Brenner: *J. Amer. Chem. Soc.* **103**, 217 (1981).
182. J. L. Bilhou, A. Theolier, A. K. Smith and J. M. Basset: *J. Mol. Catal.* **3**, 245 (1978).
183. J. L. Bilhou, V. Bilhou-Bougnol, W. F. Graydon, J. M. Basset, A. K. Smith, G. M. Zanderighi and R. Ugo: *J. Organometal. Chem.* **153**, 73 (1978).
184. R. F. Howe: *J. Catal.* **50**, 196 (1977).
185. D. B. Tkatchenko, G. Coudurier, H. Mozzanega and I. Tkatchenko: *J. Mol. Catal.* **6**, 293 (1979).
186. P. Gallezot, G. Coudurier, M. Priniet and B. Imelik: *ACS Symp. Ser.* **40**, 144 (1977).
187. R. L. Howe: *Inorg. Chem.* **15**, 486 (1976).
188. A. Brenner and D. A. Hucul: *Prepr., ACS Div. Petrol. Chem.* **22**, 1221 (1977).
189. R. L. Banks: *US Patent* 3,463,827 (1969); *Chem. Abs.* **75**, 151339 (1971).
190. R. L. Banks: *Belgian Patent* 633,418 (1963); *Chem. Abs.* **61**, 1690e (1964).
191. R. F. Howe, D. E. Davidson and D. A. Whan: *J. Chem. Soc. Faraday Trans. I* **68**, 2266 (1972).
192. A. Brenner and R. L. Burewell: *J. Amer. Chem. Soc.* **97**, 2565 (1975).
193. A. Kuzusaka and R. F. Howe: *J. Mol. Catal.* **9**, 183 (1980).
194. A. Kuzusaka and R. F. Howe: *J. Mol. Catal.* **9**, 199 (1980).

195. J. P. Van Linhoudt, L. Delmulle and G. P. van der Kelen: *J. Organometal. Chem.* **202**, 39 (1980).
196. C. L. Reichel and M. S. Wrighton: *J. Amer. Chem. Soc.* **103**, 7180 (1981).
197. K. L. Watters, R. L. Schneider and R. F. Howe: *Preprints Amer. Chem. Soc., Div. Petr. Chem.* **25**, 771 (1980).
198. A. K. Smith and J. M. Basset: *J. Mol. Catal.* **2**, 229 (1977).
199. C. U. Pittman and R. C. Ryan: *ChemTech* **8**, 170 (1978).
200. E. L. Muetterties, T. N. Rhodin, E. Band, C. F. Brucker and W. R. Pretzer: *Chem. Rev.* **79**, 91 (1979).
201. B. C. Gates and J. Lieto: *ChemTech* **10**, 195 (1980).
202. B. C. Gates and J. Lieto: *ChemTech* **10**, 248 (1980).
203. B. F. G. Johnson (Ed.): *Transition Metal Clusters*, John Wiley, Chichester, 1980.
204. A. Brenner: *JCS Chem. Commun.* 251 (1979).
205. F. Hugues, A. K. Smith, Y. B. Taarit, J. M. Basset, D. Commereuc and Y. Chauvin: *JCS Chem. Commun.* 68 (1980).
206. D. B. Tkatchenko, G. Coudurier, H. Mozzanega and I. Tkatchenko in: *Fundamental Research in Homogeneous Catalysis*, (Ed. by M. Tsutsui) Plenum Press, New York, Vol. 3, p. 257 (1979).
207. R. L. Jackson and M. R. Trusheim: *J. Amer. Chem. Soc.* **104**, 6590 (1982).
208. B. F. G. Johnson and J. Lewis: *Adv. Inorg. Radiochem.* **24**, 225 (1981).
209. J. Robertson and G. Webb: *Proc. Roy. Soc.* **A341**, 383 (1974).
210. A. F. Simpson and R. Whyman: *J. Organometal. Chem.* **213**, 157 (1981).
211. V. L. Payne: *Diss. Abs.* **B42**, 625 (1981).
212. J. G. Goodwin and C. Naccache: *Appl. Catal.* **4**, 145 (1982).
213. A. K. Smith, A. Theolier, J. M. Basset, R. Ugo, D. Commereuc and Y. Chauvin: *J. Amer. Chem. Soc.* **100**, 2590 (1978).
214. R. Psaro, R. Ugo, G. M. Zanderighi, B. Besson, A. K. Smith and J. M. Basset: *J. Organometal. Chem.* **213**, 215 (1981).
215. M. Deeba and B. C. Gates: *J. Catal.* **67**, 303 (1981).
216. M. Deeba, J. P. Scott, R. Barth and B. C. Gates: *J. Catal.* **71**, 373 (1981).
217. R. Ganzerla, M. Lenarda, F. Pinna and M. Graziani: *J. Organometal. Chem.* **208**, C43 (1981).
218. H. Knözinger and Y. Zhao: *J. Catal.* **71**, 337 (1981).
219. M. Deeba, B. J. Streusand, G. L. Schrader and B. C. Gates: *J. Catal.* **69**, 218 (1981).
220. G. Collier, D. J. Hunt, S. D. Jackson, R. B. Moyes, I. A. Pickering, P. B. Wells, A. F. Simpson and R. Whyman: *J. Catal.* **80**, 154 (1983).
221. B. Besson, A. Choplin, L. D'Ornelas and J. M. Basset: *JCS Chem. Commun.* 843 (1982).
222. D. Commerueuc, Y. Chauvin, F. Hugues, J. M. Basset and D. Olivier: *JCS Chem. Commun.* 154 (1980).
223. R. L. Schneider, R. F. Howe and K. L. Watters: *J. Catal.* **79**, 298 (1983).
224. M. Ichikawa: *J. Catal.* **56**, 127 (1979).
225. M. Ichikawa: *J. Catal.* **59**, 67 (1979).
226. J. R. Anderson, P. S. Elmes, R. F. Howe and D. E. Mainwaring: *J. Catal.* **50**, 508 (1977).
227. G. C. Smith, T. P. Chojnacki, S. R. Dasgupta, K. Iwatate and K. L. Watters: *Inorg. Chem.* **14**, 1419 (1975).
228. H. Conrad, G. Erth, H. Knözinger, J. Kuppers and E. E. Latta: *Chem. Phys. Lett.* **42**, 115 (1976).

229. P. Gelin, Y. B. Taarit and C. Naccache: *J. Catal.* **59**, 357 (1979).
230. K. L. Watters, R. F. Howe, T. P. Chojnacki, C.-M. Fu, R. L. Schneider and N.-B. Wong: *J. Catal.* **66**, 424 (1980).
231. A. K. Smith, F. Hugues, A. Theolier, J. M. Basset, R. Ugo, G. M. Zanderighi, J. L. Bilhou, V. Bilhou-Bougnol and W. F. Graydon: *Inorg. Chem.* **18**, 3104 (1979).
232. A. Theolier, A. K. Smith, M. Leconte, J. M. Basset, G. M. Zanderighi, R. Psaro and R. Ugo: *J. Organometal. Chem.* **191**, 415 (1980).
233. J. L. Bilhou, V. Bilhou-Bougnol, W. F. Graydon, J. M. Basset, A. K. Smith, G. M. Zanderighi and R. Ugo: *J. Organometal. Chem.* **153**, 73 (1978).
234. M. Ichikawa, K. Sekizawa, K. Shikakura and M. Kawai: *J. Mol. Catal.* **11**, 167 (1981).
235. M. S. Scurrell: *J. Mol. Catal.* **10**, 57 (1980).
236. J. R. Anderson and R. F. Howe: *Nature* **268**, 129 (1977).
237. K. Tanaka, K. L. Watters and R. F. Howe: *J. Catal.* **75**, 23 (1982).
238. K. Foger and J. R. Anderson: *J. Catal.* **59**, 325 (1979).
239. M. Ichikawa: *Bull. Chem. Soc. Japan* **51**, 2268 (1978).
240. G. B. McVicker and M. A. Vannice: *US Patent* 4,154,751 (1979); *Chem. Abs.* **91**, 76694 (1979).
241. N. D. Parkyns in: *Proc. Third Int. Congr. Catal.*, (Ed. by W. M. H. Sachtler, G. C. A. Schuit and P. Zweiterung) North-Holland, Amsterdam, Vol. 2, p. 914 (1965).
242. E. G. Derouane, J. B. Nagy and J. C. Vedrine: *J. Catal.* **46**, 434 (1977).
243. R. B. Bjorklund and R. L. Burwell: *J. Colloid Interface Sci.* **70**, 383 (1979).
244. A. A. Galinskii: *Poverkhn. Yavleniya Dispersnykh Sist.* **4**, 82 (1975); *Chem. Abs.* **88**, 95378 (1978).
245. A. A. Galinskii, N. P. Samchenko, P. N. Galich and A. M. Verblovskii: *Ukr. Khim. Zh.* **43**, 31 (1977); *Soviet Prog. Chem.* **43**, 30 (1977).
246. A. A. Galinskii, N. P. Samchenko and P. N. Galich: *Katal. Katal.* **14**, 61 (1976); *Chem. Abs.* **86**, 161696 (1977).
247. N. V. Pavlenko, N. P. Samchenko, P. N. Galich and A. I. Ponomarenko: *Kinet. Katal.* **14**, 66 (1976); *Chem. Abs.* **86**, 178029 (1977).
248. M. Ichikawa: *JCS Chem. Commun.* 26 (1976).
249. M. Ichikawa: *JCS Chem. Commun.* 11 (1976).
250. M. Ichikawa: *Japanese Kokai*, 77,65,201 (1977); *Chem. Abs.* **88**, 123625 (1978).
251. K. L. Watters: *Surface Interface Anal.* **3**, 55 (1981).
252. R. Bender and P. Braunstein: *JCS Chem. Commun.* 334 (1983).
253. D. D. Eley, D. A. Keir and R. Rudham: *J. Chem. Soc. Faraday Trans I* **72**, 1685 (1976).
254. J. C. W. Chien: *J. Catal.* **23**, 71 (1971).
255. K. Soga, K. Izumi, M. Teramo and S. Ikeda: *Makromol. Chem.* **181**, 657 (1980).
256. J. C. W. Chien and J. T. T. Hsieh: *J. Poly. Sci., Polym. Chem. Ed.* **14**, 1915 (1976).
257. F. Sato, H. Ishikawa, Y. Takahashi, M. Miura and M. Sato: *Tetrahedron Lett.* 3745 (1979).
258. P. Damyanov, M. Velikova and L. Petkov: *European Polym. J.* **15**, 233 (1979).
259. J. Skupinska and W. Skupinski: *React. Kinet. Catal. Lett.* **16**, 297 (1981).
260. W. Skupinski, I. Cieslowska and S. Malinowski: *J. Organometal. Chem.* **182**, C33 (1979).
261. D. Slotfeldt-Ellingsen, I. M. Dahl and O. H. Ellestad: *J. Mol. Catal.* **9**, 423 (1980).
262. O. A. Efimov, A. I. Min'kov, V. A. Zakharov and Yu. I. Yermakov: *Kinet. Katal.* **17**, 995 (1976); *Chem. Abs.* **85**, 160656 (1976).

263. D. R. Fahey and M. B. Welch: *US Patent* 4,199,475 (1980); *Chem. Abs.* 93, 47510 (1980).
264. B. E. Nasser and J. A. Delap: *US Patent* 4,188,471 (1980); *Chem. Abs.* 92, 164523 (1980).
265. J. C. W. Chien: *J. Amer. Chem. Soc.* 93, 4675 (1971).
266. A. A. Golub, V. V. Skoneko, A. A. Chuiko and V. V. Trachevskii: *Ukr. Khim. Zh.* 44, 237 (1978); *Chem. Abs.* 89, 12621 (1978).
267. J. C. W. Chien and J. T. T. Hsieh in: *Coordination Polymerisation* (Ed. by J. C. W. Chien) Academic Press, New York, p. 305 (1975).
268. W. Hanke, R. Bienert and H.-G. Jerschkewitz: *Z. Anorg. Allg. Chem.* 414, 109 (1975).
269. F. Roozeboon, T. Fransen, P. Mars and P. J. Gellings: *Z. Anorg. Allg. Chem.* 449, 25 (1979).
270. S. B. Nikishenko, A. A. Slinkin, Yu. K. Vail, I. I. Zadko, A. F. Mironov, D. A. Agievskii and A. L. Belozerov: *Kinet. Katal.* 16, 1560 (1975); *Chem. Abs.* 84, 170177 (1976).
271. W. L. Carrick, R. J. Turbett, F. J. Karol, G. L. Karapinka, A. S. Fox and R. N. Johnson: *J. Polym. Sci., A1* 10, 2609 (1972).
272. D. Gutschick, O. N. Kimkhai, B. N. Kuznetsov, A. N. Startsev, V. G. Shinkarenko and G. K. Boreskov: *Kinet. Katal.* 21, 436 (1980); *Chem. Abs.* 93, 32343 (1980).
273. V. N. Pak, A. N. Volkova, N. N. Kushakova, S. I. Koltsov and V. B. Aleskovskii: *Zh. Phiz. Khim.* 48, 2394 (1974).
274. Yu. I. Yermakov, B. N. Kuznetsov, I. A. Ovsyannikova, A. N. Startsev, S. B. Erenburg and M. A. Sheromov: *React. Kinet. Catal. Lett.* 8, 377 (1978).
275. A. N. Bashkirov, R. A. Fridman, S. M. Nosikova, L. G. Liberov and A. M. Bolshakov: *Kinet. Katal.* 15, 1607 (1974); *Chem. Abs.* 82, 155148 (1975).
276. J. R. Pearce, B. L. Gustafson and J. H. Lunsford: *Inorg. Chem.* 20, 2957 (1981).
277. G. Mercati and F. Morazzoni: *Inorg. Chim. Acta* 25, L115 (1977).
278. I. Mochida, K. Tsuji, K. Suetsugu, H. Fujitsu and K. Takeshita: *J. Phys. Chem.* 84, 3159 (1980).
279. J. C. Summers and S. A. Ausen: *J. Catal.* 52, 445 (1978).
280. A. R. Siedle, P. M. Sperl and T. W. Rusch: *Appl. Surf. Sci.* 6, 149 (1980).
281. J. J. Rooney and A. Stewart in: *Catalysis, Volume 1*, (Ed. by C. Kemball) *Chem. Soc. Spec. Per. Rep.* p. 277 (1977).
282. J. Vickerman in: *Catalysis, Volume 2*, (Ed. by C. Kemball and D. A. Dowden) *Chem. Soc. Spec. Per. Rep.* p. 107 (1978).
283. R. L. Banks in: *Catalysis, Volume 4*, (Ed. by C. Kemball and D. A. Dowden) *Chem. Soc. Spec. Per. Rep.* p. 100 (1980).
284. H. Pines: *The Chemistry of Catalytic Hydrocarbon Conversions*, Academic Press, New York, Chapter 7 (1981).
285. M. Akimoto, M. Usami and E. Echigoya: *Bull. Chem. Soc. Japan* 51, 2195 (1978).
286. S. Yoshida, Y. Magatani, S. Noda and T. Funabiki: *JCS Chem. Commun.* 601 (1981).
287. S. Yoshida, Y. Matsumura, S. Noda and T. Funabiki: *J. Chem. Soc. Faraday Trans I* 77, 2237 (1981).
288. B. Fubina, G. Ghiotti, L. Stradella, E. Garraone and C. Morterra: *J. Catal.* 66, 200 (1980).
289. D. D. Beck and J. H. Lunsford: *J. Catal.* 68, 121 (1981).
290. T. Imanaka, Y. Katoh, Y. Okamoto and S. Teranishi: *Chem. Lett.* 1173 (1981).
291. A. Andreini and J. C. Mol: *J. Colloid Interface Sci.* 84, 57 (1981).

292. R. K. Aliev, I. L. Tsitovskaya, A. A. Kadushin and O. V. Krylov: *React. Kinet. Catal. Lett.* **8**, 257 (1978).

293. A. K. Coverdale, P. F. Dearing and A. Ellison: *JCS Chem. Commun.* 567 (1983).

294. R. J. Haines and G. J. Leigh: *Chem. Soc. Rev.* **4**, 155 (1975).

295. J. C. Mol and J. A. Moulin: *Adv. Catal.* **24**, 131 (1975).

296. R. L. Banks and G. C. Bailey: *Ind. Eng. Chem. Product Res. Dev.* **3**, 170 (1964).

297. L. F. Heckelsberg, R. L. Banks and G. C. Bailey: *Ind. Eng. Chem. Product Res. Dev.* **7**, 29 (1968).

CHAPTER 4

CHARACTERISATION OF SUPPORTED CATALYSTS

The complete characterisation of supported catalysts is extremely difficult due to the inhomogeneous nature of the catalysts themselves. Accordingly a wide range of analytical and spectroscopic techniques have been applied to the problem. No one technique can do more than illuminate one facet of the structure. A typical characterisation procedure would involve a careful record of the conditions under which the supported catalyst had been prepared, microanalysis for as many of the elements present as possible followed by the use a number of spectroscopic techniques. The response of the supported catalyst to chemical reagents such as carbon monoxide, hydrogen and nitric oxide may also be used.

Clearly any attempt to characterise a supported catalyst will be severely handicapped if the nature and structure of the support itself is unknown. Although the characterisation of the supports lies outside the scope of the present book, readers are referred to [1] and [2] for polymer supports, and [3] and [4] for metal oxides.

4.1. Microanalysis

The commonest starting point for the characterisation of a supported catalyst is microanalysis. Modern microanalytical equipment requires very small amounts of material and, whilst this is clearly an advantage in terms of the amount of product consumed, care must be taken to ensure that the small sample taken is representative of the whole. Since most products are of limited homogeneity analysis of several samples may be necessary to get meaningful microanalytical data.

Ideally all the elements present should be analysed for, and this includes elements introduced as intermediates. Thus the phosphination of polystyrene, for example, is achieved in a number of steps. Each of these steps may be incomplete due to the heterogeneous nature of the substrate. Thus if phosphination has been effected by initial chloromethylation then a chlorine analysis should be undertaken in addition to carbon, hydrogen and phosphorus. When organic groups have been introduced on to the surface of metal oxides, carbon, hydrogen and nitrogen microanalyses may be undertaken in essentially the same way as for a conventional organic compound using catalytic combustion in oxygen [5].

Analysis for the metal content of supported catalysts is usually achieved by destruction of the substrate so yielding a solution of the metal. Polymeric substrates may be cleanly destroyed by oxidation using a combination of concentrated sulphuric acid and 30% aqueous hydrogen peroxide [6], which we have used consistently without any problems [7]. In contrast the use of perchloric acid instead of hydrogen peroxide is strongly discouraged as it has given rise on occasions to serious explosions [8]. A particularly common type of supported catalyst involves rhodium on a phosphinated support. It is tempting to send such samples to commercial analysts, many of whom use a gravimetric procedure in which the sample is heated to 2000 °C under a stream of hydrogen, and assume the residue to be rhodium metal. Such an assumption is invalid, as in the presence of phosphorus rhodium forms inert rhodium phosphides. Accordingly, alternative analytical techniques are necessary; of which atomic absorption spectroscopy, in the presence of excess lanthanum ions to suppress interference from phosphate and sulphate ions, is particularly convenient [7, 9–11]. Care should, however, he taken to ensure that the sample size is adequate to be representative of the total batch [7, 12], a point that has been amply demonstrated by analysing 109 different Amberlyte A–21 beads of between 0.3 and 1.0 mm diameter on to which chloroplatinic acid containing traces of palladium had been absorbed [12]. Destruction of the polymer is not always necessary prior to atomic absorption spectroscopy. Thus the direct electrothermal atomisation of a homogeneous suspension in an organic solvent of polyamide supported palladium complexes has been used to enable the palladium to be determined by atomic absorption spectroscopy [13, 13a]. A number of non-destructive methods for metal analysis have been developed, including X-ray fluorescence spectroscopy [14, 15], thermal neutron activation analysis [16, 17] and changed particle activation analysis [18]. All, however, suffer from the major drawback of being subject to many interferences necessitating the often difficult task of preparing standards in the same matrix as the materials under test.

4.2. Chromatographic Methods

4.2.1. GEL CHROMATOGRAPHY [2, 19–21]

In Section 1.2 it was emphasised that one way of promoting high selectivity is to support a metal complex within the interstices of an organic polymer. Diffusion of reactants to and products from the active site then introduces a measure of selectivity. Gel chromatography has been used to determine the maximum size and/or molecular weight of substrates that can penetrate into the polymer [22]. A conventional column is packed with the polymer supported catalyst, and the substrates under test dissolved in the solvent that will be used for the actual catalysis are added to the top of the column and then eluted using the same solvent. Those substrates excluded from the polymer are rapidly eluted from the base of the column.

4.2.2. TEMPERATURE PROGRAMMED DECOMPOSITION CHROMATOGRAPHY (TPDE) [23–27]

Temperature programmed decomposition chromatography provides a valuable technique for monitoring the progress of the chemical interaction between a metal complex and a support where that interaction gives rise to volatile by-products. It was initially developed [23] to study the chemical interaction between physically absorbed metal carbonyls and oxide surfaces (see Section 3.3.b). A solution of the metal complex is absorbed on to the support and the solvent evaporated with the aid of a stream of an inert gas such as helium. When the catalyst is dry, it is then steadily heated in a linear pre-programmed manner, with a steady helium flow. The gases evolved during this temperature programmed decomposition are passed through a cold trap at $-196\,^{\circ}C$ and then a catharometer. The cold trap removes carbon dioxide and hydrocarbons and the catharometer essentially detects the carbon monoxide present, since its response to hydrogen in a helium carrier is only 3% of its response to carbon monoxide. The effluent from this detector is then passed through a tube packed with silica at $-169\,^{\circ}C$. This removes carbon monoxide, leaving only hydrogen in the helium carrier. The hydrogen is diffused through a thin tube of palladium alloy into a stream of nitrogen. A second detector, having nitrogen as its reference gas, monitors the concentration of hydrogen. On completion of the temperature programmed decomposition the silica trap is warmed to room temperature and the gases back-flushed through a column of Spherocarb and then back through the first detector. This allows an accurate determination of carbon monoxide and methane. The first cold trap is then warmed to $-78\,^{\circ}C$ and carbon dioxide and light hydrocarbons (C_1–C_3) are analysed using a temperature programmed Spherocarb column. This system can detect gas evolution of 2×10^{-11} mol s^{-1} of carbon monoxide, 6×10^{-10} mol s^{-1} of hydrogen and 3×10^{-10} mol s^{-1} of hydrocarbons.

For a metal carbonyl absorbed on a metal oxide the decomposition reaction leading to hydrogen and carbon monoxide evolution can be written:

$$M(CO)_n(abs) + m(\sigma\text{-OH}) \xrightarrow{\Delta} (\sigma\text{-O}^-)_m M^{m+} + m/2\, H_2 + n\, CO \qquad (1)$$

Thus the evolution of hydrogen corresponds to oxidation of the metal. Some hydrocarbons are also formed, but the hydrogen in these also arises from surface hydroxyl groups which become surface oxide ions. Thus from the total hydrogen present as H_2 and hydrocarbons, the average 'oxidation number' of the metal can be calculated. Comparison of these with oxidation numbers determined independently by oxygen titration [28] shows good agreement, although the accuracy of temperature programmed decomposition oxidation numbers is only ±1 to ±1.5, whereas oxygen titration gives an accuracy of about ±0.3.

The temperature of the peak maximum, T_m varies as the heating rate, β, according to equation (2), where E_d is the activation energy for decomposition, A is the pre-exponential frequency factor, and R is the gas constant.

$$2 \ln T_m - \ln \beta = E_d/RT_m + \ln (E_d/AR) \tag{2}$$

E_d and A are evaluated by plotting $2 \ln T_m - \ln \beta$ against $1/T_m$ for runs performed at a series of heating rates, β [28]. However a more accurate value can be obtained by setting

$$A = kT_m/h \tag{3}$$

where k = Boltzmann's constant and h = Planck's constant. The value of T_m now determines E_d and a tenfold error in the value of A only shifts E_d by about 2 kcal mol^{-1} [28].

4.3. Spectroscopic Methods

A great deal of the information currently available about the nature of supported metal complex catalysts has been obtained by spectroscopic methods. Spectroscopic techniques have been described as 'sporting' techniques because they have a 'sporting chance' of giving the right answer. This must always be borne in mind when interpreting spectroscopic data especially with some inhomogeneous substrates, and the more different kinds of data that can be obtained on the same sample in the same environment the greater is the chance of reaching an accurate description of the catalyst. A brief summary of the main spectroscopic techniques is given in Table I.

4.3.1. INFRARED

Infrared spectroscopy is the most widely used technique for the study of supported metal complex catalysts, partly because of its simplicity and the relative cheapness and widespread availability of infrared spectrometers. This latter also means that a great deal of data has been accumulated about the infrared spectra of both the supports and the metal complexes on their own and it is this vast

TABLE I

Spectroscopic Technique	Availability of Equipment	Comments
Infrared	Readily available	Probably the most widely used technique. Fourier transform ir will extend its value, but at the penalty of significantly enhanced cost.
Raman	Commercially available	Rarely used, because only in occasional situations such as the study of metal-metal bonds does it have an advantage over infrared.
Inelastic Electron Tunnelling Spectroscopy	Not widely available, can readily be homemade	Potentially as valuable as infrared; results contain the same information as infrared plus Raman spectra.
Ultra-violet and Visible	Readily available	Difficult to obtain useful, precisely interpretable data. Hence not widely used.
NMR	Readily available for solution studies; equipment for high resolution solid state studies is now commercially available	Very valuable and with the availability of high resolution solid state nmr will become increasingly useful.
ESR	Fairly widely available	Only applicable to paramagnetic species; very useful for indicating presence of such species; can be used to study site separation.
Mössbauer	Not widely available	Useful for a limited number of metals; can be studied under the same temperature, pressure and sample conditions as catalyst is used under.
Mass spectrometry	Widely available	Only used to characterise functionalised polymers to date.
ESCA	Fairly widely available	Very useful within the limitation that sample must be studied *in vacuo*.
EXAFS	Ideally requires a Synchrotron Radiation Facility	Like to become more important because of its ability to bridge crystalline solid, amorphous solid, solution gap. However requires very expensive X-ray source, which severely limits availability.

infrastructure of data that enables infrared spectroscopy to be used with some confidence. The development of Fourier transform infrared spectrometers should further enhance the use of the technique particularly in situations where relatively low concentrations of species are present [29]. However, this extension will be at the expense of price and hence availability of equipment. Infrared spectroscopy has been used both to characterise the supported catalyst and to study the nature of the catalytic reaction itself. Many of the techniques developed for the study of heterogeneous catalysts are directly relevant to supported metal complex catalysts so that the reader is referred to a valuable article describing the use of infrared in that field [30a].

Infrared spectra may be recorded either by transmission or reflection of which the former is much preferred. Metal oxides such as silica and alumina may be pressed into clear, self-supporting discs of several millimetres thickness; the thickness required depends on the amount of metal complex present. The detailed techniques of oxide pretreatment, pressing load and sample thickness vary from laboratory to laboratory but representative techniques for alumina both as pressed wafers [29, 31–34] and as discs up to 1 cm thick [35], silica [36] and magnesium oxide [36] have been described. The transparency is always improved when the oxide particle size is small relative to the radiation wavelength so minimising scattering. Other things being equal discs prepared at low pressing loads are to be preferred since this largely preserves the open structure of the oxide particles. In addition to pressed discs, spectra have also been recorded on nujol suspensions of silica and alumina supported metal complexes [37] as well as in alkali metal halide discs. Nujol and hexachlorobutadiene suspensions may be prepared by placing a 'pile' of the supported catalyst on the infrared window and adding sufficient liquid from a Pasteur pipette to thoroughly wet all the oxide. A second window is placed on top and the two windows gently pressed together making sure the material under test remains in the centre of the window. This is relatively easy with the fairly viscous nujol but more difficult with the less viscous hexachlorobutadiene. However, with practice and care it can be accomplished. Lightly clamping the plates in a standard infrared sample holder helps. The same technique has been used for polymer supported complexes where the polymer cannot be either ground under liquid nitrogen or compressed into a disc without heating, as is the case with polypropylene [38].

Infrared spectra of complexes supported on alumina have been obtained by plating the alumina sample out of a chloroform slurry on to the inner surface of the window of an infrared gas cell which was jointed to provide ready access [39]. However this technique gives rise to a lot of scattering so that relatively wide slit settings must be used to obtain adequate signal levels.

Reflectance is relatively rarely used in the infrared although it has been used in the near ir using a commercial reflectance attachment [40] and in the main ir using a purpose built cell [41]. Reflectance, of course, concentrates on the species present on the surface of the material whereas transmittance provides information on the species present throughout the material.

The infrared spectra of polystyrene-supported metal complex catalysts have been recorded by grinding the material, or crushing it in a ball mill, then mixing it with KBr or polyethylene and preparing a disc [42–44] or recording the spectrum as a nujol mull [43]. As already mentioned above, a nujol suspension can be used with polymers such as polypropylene which can neither be ground nor pressed into pellets without heating [38].

As infrared approach has been used to study the reactions occurring within a polymer supported rhodium methanol carbonylation catalyst [45]. The catalyst was prepared by chloromethylation of a 7μm thick polystyrene–5% divinylbenzene copolymer which was phosphinated using $LiPPh_2$ and the rhodium introduced by phosphine exchange with $[RhCl(CO)(PPh_3)_2]$. The catalyst film was suspended within a heated infrared gas cell normal to the beam. A mixture of methanol, carbon monoxide, methyl iodide and helium was then passed through the sample infrared cell and then through a reference cell that was identical apart from having no catalyst film in it. By arranging the conditions to be such that the conversion was low the compositions of the vapours in the two cells were essentially identical, so that the resulting infrared spectra were characteristic of the catalyst membrane and the reactants chemically bonded to it [45, 46]. Although in this case the infrared cells were operated at room temperature, infrared cells are capable of operating up to 59 atmospheres [47–49].

4.3.2. RAMAN

Raman spectroscopy [30b] is complementary to infrared spectroscopy for molecules with high symmetry. However, for the low symmetry species generally encountered as supported metal complex catalysts the two techniques often give essentially the same information. The widespread availability of infrared spectrometers and the relative scarcity of laser-Raman spectrometers has resulted in relatively few Raman studies being reported, unless Raman spectroscopy has a clear cut advantage. The low intensity of the Raman scattered radiation makes such an advantage rare. However, the need to study a catalyst in the presence of water could provide such an advantage although many reactions catalysed by supported metal complex catalysts occur in non-aqueous systems. A second area where Raman has an advantage over infrared is in the detection and characterisation of metal-metal bonds and it has been used in this role to study the species formed when $[Os_3(CO)_{12}]$ and $[H_2Os_3(CO)_{10}]$ are supported on γ-alumina [50].

4.3.3. INELASTIC ELECTRON TUNNELLING [51]

Inelastic tunnelling spectroscopy is concerned with the inelastic tunnelling of electrons through an insulating barrier that is situated between two conducting electrodes when a bias voltage is applied across the two conductors. The second derivative $d^2I/d(eV)^2$ of the current (I) against voltage (eV) plot of a tunnelling junction is analogous to an optical spectrum. Since $eV = h\nu$ the base line can be plotted as wavenumber, which is typically in the range $240 - 4000$ cm^{-1}. For molecules adsorbed at the junction interface, the molecular symmetry is reduced so that infrared-active, Raman active and optically-forbidden transitions are allowed and observed in inelastic tunnelling spectroscopy [52]. Since the intensities of the infrared- and Raman-like modes are very similar in tunnelling spectroscopy it provides a powerful tool for looking at the interaction of substrate molecules with surfaces.

4.3.4. ULTRAVIOLET AND VISIBLE [30c]

Ultraviolet and visible spectroscopy have rarely been used to characterise support metal complex catalysts. There are several reasons for this. Many complexes of interest give rise to broad charge transfer absorption bands, so that UV-visible spectroscopy does not provide a very sensitive technique for determining the nature of the catalytic sites. Where ligand field bands sensitive to the nature and geometrical arrangement of the ligands about the metal are observable again the broadness of the absorption bands largely inhibits precise conclusions. The observation of the spectra themselves is generally far from easy due to the nature of the samples. Sometimes the samples can be pressed into discs and examined in the same way as for infrared although molar absorptivities which can provide valuable assistance in band assignments are then only obtained very approximately. The spectra of some samples have been obtained as nujol or chloroform slurries [53] but these solvents are only able to reduce not eliminate the scattering the occurs. Where UV-visible spectra are recorded, they are normally interpreted by comparison with the spectra of homogeneous samples.

4.3.5. NUCLEAR MAGNETIC RESONANCE

NMR has proved to be the most valuable spectroscopic technique of all for the characterisation of chemical compounds. Early high resolution NMR concentrated on protons, but with the development of Fourier transform NMR spectrometers ^{13}C and ^{31}P became readily accessible. More recent developments in multinuclear NMR spectrometers have resulted in solution studies of virtually every 'NMR-active' nucleus [54]. Initially high resolution NMR spectra could only be obtained in solution where the rapid isotropic motions of the nuclei averaged out the anisotropic interactions that give rise to broad absorptions in the solid state. Since one of the attractive features of supported metal complex catalysts was their insolubility NMR was not immediately of value in their study, although there were a number of reports of NMR studies of suspensions of polymer supported catalysts in suitable organic solvents and even in aqueous suspensions [55–57]. Thus ^{31}P NMR can be used to determine the relative amounts of free and coordinated phosphine as well as of phosphine oxide on a phosphinated styrene-2%-divinylbenzene polymer by recording the spectrum of a swollen suspension in toluene in the presence and absence of [RhCl(cyclo-octene)$_2$]$_2$ [57]. Similarly the presence of free and coordinated phosphine groups on phosphinated silica that has been treated with [Rh(acac)(CO)$_2$] has been demonstrated by recording the ^{31}P NMR spectrum of a suspension of the material in toluene [58]. The resolution is, however, very low.

Recently, the development of high-power decoupling [59], cross-polarisation [60] and magic-angle spinning [61–64] techniques has allowed 'high resolution' NMR studies of dilute nuclei to be performed on solid samples under variable temperature conditions [65–67]. The technique depends on the fact that very rapid rotation of a solid sample about an axis inclined at an angle of 54° 44′ (the 'magic angle') to the direction of the applied magnetic field can remove many sources of broadening from the NMR spectrum of a solid so enabling the finer features to be revealed. The 'magic angle' arises because the dipolar interaction Hamiltonian is proportional to $(3 \cos^2 \theta - 1)$ and this term has the value zero at $\theta = 54° 44′$.

The power and limitations of cross-polarisation/magic-angle spinning (CP/MAS) NMR spectroscopy for characterising supported metal complex catalysts are well illustrated by Figures 1 and 2. Figure 1a shows the spectrum of *cis*-[PtCl$_2$ {Ph$_2$PCH$_2$CH$_2$Si(OEt)$_3$}$_2$] which exhibits three central components due to the presence of three phosphorus environments with a single crystal. On supporting this complex on activated silica gel a single broad absorption ($\nu_{1/2} \approx 500$ Hz) is observed (Figure 1b). A broad absorption arises from the highly disordered environment of the surface-immobilised species. The *cis*-geometry of the immobilised species is confirmed by the retention of a ^{195}Pt–^{31}P coupling of 3633 Hz which contrasts with the smaller coupling of 2656 Hz observed when *trans*-[PtCl$_2$ {Ph$_2$PCH$_2$CH$_2$Si(OEt)$_3$}$_2$] was immobilised on silica gel [68, 69]. Figure 2 shows the ^{31}P NMR spectrum of [Rh(Ⓟ—PPh$_2$)-(acac)CO] where Ⓟ—PPh$_2$ is *p*-styrylPPh$_2$ γ-radiation grafted on to polypropylene [70]. The strong absorption at −5.4 ppm is due to uncomplexed phosphine groups whilst the broad absorption at +50 ppm arises from the rhodium(I) complex: *cis*-[Rh(*p*-styrylPPh$_2$)(acac)CO] has a ^{31}P NMR chemical shift in the solid state of +46.7 ppm relative to 85% phosphoric acid.

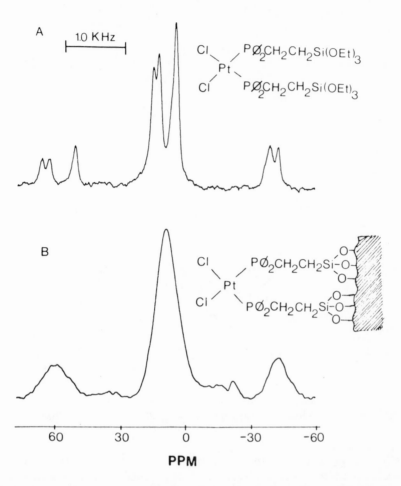

Fig. 1. Cross-polarisation – Magic Angle Spinning ^{31}P NMR spectra of *cis*-[PtCl$_2$ {Ph$_2$-PCH$_2$CH$_2$Si(OEt)$_3$}$_2$] (a) as the solid complex and (b) after immobilisation on silica gel (reproduced with permission from [69]).

^{29}Si CP/MAS NMR has been used to study the nature of the reaction of Me$_3$SiCl with silica gel [71], a reaction that can be used to remove surface hydroxyl groups (see Section 2.5). Before reaction the spectrum shows three silicon environments (Figure 3a), whereas after incomplete reaction a fourth due to —OSiMe$_3$ surface groups is clearly visible (Figure 3b) [71].

Proton NMR spectra have been recorded on solid samples such as the product formed by treating silica with [Rh(allyl)$_2$] [72, 73]. In general the resolution

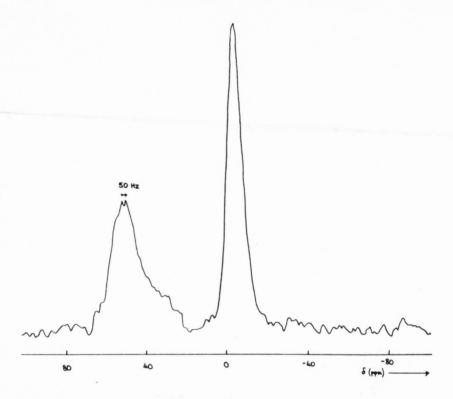

Fig. 2. Cross-polarisation – Magic-Angle Spinning solid state ^{31}P NMR spectrum of [Rh-((P)–PPh$_2$)(acac)(CO)], where (P)–PPh$_2$ is p-styrylPPh$_2$ γ-radiation grafted on to poly-propylene (reproduced with permission from [70])

is inadequate to distinguish η^1- and η^3-bonded allyls, although in favourable cases peak widths at half-height as narrow as 1.7 ppm have been observed.

4.3.6. ELECTRON SPIN RESONANCE

ESR has been used extensively to study the paramagnetic species that exist on various supported catalysts. The technique can only be applied where isolated paramagnetic molecules or ions are present. These should preferably have only one unpaired electron and have neither very short nor very long relaxation times. The technique therefore has specific, rather than general, applications. Although the extent of information available from ESR theoretically varies from simple confirmation that an unknown paramagnetic species is present to a detailed description of the bonding and orientation of a surface complex, in practice

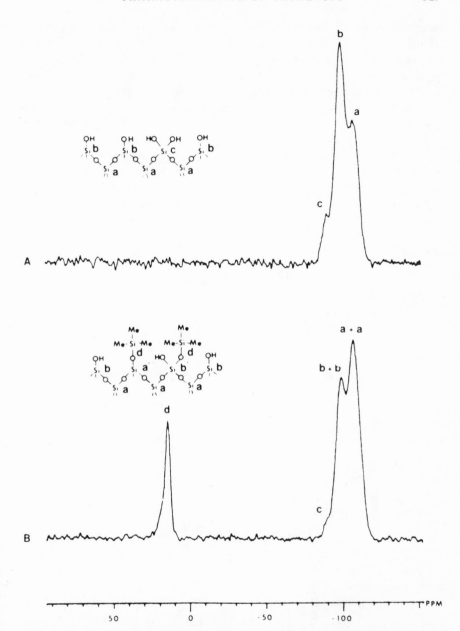

Fig. 3. Cross-polarisation – Magic Angle Spinning ^{29}Si NMR spectra of silica gel (a) before and (b) after reaction with trimethylchlorosilane. Chemical shifts are in ppm relative to liquid tms; larger numbers correspond to lower shielding (reproduced with permission from [71]).

samples of supported metal complex catalysts give spectra which are the envelope of the spectra from all possible orientations of the radical with respect to the magnetic field. To obtain meaningful data, it must be possible to extract the principal g and hyperfine values. A relatively straightforward analysis of such spectra can be made provided the resolution is adequate [30d].

ESR has been used to study copper(II) on poly-4-vinylpyridine [74] on polymer supported chelating amines and on Schiff bases [53, 75], cobalt(II) on polymer supported Schiff bases [75] and on phthalocyanines [76], titanium(IV) on 4-vinylpyridine-divinylbenzene copolymers [77], on grafted polyethylene-polymethylvinylketone [78] and on grafted polyethylene-polyacrylonitrile [78], vanadium(IV) and molybdenum(V) on grafted polyethylene-polyallyl alcohol [78]. The introduction of spin labels on to supports has been used to study the penetration of functional groups into supports as well as the distance between the supported radicals. Examples of such spin labels are shown in reactions (2–4) [78–80];

$$\text{(2)}$$

$$\text{(3)}$$

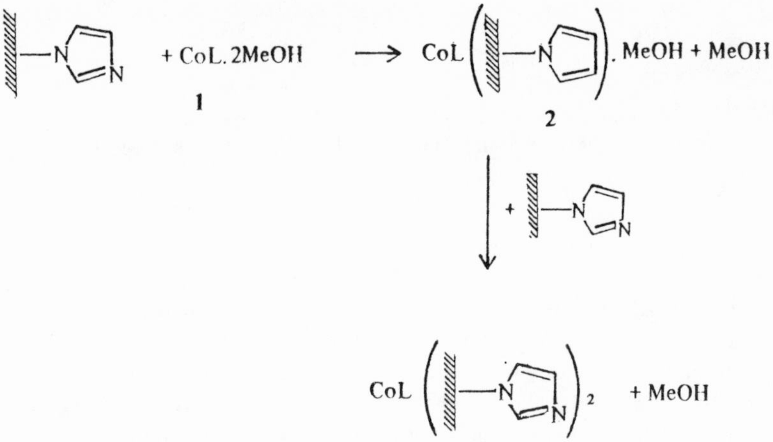

$$CoL\left(\text{❙}-N\underset{N}{\overset{}{\diagup}}\right)_2 + MeOH \qquad (4)$$

3

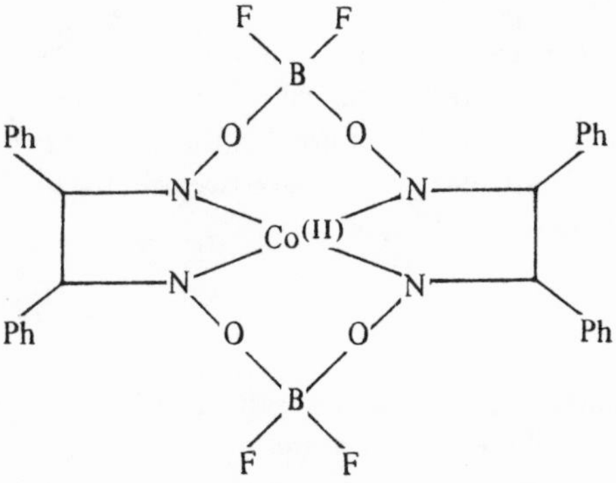

4

2 gives a three-line hyperfine splitting and **3** gives a five-line hyperfine splitting. Site separation can be demonstrated by the absence of any interaction between isolated radicals of **4** on the support [80]. However where such interaction is present then it is possible to determine the effective distances between the paramagnetic centres from the second moments of the ESR spectra [79, 81].

4.3.7. MÖSSBAUER

Although limited in the range of metal ions to which it is applicable (Figure 4), Mössbauer or nuclear gamma resonance spectroscopy is a very valuable technique for studying supported metal complex catalysts [30e, 82]. This

				Fe	Ni		Zn		Ge					Kr
			Tc	Ru						Sn	Sb	Te	I	Xe
Cs	Ba	Hf	Ta	W	Re	Os	Ir	Pt	Au	Hg				
		Pr	Nd	Pm	Sm	Eu	Gd	Tb	Dy	Ho	Er	Tm	Yb	Lu
	Th	Pa	U	Np		Am								

Fig. 4. Mössbauer Active Elements

is because its extremely high resolution enables nuclear energy levels to be determined to 1 part in 10^{14}, which is sufficiently precise to allow the weak interactions of the nucleus with the surrounding electronic environment to be readily detected. Many of the isotopes which can be observed by Mössbauer spectroscopy can be studied in microcrystalline samples under temperature and pressure conditions which reflect those under which the catalyst operates. Thus Mössbauer spectroscopy can yield data of considerable significance to the chemical state of the catalyst in operation.

Although most studies have been concerned with iron [75, 83, 84] Mössbauer has also been used to study the oxidation state and chemical nature of supported ruthenium complexes [85]. Of particular interest is the study of iron(II) supported on a polymer bound Schiff base where the susceptibility of the iron(II) to oxidation could be readily observed [75].

4.3.8. MASS SPECTROMETRY

Mass spectrometry has not been widely used in the characterisation of supported metal complex catalysts because of the obvious lack of volatility of most catalysts. However, we have used a direct insertion probe to show that γ-radiation grafting of 4-vinylpyridine on to polypropylene yields a product in which several 4-vinylpyridine units are grafted onto a single polymer backbone site [86], 5. In a similar way mass spectrometry has been used to determine the

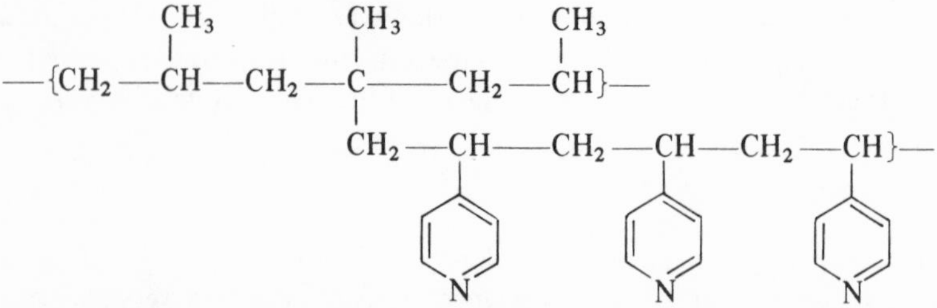

(5)

degree of ring functionalisation of polystyrene [87]. For this it was found that low molecular mass fragments ($m/z < 150$) were unsuitable because they could originate from a number of fragmentation pathways. The method was particularly useful for studying polystyrene bromination where the degree of bromination could be determined with an accuracy of about 3%, which is comparable to the accuracy of microanalysis (2%). An attempt to determine directly the degree of polystyrene substitution by $-PPh_2$ was unsuccessful because fragments containing the $-PPh_2$ give intensities which are either too strong relative to, or overlap with, bromine-containing fragments. However, the degree of $-PPh_2$ substitution can be determined conveniently by determining the degree of bromination before and after reaction with lithium diphenylphosphide. Attempts to study transition metal complexes supported on phosphinated polystyrene have so far met with little success [87].

4.3.9. ESCA

Electron spectroscopy for chemical analysis (ESCA) or X-ray photoelectron spectroscopy (XPS) has been widely used since its discovery in the characterisation of supported metal complex catalysts [30f, 88]. ESCA gives information about the binding energies of the core electrons [89] and is closely related to ultraviolet-induced photoelectron spectroscopy which is concerned with the valence electrons in gases [90].

The features which make ESCA so valuable are its ability to monitor the chemistry occurring in the outermost layer of the catalyst and, irrespective of the elements present (apart from hydrogen and helium), to detect changes in the relative concentrations of surface atoms. The extreme surface sensitivity of ESCA arises because the kinetic energy of the ejected electrons must be less than the photon energy and electrons with energies up to 1500 eV can only escape from less than 20Å below the surface. However whilst ESCA can determine relative concentrations in a series of closely related catalysts, it is rarely, if ever, possible to be quantitative. This is because of the complex range of factors which affect signal intensity including the photoelectron cross-section, atomic number and chemical state of the atom, escape depth, sample orientation with respect to the collector slit, surface roughness, surface geometry, distribution of species within the sampling depth, phase homogeneity and contamination. One of the limitations of ESCA is that samples must be studied under a vacuum of about 10^{-8} torr, so that *in situ* studies are impossible. A second limitation is that the rhodium $3d$ photoelectron spectra of supported rhodium complexes are complicated by X-ray induced decomposition effects and overlap with the carbon $1s$ inelastic scattering peak as well as various peaks from other elements [91].

ESCA has been used to study a wide variety of supported catalysts [92] including MoO_x supported on β-TiO_2 metathesis catalysts [93], $[W(\pi$-$C_4H_7)_4]$ on silica [94], ruthenium-ammine complexes on Y-type zeolites [95], cobalt complexes on both polymers and alumina [96, 97], $[Rh_6(CO)_{16}]$ on alumina [98], rhodium(I)-phosphine complexes on silica and alumina [99, 100], rhodium on phosphinated polystyrene and polyimines [101], polyamide supported $[Rh(PPh_3)_3Cl]$ [102] and palladium(II) on phosphinated [103, 104] and aminated [105, 106] supports. ESCA can be used to study the distribution of the supported metal complex on a surface [106]. In the interaction of $[Rh(PPh_3)_3Cl]$ with polyamides ESCA showed that the chloride ligand remained coordinated to rhodium(I) since the Cl_{2p} peak which lies at 196.5 eV in $[Rh(PPh_3)_3Cl]$ is only shifted to 196.6 eV in the supported complexes [102]. When $[Rh(PPh_3)_2(CO)Cl]$ supported on silica and alumina were exposed to carbon monoxide ESCA demonstrated that on alumina the initial product was $[Rh(PPh_3)(CO)_2Cl]$ with a dimeric species being formed when high carbon monoxide pressures were used, whereas on alumina the initial products were a mixture of four- and five-coordinate dicarbonyls which subsequently give two separate dimeric species [100]. ESCA has been used to resolve controversies such as whether the interaction of $[Rh_6(CO)_{16}]$ with alumina results in fragmentation accompanying the oxidation or not; the evidence suggests it does not, the Rh_6 unit being retained [98]. ESCA has been used to follow the chloromethylation of polystyrene (through the Cl_{2p} signal), its subsequent phosphination (through the decay, complete or partial, of the Cl_{2p} signal and build up of the P_{2p} signal) and later reaction with a metal complex such as rhodium(I) (through the $Rh_{2p_{3/2}}$ signal) [107].

Whilst ESCA is a very useful and powerful technique it is not always the only, or indeed the best technique to use. For example the nature of the green product formed when $[Pd(PhCN)_2Cl_2]$ is introduced on to phosphinated polystyrene was more readily shown to be palladium(0) by chemical reduction of benzoquinone than by ESCA because of the formation of $Pd(0)_{ads}$ by the palladium(0) [104].

4.3.10. EXTENDED X-RAY ABSORPTION FINE STRUCTURE

Extended X-ray absorption fine structure (EXAFS) [108] probes the local environment of a particular element. It is applicable in solution, amorphous and crystalline solids and is therefore well suited to the study of supported metal complex catalysts [109–111]. The technique depends on having an intense source of monochromatic X-rays, for which a synchrotron radiation source is ideal [112]. When an X-ray photon is absorbed by an electron in the K shell of a particular atom, the electron will be ejected. This is equivalent to spherical waves radiating from the atom. Whenever these waves encounter other atoms they are partially reflected and so interfere with the outgoing wave. The net interference pattern at the nucleus of the initially absorbing atom will then modify the cross-section for the inital process. As a result there is a sharp initial threshold corresponding to the minimum energy needed to excite the K electron. Fine structure, extending to several hundred volts higher in energy, is produced by the progressively shortening wavelength of the spherical wave, yielding additional nodes in the pattern. A Fourier transform unravelling of this fine structure can yield internuclear distances with an accuracy of about ±0.01 Å.

EXAFS has been used to determine the nature of $[Rh(PPh_3)_3X]$, X = Cl, Br on phosphinated polystyrene. $[RhP_2X]_2$, P = PPh_3 or phosphinated polystyrene, is the major species present on 2% crosslinked material whereas $[RhP_3X]$ is the major species present on 20% crosslinked material [113–115]. EXAFS has also been used to determine the nature of the species formed when trinuclear osmium clusters are grafted on to alumina and tethered to thiolated silica [116]. Undoubtedly more studies will be reported in the future – although synchrotron sources are scarce – because of the ability of the technique to bridge the solid-solution boundary [117–121]. The sensitivity may be enhanced by several orders of magnitude by detecting the X-ray fluorescence rather than absorption.

4.4. Electron Microscopy

Electron microscopy, especially when supplemented with an electron probe microanalyser, provides a very effective tool for determining the distribution of species within a polymer bead [123, 124]. The major problem is specimen preparation. A convenient technique involving embedding polystyrene beads in an epoxy resin followed by microtome section to produce 7–10 μ sections has been described [124]. Phosphorus and rhodium analyses showed that when 2% styrene-divinylbenzene is chloromethylated by Pepper's procedure [125], subsequent reaction with lithium diphenylphosphide gives even phosphination throughout 300–700 μ beads. Reaction of these beads with $[Rh(PPh_3)_3Cl]$ and $[Rh(cyclooctene)_2Cl]_2$ results in initial coordination of rhodium near the surface, but if sufficient rhodium complex is used and the reaction carried on for sufficient time rhodium(I) becomes evenly distributed throughout the polymer [124]. The evenness of phosphination does depend on the degree of cross-linking and the pore sizes within the polystyrene; large pores (~1300 Å) allow even phosphination whereas small pores (<50 Å) allow only limited penetration [126].

When small clusters, such as $[Ir_4(CO)_{12}]$, are reacted with phosphinated silica to form $[Ir_4(CO)_{14}(\{Si\}(CH_2)_3PPh_2)]$ electron microscopy can be used to detect those clusters, which are about 6 Å is diameter [127]. In particular it is possible to determine whether or not any aggregation of clusters occurs; it does not. Electron microscopy has been used in a similar way to examine aggregates in 4-vinylpyridine-grafted ethylene-propylene rubbers treated with nickel acetylacetonate [128–130].

References

1. R. B. Seymour: *Introduction to Polymer Chemistry*, McGraw-Hill, New York, Chapters 3 and 12 (1971).
2. F. W. Billmeyer: *Textbook of Polymer Science*, Wiley-Interscience, New York, Third Edition (1979).
3. H. P. Boehm: *Adv. Catal.* **16**, 179 (1966).
4. D. G. H. Ballard: *Adv. Catal.* **23**, 263 (1973).
5. G. V. Filonenko, S. B. Grinenko and N. N. Lysova, *Katal. Katal.* **18**, 98 (1980).
6. J. L. Down and T. T. Gorsuch: *Analyst* **92**, 398 (1967).
7. F. R. Hartley, S. G. Murray and P. N. Nicholson: *J. Organometal. Chem.* **231**, 369 (1982).
8. T. T. Gorsuch: *The Destruction of Organic Matter*, Pergamon Press, New York (1970).
9. W. R. Bramstedt, D. E. Harrington and K. N. Andrews: *Talanta* **24**, 665 (1977).
10. J. A. Pajares, P. Reyes, L. A. Oro and R. Sariego: *J. Mol. Catal.* **11**, 181 (1981).
11. N. M. Potter: *Anal. Chem.* **50**, 769 (1978).
12. R. Rericha, A. Vitek, D. Kolihova, V. Sychra, Z. Sir and J. Hetflejs: *Coll. Czech. Chem. Comm.* **44**, 3183 (1979).
13. S. Chiricosta, G. Cum, R. Gallo, A. Spadaro and P. Vitarelli: *At. Spectrosc.* **3**, 185 (1982).
13a. R. Rericha, A. Vitek, D. Kolihova, V. Sychra, Z. Sir and J. Hetflejs: *Coll. Czech. Chem. Comm.* **44**, 3183 (1979).
14. L. Leoni, G. Braca, G. Sbrana and E. Gianetti: *Anal. Chim. Acta* **80**, 176 (1975).
15. L. D. Hulett, H. W. Dunn and J. G. Tartar: *J. Radioanal. Chem.* **43**, 541 (1978).
16. E. L. Steele and W. W. Meincke: *Anal. Chim. Acta* **26**, 269 (1962).
17. K. C. Cambell, J. W. Reid and D. Gibbons: *Radiochem. Radioanal. Lett.* **16**, 283 (1974).
18. J. R. McGinley and E. A. Schweikert: *J. Radioanal. Chem.* **16**, 385 (1973).
19. H. Determann and J. E. Brewer in: *Chromatography* (Ed. by E. Heftmann) Van Nostrand Reinhold, New York, Chapter 14 (1975).
20. W. W. Yau, J. J. Kirkland and D. D. Bly: *Modern Size-Exclusion Liquid Chromatography*, Wiley-Interscience, New York (1979).
21. T. Kremmer and L. Boross: *Gel Chromatography*, John Wiley, Chichester (1979).
22. W. Heitz and R. Michels: *Angew. Chem. Int. Ed.* **11**, 298 (1972).
23. A. Brenner and D. A. Hucul: *Prepr., Div. Petr. Chem., Am. Chem. Soc.* **22**, 1221 (1977).
24. A. Brenner, D. A. Hucul and S. J. Hardwick: *Inorg. Chem.* **18**, 1478 (1979).
25. T. J. Thomas, D. A. Hucul and A. Brenner: *Amer. Chem. Soc., Symp. Ser.* **192**, 267 (1982).

26. J. L. Falconer and J. A. Schwarz: *Catal. Rev. Sci. Eng.* **25**, 141 (1983).
27. B. A. Matrana: *Diss. Abs.* **B43**, 1841 (1982).
28. A. Brenner and R. L. Burwell: *J. Catal.* **152**, 353 (1978).
29. M. Deeba and B. C. Gates: *J. Catal.* **67**, 303 (1981).
30. W. N. Delgass, G. L. Haller, R. Kellerman and J. H. Lunsford, *Spectroscopy in Heterogeneous Catalysis*, Academic Press, New York (1979), (a) Chapter 2, (b) Chapter 3, (c) Chapter 4, (d) Chapter 6, (e) Chapter 5, (f) Chapter 8.
31. H. Knözinger, H. Stolz, H. Bühl, G. Clement and W. Meye: *Chem. Ing. Tech.* **42**, 548 (1970).
32. H. Knözinger, *Acta Cient. Venez.* **24** (Suppl. 2), 76 (1973).
33. A. Kuzusaka and R. F. Howe: *J. Mol. Catal.* **9**, 183 (1980).
34. A. A. Olsthoorn and J. A. Moulijn: *J. Mol. Catal.* **8**, 147 (1980).
35. A. A. Olsthoorn and C. Boelhouwer: *J. Catal.* **44**, 197 (1976).
36. M. S. Scurrell: *J. Mol. Catal.* **10**, 57 (1980).
37. A. Luchetti, L. F. Wieserman and D. M. Hercules: *J. Phys. Chem.* **85**, 549 (1981).
38. F. R. Hartley, S. G. Murray and P. N. Nicholson: *J. Mol. Catal.* **16**, 363 (1982).
39. G. C. Smith, T. P. Chojnacki, S. R. Dasgupta, K. Iwatate and K. L. Watters: *Inorg. Chem.* **14**, 1419 (1975).
40. J. H. Anderson: *J. Catal.* **28**, 76 (1973).
41. J. T. Yates and D. A. King: *Surface Sci.* **30**, 601 (1972).
42. W. D. Bonds, C. H. Brubaker, E. S. Chandrasekaran, C. Gibbons, R. H. Grubbs and L. C. Kroll: *J. Amer. Chem. Soc.* **97**, 2128 (1975).
43. C. Andersson and R. Larsson: *J. Catal.* **81**, 179 (1983).
44. B. A. Sosinsky, W. C. Kalb, R. A. Grey, V. A. Uski and M. F. Hawthorne: *J. Amer. Chem. Soc.* **99**, 6768 (1977).
45. M. S. Jarrell and B. C. Gates: *J. Catal.* **40**, 255 (1975).
46. R. Thornton and B. C. Gates: *J. Catal.* **34**, 274 (1974).
47. R. Whyman: *Laboratory Methods in Infrared Spectroscopy* (Ed. by R. G. J. Miller), Heyden & Sons, London, Second Edition, p. 149 (1972).
48. D. L. King: *J. Catal.* **61**, 77 (1980).
49. J. M. L. Penninger: *J. Catal.* **56**, 287 (1979).
50. M. Deeba, B. J. Streusand, G. L. Schrader and B. C. Gates: *J. Catal.* **69**, 218 (1981).
51. W. H. Weinberg: *Vib. Spectra Structure* **11**, 1 (1982).
52. J. R. Kirtley, D. J. Scalapino and P. K. Hansma: *Phys. Rev. B* **14**, 3177 (1976).
53. G. D. Shields and L. J. Boucher: *J. Inorg. Nucl. Chem.* **40**, 1341 (1978).
54. J. A. Davies in: *The Chemistry of the Metal Carbon Bond* (Ed. by F. R. Hartley and S. Patai) Wiley, Chichester, Vol. 1, Chapter 21 (1982).
55. N. Kawata, T. Mizoroki and A. Ozaki: *Bull. Chem. Soc. Japan* **47**, 1807 (1974).
56. S. Shinoda and Y. Saito: *Inorg. Chim. Acta* **63**, 23 (1982).
57. R. H. Grubbs and S.-C. H. Su in: *Organometallic Polymers* (Ed. by C. E. Carraher, J. E. Sheats and C. U. Pittman) Academic Press, New York, p. 129 (1978).
58. S. Shinoda, K. Nakamura and Y. Saito: *J. Mol. Catal.* **17**, 77 (1982).
59. J. Schaefer and E. D. Stejskal: *J. Amer. Chem. Soc.* **98**, 1031 (1976).
60. A. Pines, M. G. Gibby and J. S. Waugh: *J. Chem. Phys.* **59**, 569 (1973).
61. E. R. Andrew: *Prog. Nucl. Mag. Reson. Spectrosc.* **8**, 1 (1971).
62. J. Schaefer, E. O. Stejskal and R. Buchdahl: *Macromolecules* **10**, 384 (1977).
63. E. R. Andrew: *Phil. Trans. Roy. Soc. London* **A299**, 505 (1981).
64. E. R. Andrew: *Int. Revs. in Phys. Chem.* **1**, 195 (1981).
65. W. W. Fleming, C. A. Fyfe, R. D. Kendrick, J. R. Lyerla, H. Vanni and C. S. Yannoni: *Amer. Chem. Soc., Symp. Ser.* **142**, 194 (1980).

66. C. A. Fyfe, L. Bemi, R. Childs, H. C. Clark, D. Curtin, J. A. Davies, D. Drexler, R. L. Dudley, G. C. Gobbi, J. S. Hartman, P. Hayes, J. Klinowski, R. E. Lekinski, C. J. L. Lock, I. C. Paul, A. Rudin, W. Tchir, J. M. Thomas and R. E. Wasylichen: *Phil. Trans. Roy. Soc. London* **A305**, 591 (1982).

67. H. C. Clark, J. A. Davies, C. A. Fyfe, P. J. Hayes and R. E. Wasylichen: *Organometallics* **2**, 197 (1983).

68. L. Bemi, H. C. Clark, J. A. Davies, D. Drexler, C. A. Fyfe and R. E. Wasylichen: *J. Organometal. Chem.* **224**, C5 (1982).

69. L. Bemi, H. C. Clark, J. A. Davies, C. A. Fyfe and R. E. Wasylichen: *J. Amer. Chem. Soc.* **104**, 438 (1982).

70. F. R. Hartley, S. G. Murray and P. N. Nicholson: *J. Mol. Catal.* **16**, 363 (1982).

71. D. W. Sindorf and G. E. Maciel: *J. Amer. Chem. Soc.* **103**, 4263 (1981).

72. S. J. DeCanio, H. C. Foley, C. Dybowski and B. C. Gates: *JCS Chem. Commun.* 1372 (1982).

73. H. C. Foley, S. J. DeCanio, K. D. Tau, K. J. Chao, J. H. Onuferko, C. Dybowski and B. C. Gates: *J. Amer. Chem. Soc.* **105**, 3074 (1983).

74. A. I. Kokorin and K. I. Zamaraev: *Proc. 17th Int. Coord. Chem. Conf.*, Hamburg, 238 (1976).

75. R. S. Drago, J. Gaul, A. Zombeck and D. K. Straub: *J. Amer. Chem. Soc.* **102**, 1033 (1980).

76. J. Zwart and J. H. M. C. Van Wolput: *J. Mol. Catal.* **5**, 235 (1979).

77. Y. Chimura, M. Beppu, S. Yoshida and K. Tarama: *Bull. Chem. Soc. Japan* **50**, 691 (1977).

78. F. S. Dyachovskii and A. D. Pomogailo: *J. Poly. Sci., Poly Symp.* **68**, 97 (1981).

79. F. S. Dyachovskii, A. D. Pomogailo and N. M. Bravaya: *J. Poly. Sci., Poly. Chem. Ed.* **18**, 2615 (1980).

80. R. S. Drago and J. H. Gaul: *Inorg. Chem.* **18**, 2019 (1979).

81. A. I. Kokorin: Ph.D. Thesis, Moscow (1973).

82. W. Jones in: *Characterisation of Catalysts* (Ed. by J. M. Thomas and R. M. Lambert) Wiley, Chapter 8 (1980).

83. C. R. F. Lund and J. A. Dumesic: *J. Phys. Chem.* **85**, 3175 (1981).

84. K. Lazar, Z. Schay and L. Guczi: *J. Mol. Catal.* **17**, 205 (1982).

85. C. A. Clausen and M. L. Good: *Mössbauer Effect Methodol.* **10**, 93 (1976); *Chem. Abs.* **91**, 112958 (1979).

86. F. R. Hartley, D. J. A. McCaffrey, S. G. Murray and P. N. Nicholson: *J. Organometal. Chem.* **206**, 347 (1981).

87. N. J. Coville and C. P. Nicolaides: *J. Organometal. Chem.* **219**, 371 (1981).

88. T. Edmonds in: *Characterisation of Catalysts* (Ed. by J. M. Thomas and R. M. Lambert) Wiley, Chichester, p. 30 (1980).

89. K. Siegbahn, C. Nordling, A. Fahlman, R. Nordberg, K. Hamrin, J. Hedman, G. Johansson, T. Bergmark, S.-E. Karlsson, I. Lindgren and B. Lindberg. *Nova Acta Regiae Sci. Ups.* **20**, 4 (1967).

90. D. W. Turner, A. D. Baker, C. Baker and C. R. Brundle: *Molecular Photoelectron Spectroscopy; a Handbook of He 584Å Spectra*, Wiley Interscience, New York (1970).

91. S. Lars and T. Andersson: *J. Microscop. Spectrosc. Electron.* **7**, 159 (1982).

92. F. F.-L. Ho: *Chem. Anal.* **63**, 135 (1982).

93. K. Tanaka, K. Miyahara and K. Tanaka: *Bull. Chem. Soc. Japan* **54**, 3106 (1981).

94. Y. M. Shulga, A. A. Startsev, Yu. I. Ermakov, B. N. Kuznetsov and Y. G. Borod'ko: *React. Kinet. Catal. Lett.* **6**, 377 (1977).

95. M. D. Patel: *Diss. Abs.* **B42**, 1882 (1981).
96. I. N. Ivleva, A. D. Pomogailo, S. B. Echmaev, M. S. Ioffe, N. D. Golubeva and Y. G. Borod'ko: *Kinet. Katal.* **20**, 1282 (1979).
97. M. Howalla and B. Delmon: *Appl. Catal.* **1**, 285 (1981).
98. S. L. T. Andersson, K. L. Watters and R. F. Howe: *J. Catal.* **69**, 212 (1981).
99. S. Lamalle, A. Mortreux, M. Evrard, F. Petit, J. Grimblot and J. P. Bonnelle: *J. Mol. Catal.* **6**, 11 (1979).
100. D. M. Hercules: *Report* ARO-14839.3-C (1980); *Chem. Abs.* **94**, 163248 (1981).
101. T. Uematsu and H. Hashimoto: *Kogakubu Kenkyu Hokoku* **33**, 99 (1981); *Chem. Abs.* **96**, 110836 (1982).
102. T. H. Kim and H. F. Rase: *Ind. Eng. Chem., Prod. Res. Dev.* **15**, 249 (1976).
103. V. A. Semikolenov, V. A. Likholobov, P. A. Zhdan, A. I. Nizovskii, A. P. Shepelin, E. M. Moroz, S. P. Bogdanov and Yu. I. Yermakov: *Kinet. Katal.* **22**, 1247 (1981).
104. C. Andersson and R. Larson: *J. Catal.* **81**, 194 (1983).
105. M. Terasawa, K. Sano, K. Kaneda, T. Imanaka and S. Teranishi: *J. C. S. Chem. Commun.* 650 (1978).
106. D. Wang, Y. Li and Y. Jiang: *Cuihua Xuebao* **3**, 78 (1982); *Chem. Abs.* **97**, 169666 (1982).
107. F. F. L. Ho: *Applied Electron Spectroscopy for Chemical Analysis* (Ed. by H. Windawi and F. F. L. Ho) Wiley, New York, p. 135 (1982).
108. B. K. Teo and D. C. Joy (Eds.): *EXAFS Spectroscopy, Techniques and Applications*, Plenum Press, New York (1980).
109. R. W. Joyner in: *Characterisation of Catalysts* (Ed. by J. M. Thomas and R. M. Lambert) Wiley, Chichester, p. 238 (1980).
110. R. F. Pettifer in: *Characterisation of Catalysts* (Ed. by J. M. Thomas and R. M. Lambert) Wiley, Chichester, p. 264 (1980).
111. B. R. Stults, R. M. Friedman, K. Koenig and W. Knowles: *J. Amer. Chem. Soc.* **103**, 3235 (1981).
112. I. H. Munro: *Chem. in Britain* **15**, 330 (1979).
113. J. Reed, P. Eisenberger, B.-K. Teo and B. M. Kincaid: *J. Amer. Chem. Soc.* **100**, 2375 (1978).
114. J. Reed, P. Eisenberger, B.-K. Teo and B. M. Kincaid, *J. Amer. Chem. Soc.* **99**, 5217 (1977).
115. J. Reed and P. Eisenberger: *JCS Chem. Commun.* 628 (1977).
116. S. L. Cook, J. Evans and G. N. Greaves: *JCS Chem. Commun.* 1287 (1983).
117. B. R. Stults, R. M. Friedman, W. S. Knowles, K. Koenig, F. W. Lytle and R. B. Greegor: *Stanford Synchrotron Radiation Laboratory Report* **9**, 52 (1978).
118. J. Garmendia, J. Lieto, J. Rafalko, J. Katzer, D. Sayers and B. Gates: *Stanford Synchrotron Radiation Laboratory Report* **9**, 46 (1978).
119. R. B. Greegor, F. W. Lytle, R. L. Chin and D. M. Hercules: *J. Phys. Chem.* **85**, 1232 (1981).
120. P. Gallezot, R. Weber, R. D. Betta and M. Boudart: *Stanford Synchrotron Radiation Laboratory Report* **9**, 44 (1978).
121. A. D. Cox in: *Characterisation of Catalysts* (Ed. by J. M. Thomas and R. M. Lambert) Wiley, Chichester, p. 254 (1980).
122. J. Jaklevic, J. A. Kirby, M. P. Klein and A. S. Robertson: *Solid State Commun.* **23**, 679 (1977).
123. J. Szymura and T. Paryjaczak: *Wiad. Chem.* **32**, 175 (1978); *Chem. Abs.* **89**, 65924 (1978).
124. R. H. Grubbs and E. M. Sweet: *Macromolecules* **8**, 241 (1975).

125. K. W. Pepper, H. M. Paisley and M. A. Young: *J. Chem. Soc.* 4097 (1953).
126. D. Tatarsky, D. H. Kohn and M. Cais: *J. Poly. Sci., Poly. Chem. Ed.* 18, 1387 (1980).
127. T. Castrillo, H. Knözinger, M. Wolf and B. Tesche: *J. Mol. Catal.* 11, 151 (1981).
128. V. A. Kabanov, I. A. Litvinov, T. V. Budantseva and V. I. Smetanyuk: *Dokl. Akad. Nauk SSSR* 262, 1169 (1982).
129. T. V. Budantseva, I. A. Litvinov and V. A. Kabanov: *VINITI Deposited Document* 3879 (1980); *Chem. Abs.* 96, 104819 (1982).
130. T. V. Budantseva, I. A. Litvinov, S. K. Pluzhnov, V. I. Smetanyuk and V. A. Kabanov: *VINITI Deposited Document* 3880 (1980); *Chem. Abs.* 96, 104820 (1982).

CHAPTER 5

THE USE OF SUPPORTED METAL COMPLEX CATALYSTS

5.1. Introduction

In the present Chapter we shall examine briefly the practical aspects of using supported metal complex catalysts as a way of introducing Chapters 6 to 10 which will consider their use in specific reactions.

When a supported metal complex catalyst is used to promote a chemical reaction that takes place in solution, the sequence of events can be broken down into a series of sequential steps.

1. Mass transfer of the reactants from the bulk solution to the exterior surface of the catalyst particle.
2. Diffusion of the reactants from the surface to the active catalytic site.
3. The catalytic reaction itself.
4. Diffusion of the products away from the active site to the catalyst particle surface.
5. Mass transfer of the products from the surface into the bulk solution.

Any of these 5 steps can, in theory, be rate-limiting. By altering such reaction conditions as temperature, pressure, nature of the solvent, rate of stirring and so on it is possible to alter the rate limiting step. In this way it is possible to alter the specificity of the catalyst to promote the reaction of one substrate in preference to others that may be present in a mixture.

For very fast reactions involving very active catalysts the rate of mass transfer of reactants to the catalyst surface, step 1, is rate limiting [1, 2]. In general this mass transfer can be enhanced by increasing the stirring or shaking speed [3–5]. In any kinetic study it is essential to operate under conditions which are not mass transfer limited. Accordingly the reaction rate should be monitored as a function of the stirring or shaking speed and that speed adjusted so that it is above the level at which the rate of reaction becomes independent of speed. In the case of fixed-bed, flow reactors it is not possible to vary an equivalent parameter to stirring speed, although the rate of flow and reactor design will influence the residence time of a molecule of reactant on the catalyst surface [6].

However, the rate of mass transfer of material through the medium can be very dramatically enhanced from a fixed to a fluidised bed [7, 8].

The effect of the rate of diffusion of the reactants within the support on the overall rate will depend on the distance over which diffusion has to take place [9, 10]. Thus if the diffusion rate is modest, relative to the rate of reaction at the catalytic site, the overall rate of reaction will depend on the size of the catalyst particle. By decreasing the size of the catalyst particle it is sometimes possible to reach a size below which the overall rate of reaction is independent of particle size [11]. The point at which this is reached will depend on the concentration of reactant if a liquid or its partial pressure if a gas. As the concentration or partial pressure of the reactant is increased so the size of particle which becomes rate-limiting increases.

The same criteria apply to steps 4 and 5 which concern removal of the products from the catalytic site, as affect steps 1 and 2. The ideal rate will be that at which the rate of the reaction at the catalytic site is rate-limiting unless one, or a combination of steps 1, 2, 4 and 5, is being used to enhance catalyst specificity.

5.2. Optimisation of Conditions

Clearly in order to optimise both the rate and specificity of a particular catalyst system it will be necessary to consider many factors. These will include:

Solvent (if used)
Temperature
Pressure (if gases are involved)
Rate of stirring
Particle size
Nature of the support (which organic polymer or inorganic support)
Degree of cross-linking (in a cross-linkable polymer)
Pore size (in an inorganic support)
Number of donor groups (L) on the support
Distribution of donor groups on the support (whether throughout the support or in localised pockets)
Nature of the metal and the other ligands surrounding it
Ratio of donor groups (L) to metal
Nature and concentration of any other added groups (such as ligands added to the solution).

Clearly optimisation of so many variables can present a massive problem involving a major expenditure of research effort. An excellent way to expedite this process is to use a Plackett-Burman matrix approach [12, 13]. Thus to optimise an experiment involving 15 variables to the extent that one variable such as temperature is investigated at two values such as high and low, whilst the other 14 variables are held constant and then examine each of the other variables in sequence would require 2^{15} or 32 768 experiments. However, by using a matrix approach, the number of experiments necessary to optimise n variables between high and low values can be reduced to $n + 1$ experiments. For 15 variables these 16 experiments would be organised as in Table I. The order in which the experiments would be undertaken is shown in the column 'random order'.

TABLE I

Plackett-Burman matrix which enables 16 experiments to be carried out in order to optimise 15 variables between two values, high (+) and low (−).

Run No.	Random Order	Variable														
		A	B	C	D	E	F	G	H	I	J	K	L	M	N	O
1	1	+	+	+	−	−	+	−	+	+	−	−	+	−	−	−
2	4	+	+	+	−	+	−	+	+	−	−	+	−	−	−	+
3	7	+	+	+	−	−	+	+	−	−	+	−	−	−	+	+
4	5	+	−	+	−	+	+	−	−	+	−	−	−	+	+	+
5	16	−	+	−	+	+	−	−	+	−	−	−	+	+	+	+
6	10	+	−	+	+	−	−	+	−	−	−	+	+	+	+	−
7	12	−	+	+	−	−	+	−	−	−	+	+	+	+	−	+
8	15	+	+	−	−	+	−	−	−	+	+	+	+	−	+	−
9	6	+	−	−	+	−	−	−	+	+	+	+	−	+	−	+
10	3	−	−	+	−	−	−	+	+	+	+	−	+	−	+	+
11	2	−	+	−	−	+	+	+	+	−	+	−	+	+	−	
12	13	+	−	−	−	+	+	+	+	−	+	−	+	+	−	−
13	11	−	−	−	+	+	+	+	−	+	−	+	+	−	−	+
14	9	−	−	+	+	+	+	−	+	−	+	+	−	−	+	−
15	14	−	+	+	+	+	−	+	−	+	+	−	−	+	−	−
16	8	−	−	−	−	−	−	−	−	−	−	−	−	−	−	−

Inspection of variable A shows that during the 16 runs it is at a high level 8 times and at a lower level 8 times. This is also true for all the other variables. The effect of a given variable A (E_A), is simply the difference between the average value of the response for the 8 runs at high level (R at +) and the average value of the response for the 8 runs at low level (R at −) as in equation (1).

$$E_A = \frac{R \text{ at } (+)}{8} - \frac{R \text{ at } (-)}{8} \tag{1}$$

Although this seems reasonable, how is it possible to distinguish the effect of one variable when all the others are changing at the same time? This arises, as inspection of Table I will reveal, because when variable A is at its high level, variable B is high four times and low four times. The same applies when A is at its low level, on four occasions B is high and on four occasions B is low. Thus, the net effect of changing variable B on the response due to A cancels out. Exactly the same occurs with variables $C-O$. The great strength of the Plackett-Burman approach is the simplicity with which it yields results that are mathematically and numerically equivalent to those that can be obtained using a complete multiple regression analysis program.

The Plackett-Burman method can be further refined to enable confidence limits to be placed on the effect of each variable. This is done by making some of the variables dummies; that is, the experiments listed under these variables at + and − are in fact identical. The effects of the dummy variables are calculated using equation (1) in exactly the same way as the effects of real variables. If there are no interactions between variables and all levels are reproduced perfectly with no errors in measuring the responses the effect shown by a dummy variable will be zero. Any difference from zero arises from a combination of lack of precision in setting the experimental conditions and analytical errors in measuring the response. Typically three such estimates of experimental error with three dummy variables will provide adequate confidence. The variance of an effect (V_{eff}) is given by equation (2) in which E_d = effect of a dummy and n = number of dummies

$$V_{\text{eff}} = \frac{\Sigma(E_d)^2}{n} \, . \tag{2}$$

Thus if C, F and M are the three dummy variables,

$$V_{\text{eff}} = \frac{(E_C)^2 + (E_F)^2 + (E_M)^2}{3} \, . \tag{3}$$

The standard error of an effect (S) is given by

$$S = (V_{\text{eff}})^{1/2} / E_{\text{eff}}. \tag{4}$$

The significance of each effect (t_A) can then be determined using the standard t-test (equation 5).

$$t_A = E_A / (S \cdot E_{\text{eff}}) \tag{5}$$

Since three dummy variables have been used in calculating t_A, the significance of effect A is found by entering the tabulated t-values under three degrees of freedom [14]. Once a Plackett-Burman matrix approach has been used to identify the important variables in a catalytic reaction it is possible to refine these variables through a series of detailed experiments.

5.3. Laboratory Application

Most laboratory applications of supported metal complex catalysts will be carried out in a batchwise manner rather than using flow systems. Under batch conditions a small amount of leaching of the metal complex from the support is often tolerable whereas in a flow system the leaching requirements are more stringent. There are four reasons for this:

1. If leaching occurs in a flow system it will inevitably lead to a downstream migration of the metal complex, whereas in a batch process the metal complex may detach and reattach itself many times during the catalytic run.

2. In a batch process it is possible to cool the reaction mixture to room temperature before separating the catalyst. Leaching is very often reversible, and lower at lower temperatures.

3. A small amount of leaching in a batch-process can result in relatively small costs relative to the cost of working up the reaction mixture.

4. Following catalyst separation it is often necessary to separate and purify the product. During this separation and purification very small traces of leached material will be readily removed from the product.

Although most supported catalysts are insoluble in the reaction medium this is not automatically an essential prerequisite when a batchwise process is used. Soluble supported catalysts can be used and may be separated at the end of the reaction in one of two ways. Either the solvent used in the reaction may be changed at the end to one in which the polymer is insoluble so enabling it to be filtered off conventionally, or the soluble polymer may be separated from solution by gel or membrane filtration [15]. Thus, for example, phosphinated polystyrene-rhodium(I) complexes that are soluble in benzene can be either membrane filtered using a polyamide membrane or can be precipitated by the addition of n-hexane.

5.4. Industrial Application

Supported metal complex catalysts, except the supported Ziegler-Natta catalysts (see Section 9.5.1), are not generally used in industry at present. Accordingly it is impossible to give specific instances of the way in which processes have been developed. However the problems that would need to be solved are basically very similar to those encountered with heterogeneous catalysts. We refer readers elsewhere for such information [2, 16–24] and concentrate in this Section on the additional problems likely to be encountered with supported metal complex catalysts.

Industrial applications may be conducted in either a batchwise or a flow manner. If a flow process is adopted then a critical parameter will be the degree of leaching. This must either be zero, or if it is not there must be some very compelling selectivity or specificity advantages accruing from the supported catalyst. If those were to be present then the chemical engineering would have to be undertaken using essentially the same techniques as used in the case of a homogeneous catalyst [17, 25]. A mathematical analysis of the chemical engineering implications of using 'leaky' polymer supported catalysts in packed bed reactors has been reported [26].

A major problem likely to be encountered with polymer supported metal complex catalysts is the high local temperatures developed [27, 28]. Many of the reactions of interest are highly exothermic and location of the active sites within poor thermally conducting materials will result in poor removal of the heat of reaction from the catalytic site. If the heat is not removed then it may result in local structural changes within the polymer or even thermal degradation of either the polymer or the metal complex that it is supporting, or both. Whilst inorganic supports themselves are less susceptible than organic polymers to thermal damage, the metal complex will still potentially be susceptible to thermal degradation. Fluidised beds have a great advantage over fixed beds in enabling the heat to be transported readily within the bed, so enabling close temperature control in the reaction zone to be maintained [29].

The mechanical properties of the catalyst particles are very important, particularly the strength of particles used in large fixed catalyst beds and the resistance to attrition of particles used in fluidised beds. If the strength of the particles is insufficient to carry the load acting on them, then they will break and as a result increase the resistance to flow across the reactor bed. Uneven distribution of the reactants to the catalyst particles will result which will lead to poor overall performance of the reactor system. If exothermic reactions are involved uneven temperature distributions may also result and may damage the reactor. In a fluidised bed attrition of catalyst particles upsets the even flow of material through the bed and will eventually lead to removal of fine particles from the bed down the output.

Whilst polymer swelling is a very useful parameter to control both specificity and selectivity in a batch reactor, in a fixed bed reactor bead swelling is highly undesirable because it leads to blocking of the inter-bead channels and so prevents flow through the reactor bed. This phenomenon has been claimed to be of such practical engineering importance that it has sometimes been stated that polymer supported catalysts will never be useful in large-scale industrial applications [30]. Whilst such a statement is clearly an exaggeration since if it were true polymers would not be used on the large scale they are used for ion-exchange purposes, it clearly indicates that polymers will be of strictly limited value in situations where they will be swollen by either the solvents, or the reactants, or the products of reaction.

References

1. J. R. Katzer in: *Chemistry and Chemical Engineering of Catalytic Processes* (Ed. by R. Prins and G. C. A. Schuit) Sijthoff and Noordhoff, The Netherlands, p. 49 (1980).
2. J. M. Thomas and W. J. Thomas: *Introduction to the Principles of Heterogeneous Catalysis*, Academic Press, London (1967).
3. S. J. Thomson and G. Webb: *Heterogeneous Catalysis*, Oliver and Boyd, Edinburgh (1968).
4. G. C. Bond: *Heterogeneous Catalysis*, Oxford (1974).
5. M. Chanda, K. F. O'Driscoll and G. L. Rempel: *J. Mol. Catal.* **7**, 389 (1980).
6. G. J. K. Acres, A. J. Bird and P. J. Davidson: *Chem. Eng. (London)* **283**, 145 (1974).
7. H. W. Flood and B. S. Lee: *Scientific American* **219** (July), 94 (1968).
8. T. Miyauchi, S. Furaski, S. Morooka and Y. Ikeda: *Adv. Chem. Eng.* **11**, 276 (1981).
9. R. Aris: *The Mathematical Theory of Diffusion and Reaction of Permeable Catalysts*, Oxford Univ. Press, Volumes 1 and 2 (1975).
10. D. Ryan, R. G. Carbonell and S. Whitaker: *Amer. Inst. Chem. Eng., Symp. Ser.*, **77** (202), 46 (1981).
11. M. Chanda, K. F. O'Driscoll and G. L. Rempel: *J. Mol. Catal.* **11**, 9 (1981).
12. R. A. Stowe and R. P. Mayer: *Ind. Eng. Chem.* **58** (2), 36 (1966).
13. R. L. Plackett and J. P. Burman: *Biometrika* **33**, 305 (1946).
14. R. A. Fisher: *Statistical Methods for Research Workers*, Oliver and Boyd, Edinburgh, 14th edition (1970).
15. E. L. Bayer and V. Schurig: *Angew. Chem. Int. Ed.* **14**, 493 (1975).
16. R. Pearce and W. R. Patterson: *Catalysis and Chemical Processes*, Blackie and Sons, Glasgow (1981).
17. B. C. Gates in: *Chemistry and Chemical Engineering of Catalytic Processes* (Ed. by R. Prins and G. C. A. Schuit) Sijthoff and Noordhoff, Netherlands, p. 437 (1980).
18. J. M. Coulson and J. F. Richardson: *Chemical Engineering*, Pergamon, Oxford, Vols 1–3 (1977, 1978 and 1979).
19. D. G. Jordan: *Chemical Process Development*, Wiley-Interscience, New York, Parts 1 and 2 (1968).
20. A. W. Westerberg, H. P. Hutchinson, R. L. Motard and P. Winter: *Process Flowsheeting*, Cambridge University Press (1979).
21. E. V. Thompson and W. Ceckler: *Introduction to Chemical Engineering*, McGraw-Hill, London (1977).
22. R. H. Perry (Ed.): *Chemical Engineering Handbook*, McGraw-Hill, London (1973).
23. J. J. Carberry: *Chemical and Catalytic Reaction Engineering*, McGraw-Hill, London (1976).
24. J. Horak and J. Pasek: *Design of Industrial Chemical Reactors from Laboratory Data*, Heyden, London (1978).

25. B. C. Gates, J. R. Katzer and G. C. A. Schuit: *Chemistry of Catalytic Processes*,
     McGraw-Hill, New York, Chapter 2 (1979).
26. H. W. Altmann: *Diss. Abs.* **39B**, 5463 (1979).
27. P. E. Starkey in [16], chapter 3.
28. J. H. Russell, L. L. Oden and P. E. Sanker: *U.S. Patent Appl.* 778,272 (1977); *Chem.
     Abs.* 88, 75945 (1978).
29. G. A. Mills in: *Kirk-Othmer Enclyclopaedia of Chemical Technology* (Ed. A. Standen)
     Wiley-Interscience, New York, Second Edition, Vol. 4, p. 580 (1964).
30. N. L. Holy in: *Fundamental Research in Homogeneous Catalysis* (Ed. by M. Tsutsui)
     Plenum Press, New York, 3, 691 (1979).

CHAPTER 6

HYDROGENATION

6.1. Introduction

One of the most widely studied catalytic reactions is hydrogenation. All the early work on this involved using heterogeneous catalysts [1], but more recently, spurred on by the discovery of Wilkinson's catalyst [2], a great deal of work has been done on homogeneous catalysts [3–6]. The success of the homogeneous catalysts has led to great interest in supporting them in order to combine the advantages inherent in their molecular nature with the ease of separation at the end of the reaction that should be possible if an insoluble support is used. Not surprisingly therefore many supported metal complex hydrogenation catalysts are little more than the corresponding homogeneous catalysts linked to a support. Whilst many of these have had inferior properties to their homogeneous analogues, a significant number have demonstrated enhanced activity and selectivity as well as greater stability and resistance to deactivation and decay. In the following sections there are many reports of such catalysts.

A number of workers have taken a different approach and decided that in designing a supported metal complex for use as a catalyst it may not be best to attempt to simply mimic the ligands found in homogeneous catalysts. Rather, because of the uniquely different nature of a supported metal complex catalyst a quite different type of ligand may be more effective [7]. For example, when choosing a ligand for a homogeneous catalyst it is usually necessary to select one which does not dissociate from the metal too readily, because if this were to occur reduction of the metal ion all the way to the free metal is likely to take place in the strongly reducing conditions that accompany hydrogenation. In the interstices of a polymer, however, ligand dissociation is suppressed by spatial restrictions. Consequently ligands that have not been particularly successful in homogeneous catalysis may prove to be effective in polymer supported catalyses. Examples of ligands that are rarely used in homogeneous catalysts, but which have proved effective in polymer-supported catalysts include anthranilic acid, which has been used to support rhodium(I) and palladium(II) [7–12]. Such rhodium(I) catalysts (see Section 6.9.5.6) are exceptionally active hydrogenation catalysts that have long term stability and are fairly insensitive to poisons. The palladium(II)-supported anthranilic acid complex catalyses the hydrogenation of benzene and nitrobenzene while the homogeneous N-benzylanthranilic acid complex is inactive [10, 13, 14].

The ideal support will generally be one that promotes coordinative-unsaturation on the metal complex, whilst at the same time binding it sufficiently tightly to prevent either leaching or reduction to the free metal. Although reduction to the free metal is generally undesirable there are many examples of successful nickel, palladium and platinum supported catalysts where the initial metal(II) complex is reduced during the catalysis to the free metal which is stabilised in a very finely dispersed, highly active form (see Sections 6.9.8 and 6.9.9).

In order to prepare the optimum catalyst for a particular hydrogenation reaction there are a number of parameters which can be varied. We consider these in turn.

6.2. Nature of the Support

The nature of the support can have a very profound influence on the catalyst activity. Thus phosphinated polyvinylchloride supports are fairly inactive [15], phosphinated polystyrene catalysts are considerably more active [16], but rather less active particularly when cyclic olefins are the substrates than phosphinated silica supports [17]. The silica supported catalysts may be more active because the rhodium(I) complexes are bound to the outside of the silica surface and are therefore more readily available to the reactants than in the polystyrene based catalysts where the rhodium(I) complex may be deep inside the polymer beads. If this is so, the polystyrene based catalysts should be more valuable when it is desired to selectively hydrogenate one olefin in a mixture of olefins whereas the silica-based catalysts should be more valuable where a rapid hydrogenation of a pure substrate is required. This has been demonstrated on many occasions (see Sections 6.9.1, 6.9.3, 6.9.5.1a, 6.9.5.1c, 6.9.5.1d and 6.9.9) [16, 18–32]. Polymeric supports may undergo structural changes on heating or during reaction. When this occurs sharp changes in reaction rate [33] or even complete deactivation may occur [10, 34]. How reversible these are will depend, in part, on the degree of reversibility of the structural change.

Many of nature's enzymes use the area around the active catalytic site to select and align the substrate. In this way the enzymes achieve their remarkable powers of discrimination between substrates as well as their regio- and stereoselectivities. The support of a supported metal complex catalyst should enable these catalysts to do the same and therefore to become significantly more selective and specific than their homogeneous analogues. Although we have a very long way to go, some progress has already been made in the field of asymmetric hydrogenation (Section 6.9.6). Thus, when an asymmetric alcoholic functional group is introduced next to an asymmetric rhodium(I)-DIOP catalytic site the optical yield in a non-protic solvent is altered as the chirality of the neighbouring functional group is altered [35]. A supported catalyst that is even closer to nature has been described in which an achiral rhodium(I) complex is bound within an enzyme which has a chiral cavity. This has been used to catalyse the asymmetric hydrogenation of α-acetamidoacrylic acid to yield enantiomeric excesses of between 34% and 41%. Thus the asymmetric environment of the enzyme cavity induces asymmetry in the archiral rhodium(I) complex catalysed hydrogenation [36].

6.3. Effect of Cross-Linking

In the organic polymer supports it is possible to vary very considerably the degree of cross-linking present in the support – in the polystyrene based catalysts this is usually achieved by varying the amount of divinylbenzene which forms the cross links. The influence of cross-linking can be very profound. For example largish beads (74–149μ diameter) of 2% cross-linked phosphinated polystyrene exchanged with $[Rh(PPh_3)_3Cl]$ are only 0.06 times as active as the corresponding homogeneous catalyst [16, 37]. By contrast, smaller (37–74μ diameter) beads of lower (1%) cross-linking are 0.8 times as active as the homogeneous catalyst, an improvement by a factor of 13 [38]. In addition to altering the activity, varying the degree of cross-linking alters the specificity – the greater the cross-linking the tighter the polymer chains are bound and hence the more difficult it becomes for any but the smallest substrate molecules to enter the polymer. Thus the rates of hydrogenation of a series of olefins in benzene using 100–200 mesh 2% cross-linked polystyrene were found to decrease in the order 1-hexene $\gg$ cyclohexene $\gg$ cyclooctene $>$ cyclododecene $\gg$ Δ^2-cholestene [23–27].

6.4. Nature of the Solvent

The nature of the solvent is an important factor that can be varied in order to control the activity and selectivity of polymer supported hydrogenation catalysts. Thus coordinating solvents often take part in the intimate mechanism of the reaction being coordinated to and displaced from the active site during the catalytic cycle. The second way in which a solvent can influence a reaction is through its ability to swell the support. Thus polymeric supports such as polystyrene are swollen by non-polar solvents such as benzene. By contrast the silicates hectorite and montmorillonite (see Section 1.4.2) are swollen by polar solvents such as water and alcohols (see Section 6.9.5.2). Thirdly the polarity of the solvent will lead to a polarity gradient between the bulk solvent and the local environment of the catalytic site. By suitably adjusting the solvent polarity it is possible to enhance or inhibit substrate migration to the active site. With these three often conflicting effects of solvents it is not surprising to find that many reactions catalysed by supported metal complex catalysts are very sensitive to solvent.

One of the best examples of the influence of solvent polarity is described in Section 6.9.5.1.b (Figure 2) in which a change from benzene to ethanol decreases the ability of phosphinated polystyrene containing rhodium(I) to catalyse the hydrogenation of polar olefins whilst enhancing its ability to catalyse the hydrogenation of non-polar olefins [23]. Furthermore, the influence of solvent can itself depend on the substrate concentration.

The effect of solvent on selectivity is very complex. Thus when the relative rates of hydrogenation of 1-hexene and cyclohexene in the presence of [Rh(PPh$_3$)$_3$Cl] supported on 1%, 2% and 4% cross-linked polystyrene are studied, the influence of solvent depends on substrate concentration [23, 39]. On changing from benzene to 1 : 1 benzene : ethanol there is a significant enhancement in the hydrogenation rate at low substrate concentrations, whereas at high substrate concentrations there is relatively little change in rate; furthermore, for higher cyclohexene concentrations, the rate decreases as benzene is replaced by ethanol. These complex observations arise from the mutually conflicting effects of an increase in substrate concentration within the resin beads when the polarity of the solvent is increased and the lower resin swelling power of ethanol than benzene. These effects are only significant under conditions when the amount of substrate is rate limiting [23, 39]. Clearly the optimum solvent would be one that combines good swelling ability and high polarity. Tetrahydrofuran meets both requirements and results in enhanced hydrogenation rates for the sterically more demanding cyclohexene than benzene [39, 40].

6.5. Nature of the Metal Complex

The nature of the metal complex can have a profound influence on the rate of hydrogenation and even in cases where similar complexes have been prepared by different routes considerable differences in reactivities can be observed. Thus the effectiveness of rhodium(I) complexes on phosphinated polystyrene decreases in the order [Rh(PPh$_3$)$_3$(CO)H] > RhCl$_3$ > RhCl$_3$ + PPh$_3$ > RhCl$_3$ + PHPh$_2$ > RhCl$_3$ + C$_2$H$_4$ > [Rh(PPH$_3$)$_3$Cl] > [Rh(PHPh$_2$)$_3$Cl] [17, 41].

6.6. Activity of Supported as Compared to Homogeneous Catalysts

Although supported catalysts have advantages of selectivity with respect to substrate and ease of separation of the catalyst after reaction, as compared to homogeneous catalysts, most industrialists still tend to ask about the relative activities of the two types of catalysts. At present supported catalysts are generally less active than their homogeneous analogues. This is because some of their metal centres are bound in substrate-inaccessible sites within the pores of the catalyst; accordingly the less of these sites there are, as in silica, the greater is the activity.

A special situation arises where the active homogeneous catalyst readily dimerises, because once the monomeric units have been bound to the support dimerisation is prevented. Thus titanocene is a reactive intermediate in the reduction of olefins by mixtures of [Cp$_2$TiCl$_2$] and phenylmagnesium bromide [42], but it is rapidly converted into an inactive dimeric compound [43]. However when [Cp$_2$TiCl$_2$] was bound to a polymeric support as in reaction (2) it gave, on treatment with butyllithium, a catalyst that was 6.7 times more active as a homogeneous hydrogenation catalyst than unsupported [Cp$_2$TiCl$_2$] also treated with butyllithium [18]. The greater reactivity of the supported catalyst was ascribed to the prevention of dimerisation, a suggestion that was partly based on ESR evidence. Dimerisation provides the main deactivation mechanism for homogeneous rhodium(I)-phosphine complexes. Accordingly, where support on a polymer prevents this without significantly impeding substrate access, higher activity is found in the supported catalyst [39, 44]. When rhodium(I) complexes are supported by long alkyl chains anchored to a polymer the complexes have essentially the same activity as their homogeneous analogues [45].

Polymer bound [(ⓟ—PPh$_2$)$_2$Ir(CO)Cl] was found to be more active as a hydrogenation catalyst for 4-vinylcyclohexene than its homogeneous counterpart because equilibrium (1) lay further to the right for the polymeric phosphine

$$[(ⓟ—PPh_2)_2 Ir(CO)(olefin)Cl]$$
$$\rightleftharpoons [(ⓟ—PPh_2)Ir(CO)(olefin)Cl] + ⓟ—PPh_2 \qquad (1)$$

than for triphenylphosphine because of steric constraints within the polymer that retard the reverse reaction [46]. Similarly palladium(II) phosphine complexes are more active hydrogenation catalysts when supported than in homogeneous solution because of having less than two phosphine ligands per palladium in the supported case [47].

6.7. Selectivity [47a]

One of the major advantages offered by supported catalysts is greater selectivity than their homogeneous counterparts. We have already noted the role that the degree of cross-linking, choice of solvent and neighbouring groups can play in determining the degree of selectivity. A particularly selective catalyst for the hydrogenation of small olefins has been prepared by swelling a phosphinated polystyrene, supporting rhodium(I) and then, after removing the solvent to contract the beads, poisoning the surface catalytic sites. On reswelling, the only catalytic sites that remain are those deep within the polymer and these are only accessible to small olefins [16]. Selectivity can be promoted by increasing the loading of catalytic centres on a support [29], because the polymer surrounding a catalytic centre imposes a diffusional barrier between the bulk solution and the active site. The more active the polymer bead the greater will be its demand on the bulk solution to supply substrate which will result in an enhancement of the differences in the diffusion rates of the substrates.

6.8. Stability

Many supported catalysts are more stable than their homogeneous analogues. We have already referred to cases where this is due to the support preventing a deactivating dimerisation. There are many examples where polymer supported rhodium(I) complexes are less sensitive to air than their homogeneous analogues. However, in practical terms, supported catalysts are usually best handled in the absence of air. Although many rhodium(I) homogeneous catalysts are sensitive to poisoning by thiols, n-butylthiol reacts with silica-supported Wilkinson's catalysts, $[(\{Si\}\text{-}CH_2CH_2PPh_2)_3RhCl]$ to reduce their activity but enhance their thermal stability [48–50]. Similarly, rhodium(I)-anthranilic acid hydrogenation catalysts supported on chloromethylated polystyrene have long term thermal stability and are fairly insensitive to poisoning [7, 8].

6.9. Survey of Supported Hydrogenation Catalysts

In Section 6.9.1–6.9.10 we survey supported hydrogenation catalysts. We have not attempt to be comprehensive but to include more important and more recent papers.

6.9.1. TITANIUM, ZIRCONIUM AND HAFNIUM

Titanocene is a reactive intermediate in the reduction of olefins by mixtures of $[TiCp_2Cl_2]$ and phenylmagnesium bromide [42], but it is rapidly converted into an inactive dimeric compound [43]. However, when $[TiCp_2Cl_2]$ is formed on a polymeric support as in reaction (2),

$$(2)$$

it gives, on treatment with butyllithium, a catalyst that is 20 times more active than its homogeneous counterpart [18–20]. The enhanced reactivity of the catalysts supported on a styrene–20% divinylbenzene copolymer is ascribed to the rigidity of the support which inhibits dimerisation, a suggestion that is supported both by ESR evidence [18] and by the observation that the rate of hydrogenation reaches a maximum as the amount of titanium on the support is increased [19a, 51]. This latter observation is consistent with a monomer-dimer equilibrium. In addition to enhanced rate, the supported catalyst also has enhanced selectivity towards substrates of different degrees of bulkiness. This arises because of the dispersion of the titanium throughout the polymer beads, as shown by microtome sectioning and electron microprobe X-ray fluorescence analysis to determine both the metal and chloride content across the bead [52]. Although the zirconium and hafnium analogues of these catalysts have been prepared they have little activity as hydrogenation catalysts [51–53].

An alternative route for supporting $[TiCp_2Cl_2]$, shown in reaction (3) has the triple advantage of eliminating the potentially reactive benzyl group, not

$$(3)$$

requiring potentially carcinogenic chloromethyl methyl ether and yielding a hydrogenation catalyst that is 25–120 times as active as its homogeneous counterpart [54]. When $[TiCp_2Cl_2]$ was supported on silica (reaction (4)) instead of polystyrene, enhanced olefin hydrogenation was achieved for terminal

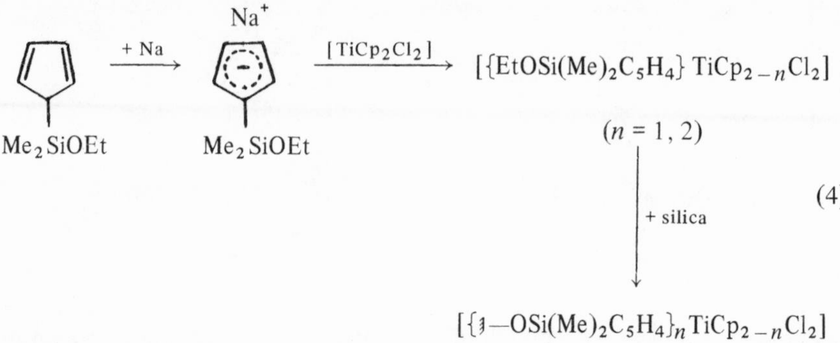

$$\text{(cyclopentadiene)} \xrightarrow{+ \text{Na}} \text{Na}^+ \text{(cyclopentadienide)} \xrightarrow{[TiCp_2Cl_2]} [\{EtOSi(Me)_2C_5H_4\}\, TiCp_{2-n}Cl_2]$$

Me$_2$SiOEt   Me$_2$SiOEt

$(n = 1, 2)$

$\Big\downarrow$ + silica   (4)

$$[\{ \text{]}-OSi(Me)_2C_5H_4\}_n TiCp_{2-n}Cl_2]$$

olefins but not for internal olefins whereas on polystyrene all olefins were hydrogenated faster than by the corresponding homogeneous catalyst [55]. The source of inhibition by silica for internal olefins is not obvious.

6.9.2. CHROMIUM, MOLYBDENUM AND TUNGSTEN

The $Cr(CO)_3$ moiety η^6-bonded to the phenyl rings of polystyrene gives a selective catalyst for the hydrogenation of methyl sorbate to Z-methyl-3-hexenoate with 96–97% selectivity (reaction (5)) [56].

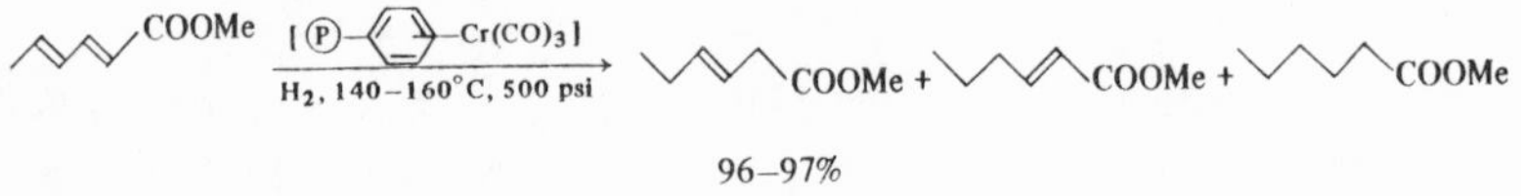

$$\text{COOMe} \xrightarrow[\text{H}_2, 140-160°C, 500 psi]{[\textcircled{P}-\langle\rangle-Cr(CO)_3]} \text{COOMe} + \text{COOMe} + \text{COOMe}$$

96–97%   (5)

The product distribution is sensitive to both solvent and temperature. Greater activity can be obtained using poly(vinylbenzoate) as the support in line with the relative activities of the homogeneous $[Cr(arene)(CO)_3]$ catalysts, which decrease in activity as the arene is altered in the order arene = PhCOOMe > PhH > PhMe [57]. Although several of the supported catalysts were very active when freshly prepared they all lost activity on recycling due to loss of chromium. Thermal analysis showed that this was due to polymer-complex instability at the temperatures necessary for hydrogenation. A phosphine supported $Cr(CO)_3$ complex that is a mildly active photohydrogenation catalyst at room temperature and atmospheric pressure is obtained by reaction (6).

$$(P)\text{—}\langle\quad\rangle\text{—}CH_2PPh_2 + [Cr(nbd)(CO)_4] \xrightarrow[h\nu]{thf} [((P)\text{—}\langle\quad\rangle\text{—}CH_2PPh_2)Cr(CO)_3(nbd)] \quad (6)$$

$(P)\text{—} = $ polystyrene/divinylbenzene

A 3% photohydrogenation of a 0.1M norbornadiene solution was achieved under 1 atmosphere of hydrogen [58].

Molybdenum catalysts prepared by anchoring $[Mo(\eta^3\text{-allyl})_4]$ on silica and alumina are more active in thiophene hydrogenolysis than catalysts prepared by impregnating the same supports with aqueous $(NH_4)_6[Mo_7O_{24}]$ [59]. The same has been observed for the corresponding tungsten catalysts where the enhanced activity of the catalysts prepared from $[W(\eta^3\text{-allyl})_4]$ has been shown to be due to the smaller tungsten sulphide aggregates [60].

Molybdenum trioxide supported on magnesium oxide is a remarkably active and selective catalyst for the hydrogenation of styrene to ethylbenzene that resists carbon disulphide poisoning, even in the absence of additives. Photo-electron emission spectroscopy demonstrated that tetrahedral Mo(IV) species formed by hydrogen reduction of the octahedral Mo(VI) species are the active centres [61]. A binary system consisting of molybdenum complexes supported on partially p-mercaptomethyl substituted polystyrene and the ferredoxin model compound $[Fe_4S_4(SPh_4)]^{2-}$ had a much higher activity for the catalytic reduction of acetylene than the corresponding unsupported molybdenum complexes [61a].

6.9.3. IRON, RUTHENIUM AND OSMIUM

Iron and ruthenium cluster carbonyls supported on alumina catalyse the hydrogenation of ethylene; their effectiveness decreases by two orders of magnitude across the series $[Ru_3(CO)_{12}] > [FeRu_2(CO)_{12}] > [Fe_2Ru(CO)_{12}] > [Fe_3(CO)_{12}]$ [62]. $[HFe_3(CO)_{11}]^-$ supported on 11 μm thick polystyrene membranes carrying $-CH_2NEt_3^+$ functional groups catalyses the reduction of nitrobenzene to aniline [62a], and iron(III) anchored to polystyrene through acetylacetonate groups, $[Fe((P)\text{—}acac)_3]$, catalyses the hydrogenation of cyclohexene [62b].

$[Ru_3(CO)_{12}]$ supported on silica was less active as a hydrogenation catalyst than the material prepared by reducing the ruthenium in this material to the metal, but more active than homogeneous $[Ru_3(CO)_{12}]$. This suggests that the silica-supported $[Ru_3(CO)_{12}]$ reacts with hydrogen and olefins predominantly by surface absorption rather than by ligand exchange. As a consequence the solvent has only a minor influence on its activity [63, 64]. A study of the hydrogenation of ethylene over $[H_4Ru_4(CO)_{12-n}(PPh_3)_n]$ showed that the Ru_4 framework provides the catalytic sites for hydrogenation perhaps by reversible $Ru-Ru$ bond breaking to form coordinatively unsaturated metal centres [65]. When $[Ru_3(CO)_{12}]$ absorbed on silica is irradiated at 25 °C, the $Ru-Ru$ bonds are cleaved and the $Ru(CO)_4$ fragments stabilised by coordination to oxygen, whereas when $[H_4Ru_4(CO)_{12}]$ absorbed on silica is irradiated the carbonyl ligands are lost but the tetraruthenium cluster H_4Ru_4 remains intact [65a].

A comparison of $[\{Si\}-(CH_2)_2PPh_2\}H_2Os_3(CO)_9]$ and $[\{Si\}-(CH_2)_2-PPh_2\}Os_3(CO)_{11}]$ where $\{Si\}-$ = silica, with their PPh_2Et homogeneous analogues showed that the supported catalysts gave lower turnover rates in ethylene hydrogenation [66, 67]. Deactivation occurred accompanied by formation of $[\{Si\}-(CH_2)_2PPh_2\}Os_3(CO)_8(\mu\text{-}H)_3(\mu\text{-}CCH_3)]$ as well as cluster decomposition. Other workers found that the same triosmium species supported on phosphinated polystyrene are inactive for ethylene hydrogenation but active for 1-hexene isomerisation [68, 68a]. $[Os_3(CO)_{12}]$ and $[Os_3(CO)_{10}(MeCN)_2]$ react with the silanol groups of silica at 423 K to form the grafted cluster 1, which is stable in air at room temperature and which is an active catalyst for ethylene hydrogenation at 353 K [68b].

1

[Ru(cot)(cod)] (cot = cycloocta-1,3,5-triene) reacts with polystyrene under 1 atmosphere of hydrogen in tetrahydrofuran at room temperature to form a supported catalyst in which the ruthenium is believed to be bound to the phenyl rings in a similar manner to $[Ru_2(PhCH_2CH_2CH_2Ph)]_n$ [69]. Although the detailed ruthenium environment is unknown, electron microscopy excludes the presence of metallic ruthenium. The supported ruthenium catalysts are active under mild conditions (25–80 °C and 50 atmospheres of hydrogen) for the hydrogenation of olefins, aromatic compounds, ketones, oximes and nitrobenzene; at 120–140 °C with 50 atmospheres of hydrogen, alkyl and aryl cyanides can be reduced to amines [69].

Ruthenium(II) complexes supported in linear polymeric ligands with carboxylic acid side groups such as poly(acrylic acid) and alternate copolymers of maleic acid and vinyl monomers are active olefin hydrogenation and isomerisation catalysts [70, 71]. Five different ruthenium species are possible of which those with bidentate carboxylate binding predominate (reaction (7)) [72].

$$\tag{7}$$

On reusing these catalysts their activity decreases slightly due to some leaching of the ruthenium [71], and some of the ruthenium(II) groups becoming bonded to more than one carboxylate group [72]. This latter can be prevented by neutralising or esterifying the uncoordinated carboxyl groups. By using alternating copolymers of maleic acid and different alkyl and aryl vinyl ethers the hydrophobicity, electronegativity and accessibility of the sites could be varied enabling the catalytic properties to be studied as a function of the secondary structure. The resulting catalytic activity then depended on a combination of: (i) the swelling ability of the solvent; a larger swelling facilitated access of the substrate to the active site but also favoured deactivation by leaching and coordination of extra carboxylate ligands; (ii) ligand dissociation and stabilisation of solvated reaction intermediates by polar solvents; (iii) the formation of active hydrido-species by hydrogen abstraction from hydroxylic solvents. The most active 1-pentene hydrogenation catalyst was that bound on 2- ethylhexyl vinyl ether–maleic acid using n-octane as the solvent and that bound on sodium polyacrylate in n-octane/ethanol (24/1) [72]. Poly(β-diketones) such as poly(methacroyl acetone) form similar complexes which are coordinatively unsaturated due to the facile loss of triphenylphosphine (reaction (8)); they are accordingly very active olefin hydrogenation and isomerisation catalysts [71].

$$
\underset{\substack{\text{CH}_3-\overset{\displaystyle\overset{\text{CH}_2}{|}}{\underset{\text{H}_2\text{C}}{C}}-C=O \\ \quad \underset{\text{CH}_3}{\overset{\diagdown}{C}=O}}}{} + [\text{RuH}_2(\text{CO})(\text{PPh}_3)_3] \xrightarrow[n=1,2]{} \left[\begin{array}{c} \text{CH}_2 \\ \text{CH}-C \\ \quad \text{C}-\text{O} \\ \text{HC} \qquad \text{RuH(CO)(PPh}_3)_n \\ \quad \text{C}-\text{O} \\ \text{CH}_3 \end{array} \right] \quad (8)
$$

[Ru(PPh$_3$)$_3$Cl$_2$] supported on phosphinated polystyrene [73] gives some selectivity in olefin hydrogenation [21]. Short chain terminal olefins are hydrogenated more rapidly than their long chain counterparts. The ability of these catalysts to discriminate between styrene and 1-hexene decreases as the degree of metal loading increases and as the percentage of benzene in the benzene-ethanol solvent increases. As the proportion of benzene is increased, so the polymer swells and, accordingly, becomes less selective. Under conditions of very low metal loading the catalysts become almost as active as their homogeneous analogues [21].

[Ru(PPh$_3$)$_2$(CO)$_2$Cl$_2$] supported on diphenylphosphinated polystyrene-divinylbenzene gave, in the presence of 18–20 mol. excess of triphenylphosphine a highly selective catalyst for the hydrogenation of the dienes 2–5 to the monenes 6–9 [38, 38a].

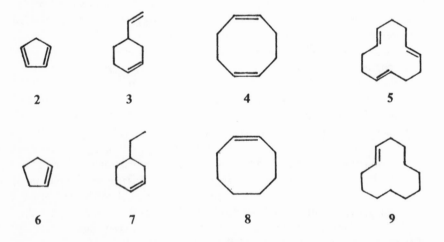

The catalyst could be recycled without detectable loss of the ruthenium in spite of the large excess of free triphenylphosphine, which is in sharp contrast to the case with supported rhodium(I) complexes (see below). An attempt to determine whether a large P : Ru ratio in the catalyst itself could effect the same selective hydrogenation as achieved using added triphenylphosphine showed initially that it could not. The coordinated phosphine moieties did not appear to encounter the bound ruthenium fast enough to influence the selectivity in the same manner as added triphenylphosphine can [38]. This result is in direct contrast to our own observations on the influence of bound phosphine groups on the selectivity of rhodium(I)-catalysed hydroformylation (see Section 8.2.b) [74]. A more detailed study of the ruthenium(II) demonstrated that the degree of phosphine loading was very important in determining selectivity [22]. At all phosphine loadings the selectivity increased as the P : Ru ratio increased (Figure 1). However examination of Figure 1 indicates that the degree of selectivity

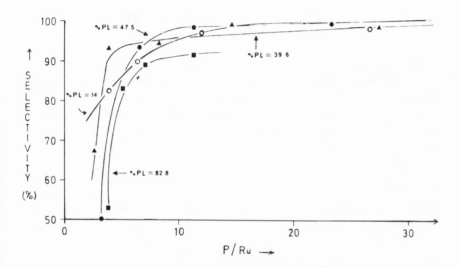

Fig. 1. The selectivity of phosphinated polystyrene supported ruthenium(II) catalysts as a function of phosphine loading (PL) and P/Ru ratio in the hydrogenation of 1,5-cyclo-octadiene. (Reproduced with permission from [22]).

achievable is a complex function of both phosphine loading (PL) and P : Ru ratio. This is because chain mobility restrictions influence the dynamic ligand-metal equilibrium within the polymer, which in turn affects the selectivity. The selectivity is particularly low for highly phosphine loaded polymers at low P : Ru ratios because these polymers have the lowest swelling factors. Low swellability results in decreased chain mobility. Therefore, despite the high concentration of free phosphine sites, the effective local concentration of free phosphine is actually quite low due to chain mobility restrictions. As the phosphine loading is decreased so swellability and hence chain mobility increase due to few phosphine-metal-phosphine cross-links and so the locally available phosphine concentration rises and the selectivity increases [22]. $[Ru(PPh_3)_2(CO)_2Cl_2]$ has been shown to be an active catalyst for the hydrogenation of aldehydes and ketones at 15 atmospheres and 160 °C in the gas phase and under supported liquid phase conditions in a gas chromatograph (see Section 1.8.b). In the gase phase, reactants with boiling points near the reaction temperature can be readily hydrogenated [75].

$RuCl_3$ supported on poly-L-methylenimine catalyses the asymmetric hydrogenation of methylacetoacetate (reaction (9)) to give a slight (5.3% maximum) optical yield. The yield drops if the ligand : ruthenium ratio is reduced below 10 : 1 [76].

$$CH_3CCH_2COOCH_3 \xrightarrow{H_2} CH_3\overset{*}{C}HCH_2COOCH_3 \tag{9}$$
$$\underset{O}{\|} \qquad\qquad\qquad \underset{OH}{|}$$

$[Ru(PPh_3)_2CO(CF_3COO)_2]$ catalyses the dehydrogenation of alcohols to aldehydes. When this complex was supported on phosphinated and carboxylated polystyrene it was found that the complex on the phosphinated polystyrene was the more active indicating that carboxylate lability is the factor that determines catalytic activity. The supported catalysts were all less active than their homogeneous counterparts [77].

6.9.4. COBALT

When cobalt carbonyl was bound onto bituminous coal, as shown in reaction (10), better liquefaction of the coal with less hydrogen consumption was

$$coal + CH_3OCH_2Cl \xrightarrow{SnCl_4} \boxed{coal}-CH_2Cl \xrightarrow{LiPPh_2} \boxed{coal}-CH_2PPh_2$$
$$\downarrow Co_2(CO)_8 \tag{10}$$
$$[(\boxed{coal}-CH_2PPh_2)_nCo_2(CO)_{8-n}]$$

achieved at 390°C under 2000 psi of hydrogen than when the metal was merely absorbed on to the coal [78]. $[Co_2(CO)_8]$ supported on phosphinated silica, $\{Si\}-O\dot{S}iCH_2CH_2CH_2PR_2$ (R = Ph, Bu, cyclohexyl), selectively catalysed the reduction of cis,trans,trans-1,5,9-cyclododecatriene to the monoene [79]. The use of alumina, silica and NaX zeolites as carriers for $[Co(acac)_3]-AlEt_3$ catalysts for the hydrogenation of cyclohexene and phenylacetylene increased their stability and selectivity and decreased polymer formation [80].

6.9.5. RHODIUM

Supported rhodium complexes have been so extensively used for hydrogenations that it makes sense to subdivide the present Section into ten parts covering analogues of [Rh(PPh$_3$)$_3$Cl], other rhodium(I)-phosphine systems, rhodium(I)-carbonyl complexes, other rhodium catalysts and rhodium(I) asymmetric hydrogenation catalysts.

6.9.5.1. *Analogues of [Rh(PPh$_3$)$_3$Cl]*
Wilkinson's catalyst is now so extensively used that 1.75% of all papers currently recorded in *Chemical Abstracts* are concerned with [Rh(PPh$_3$)$_3$Cl] [81]. Accordingly it is not surprising to find that over 50 papers have described attempts to support it on a range of polymeric and inorganic substrates. Although most papers are concerned with supports that bond to the rhodium, two interesting papers report the use of a functionalised polymer, the silver salt of sulphonated polystyrene, which enhances the activity of homogeneous [Rh(PPh$_3$)$_3$Cl] by pulling a triphenylphosphine ligand off the rhodium(I) [82, 83].

6.9.5.1.a. *Nature of support.* Phosphinated polystyrene, both soluble and insoluble [84, 85], and phosphinated silica have been most widely used to support Wilkinson's catalyst. However a range of other supports have been used, occasionally including aromatic polyamides [86], phosphinated polydiacetylenes [87], and phosphinated polyvinylchloride, although the latter is fairly inactive [15, 88]. Since the phosphine groups on silica are largely on the surface, silica supports tend to give more active catalysts than polymer supports, in which many of the catalytically active sites are buried deep within the polymer, necessitating slow diffusion of reactants and products to and from these active sites [17]. However, whilst inhibiting activity, this diffusion can be used to enhance selectivity, since bulky reactants will have more difficulty in diffusing into the polymer [16].

The importance of the nature of the support is particularly well illustrated by a series of supported catalysts in which [Rh(nbd)Cl]$_2$ was supported on phosphinated diacetylene, **10**, phosphinated silica, {Si}—CH$_2$CH$_2$CH$_2$PPh$_2$, and phosphinated polystyrene [88a].

10

The catalysts supported on diacetylene and silica are active for the reduction of arenes to cyclohexane derivatives, whereas those supported on polystyrene are not. This is believed to be due to the greater rigidity of the first two supports which stabilise the coordinatively unsaturated rhodium(I) centres more effectively than the more flexible polystyrene [88a].

6.9.5.1.b. *Activity*. The activity of a supported catalyst depends on both the number and accessibility of the active sites. Thus when very fine beads of a lightly cross-linked polystyrene are used [38] the activity is very much greater than when larger beads or more highly cross-linked material is used [16, 39]. Indeed it has been claimed that rhodium(I) complexes supported on completely uncrosslinked polystyrene and poly(iminoethylene) are more active and more stable than their homogeneous counterparts [39]. The higher activities and stabilities of these and other rhodium(I) systems have been ascribed to the role of the polymer support in anchoring the rhodium(I) firmly, so reducing its mobility and so preventing dimerisation to form catalytically inactive $[Rh(PPh_3)_2Cl]_2$ species [39]. This explanation of the increase in activity has been questioned, following a detailed analysis of the mechanism of the homogeneously catalysed reaction, which showed that reactions at monomeric $[Rh(PPh_3)_3Cl]$ sites are very fast and not rate determining, so that increasing the amount of this monomer will not markedly increase the rate. Thus a near optimisation of rate is achieved in the homogeneous case so that no significant rate advantage remains to be realised by anchoring rhodium(I) onto a support [89]. Whilst this suggestion is probably true as far as it goes, it neglects the important influence on reaction rate of the solvent which, in the polymer supported catalysts, will include the polymer itself. Such a suggestion is consistent with the conflicting results in the literature, since another group found that the activity of homogeneous and soluble polystyrene supported catalysts was essentially identical and significantly greater than of their insoluble polystyrene supported analogues [85]. Dimerisation as a cause of deactivation has been suggested to account for the increasing rate of deactivation with increasing chain length (n) of rhodium(I) supported on silica containing $-Si(CH_2)_nPPh_2$ functional groups [90].

When rhodium trichloride is reacted with silica phosphinated with $Ph_2-P(CH_2)_3Si(OMe)_3$ and the product reduced with hydrogen the resulting catalyst is 280 times as active in the hydrogenation of cyclohexene at 25 °C and 1 atmosphere of hydrogen as homogeneous $[Rh(PPh_3)_3Cl]$ [44]. Although the detailed structure of the supported catalyst has not been determined the authors are confident that metallic rhodium is not present. Very active catalysts are formed by covering silica with a pre-polymer of a polyphenylsiloxane, chloromethylating, phosphinating and finally treating with $[Rh(C_2H_4)_2Cl]_2$ [91].

The high activity may be due to clusters of rhodium atoms, intermediate in oxidation state between Rh(I) and Rh(0), surrounded by ligands [92, 92a]. In support of this, when $[Rh(CO)_2Cl]_2$ is supported on phosphinated silica treatment with hydrogen for a long time yields metallic rhodium, whereas when $[Rh(PPh_3)_3Cl]$ is used as the source of rhodium(I) the phosphine groups prevent reduction to metallic rhodium [93].

The phosphorus : rhodium ratio in the catalyst influences activity. Since a phosphine must be released from rhodium(I) in order to activate the catalyst, local P : Rh ratios of $< 3 : 1$ are desirable, although dimerisation must be prevented [34, 94]. Reduction to metallic rhodium may occur if sufficient phosphine groups are present [95]. The optimum phosphorus : rhodium ratio depends on both the support and the method of preparation of the catalyst itself, because many phosphine ligands bound to the polymer will be remote from rhodium sites. X-ray absorption of synchrotron radiation (EXAFS) has shown that at least one active catalyst prepared by treating phosphinated polystyrene with $[Rh(PPh_3)_3Cl]$ contains $trans$-$[Rh((P)-PPh_2)(PPh_3)(\mu\text{-}Cl)]_2$ groups which react with hydrogen to form the coordinatively unsaturated rhodium(III) complex $[Rh((P)-PPh_2)(PPh_3)H_2Cl]$ [96]. The same technique, EXAFS, has been used to show that the nature of the active sites when $[Rh(PPh_3)_3Br]$ is refluxed with phosphinated polystyrene depends on the degree of cross-linking. At low (2%) cross-linking the predominant species present was the dimer $trans$-$[Rh((P)-PPh_2)(PPh_3)(\mu\text{-}Br)]_2$ whereas at higher (20%) cross-linking the predominant species was monomeric $[Rh((P)-PPh_2)(PPh_3)_2Br]$ [97].

The influence of the phosphorus : rhodium ratio on the activity and selectivity of the polymer-supported catalyst is likely to depend on the detailed nature of the link between the polymer and the phosphine ligand since in the homogeneous analogues prepared by treating $[Rh(cyclooctene)_2Cl]_2$ with p-$C_nH_{2n+1}C_6H_4PPh_2$ the optimum P : Rh ratio depends on n [98]. It also depends on solvent, with more powerfully coordinating solvents giving rise to higher optimum P : Rh ratios as expected due to phosphine-solvent competition [98].

The nature of the solvent can have a profound effect on the activity of a supported catalyst. Thus the effect of changing the solvent from benzene to ethanol on the rate of hydrogenation depends on the nature of the olefin. Thus benzene is a much better solvent than ethanol for swelling polystyrene. Conversely, the solubilities of non-polar olefins are much greater in non-polar polymers such as polystyrene, than they are in ethanol. The complex and normally mutually conflicting effects of substrate solubility and polymer swelling as a function of solvent polarity give rise to the complex selectivity pattern shown in Figure 2 [23]. Whilst non-polar olefins show initial rate enhancements at low olefin concentration when the percentage of ethanol is increased due to

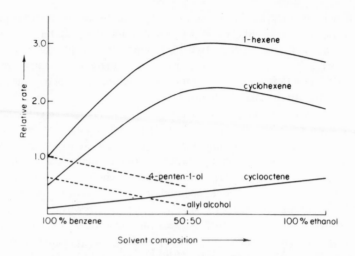

Fig. 2. The influence of solvent polarity on the relative rates of hydrogenation of non-polar
(—) and polar (— — — —) olefins. (Reproduced with permission from [23]).

enhanced solubility in the resin, polar olefins suffer just the opposite effect due
to reduced resin solubility. For polar olefins the reduction in polymer swelling is
dominant, as it is for some of the non-polar olefins at high ethanol concentrations [23]. Tetrahydrofuran is a good swelling solvent for polystyrene that also
has a high polarity. As a result it significantly enhances the rate of hydrogenation
of non-polar olefins [94]. However the rates of such hydrogenations are inversely
dependent on olefin concentration, because the swelling of the polymer in
tetrahydrofuran is suppressed by the olefin [94].

An important deactivation mechanism is leaching. Leaching is always likely
to occur with rhodium(I) phosphine complexes, because the mechanism of
catalytic hydrogenation involves cleavage of a rhodium-phosphorus bond to
create a coordinately unsaturated species to which the olefin may coordinate
[81]. If the Rh-P bond cleaved is that linking rhodium to the support leaching
will occur unless recoordination either at the original or a fresh site takes place.
In a flow system recoordination at the fresh site will inevitably occur downstream so that leaching will result; leaching may not be detectable in a batch
process. However, in a batch process great care must be taken to prevent the
ingress of oxygen at the end of a run, since polymer supported phosphines are
very sensitive to oxidation [74, 99]; the resulting phosphorus(V) compounds
do not bind to rhodium(I). To prevent leaching, at least two bonds from
rhodium(I) to polymer bound phosphine ligands are recommended [100].
Both phosphine groups may be attached to a single site, as in $\{Si\}-O-SiMe_2$-

$CH_2CH(CH_2PPh_2)_2$ [101–103]. Alternatively the polymer-rhodium(I) link may be made through a group such as a carborane, which remains coordinated throughout the hydrogenation [104, 105].

Many rhodium(I) catalysts are sensitive to poisoning by mercaptans. Although Wilkinson's complex shows some resistance to such poisoning [106], Wilkinson himself isolated $[Rh(PPh_3)_2HCl(SR)]$ from solutions of $[Rh(PPh_3)_3Cl]$ and thiols [107]. Thiols such as n-butylthiol react with Wilkinson's complexes supported on phosphinated silica, $[\{Si\}-O-CH_2CH_2PPh_2)_3RhCl]$ to reduce the activity of the catalyst, but to enhance its thermal stability [48–50]. The reasons for the enhanced stability are far from clear, since the rhodium(I) remains bound to the silica throughout indicating that at least one tertiary phosphine group remains bound to the silica. In addition the steady trend in activity with change of halide in the order $Cl < Br \leqslant 1$ strongly implies that the halide also remains bound. Other sulphur compounds are more deleterious. Carbon disulphide, thiophene and ethane-1-2-diol render the supported catalysts inactive; di-n-butyl sulphide reduces the activity but fails to enhance the thermal stability.

The activity of supported catalysts depends heavily on the nature of the support. Thus the ethylene hydrogenation activity of $[Rh(PPh_3)_3Cl]$ supported on a ternary copolymer of styrene, divinylbenzene and p-styryldiphenylphosphine changed abruptly at 68 °C [33]. This was due to a reversible discontinuous change in the pre-exponential factor due to the glass transition of the polymer. Attempts to measure the glass transition temperature by differental scanning calorimetry showed it occurred at 80 °C. Although the authors ascribed the difference between 68 °C and 80 °C to the different heating rate in the calorimeter it may well have arisen from local heating within the polymer during the catalysis giving rise to an apparently lower temperature for the glass transition. Not all changes in polymer structures are reversible and irreversible changes may well give rise to catalyst deactivation [34].

6.9.5.1.c. *Selectivity*. Polymer supported catalysts can give rise to selectivity by reducing the hydrogenation rate of bulky olefins which have difficulty in diffusing to the active sites within the polymer. Thus the relative rates of hydrogenation of a series of olefins in benzene using 100–200 mesh 2% crosslinked polystyrene were found to decrease in the order 1-hexene $\gg$ cyclohexene $>$ cyclododecene $\gg \Delta^2$-cholestene [23–27]. A first attempt to calculate the effective pore size within the polymer from the substrate selectivity has not been over-successful since calculated pore sizes of about 7 Å were obtained in polymers with measured pore sizes in the region of 150–300 Å. Although, of course, the substrate will be solvated and hence appear larger than the simple molecule, this is unlikely to completely account for the discrepancy. A supported catalyst with a very high selectivity for small olefins can be prepared by swelling the phosphinated polystyrene, treating with Wilkinson's catalyst, removing the solvent to contract the beads and then poisoning the surface catalytic sites. On reswelling the only catalytic sites available are those deep within the polymer, so that the catalyst only hydrogenated small olefins [16]. Such experiments readily account for the differences in selectivity of polymer supported catalysts prepared under different conditions, since alteration of the preparative conditions results in an alteration of the location of the catalytic sites within the polymer [108–110].

The importance of site location on selectivity is well illustrated by a comparison of the hydrogenation of 1,7-p-menthene to a mixture of cis- and trans-p-menthane. The homogeneous catalyst $[Rh(PPh_3)_3Cl]$ gave rise to a cis : trans ratio of 2.0 ; 1 whilst supporting on polystyrene gave a ratio of 3.1 : 1 and on silica a cis : trans ratio of 9.2 : 1 was obtained [28]. It appears that the lowest steric demands are created by supporting on silica which is presumably effective at promoting and stabilising coordinative unsaturation without imposing its own steric demands. An analogous effect has been achieved by introducing polar groups on the polymer support and demonstrating that these influence the selectivity of the rhodium(I)-catalysed hydrogenation of $CH_2=CH(CH_2)_n COOH$ where $n = 0, 1, 2, 4$ and 8 [11].

The effect of solvent on selectivity is very complex. Thus when the relative rates of hydrogenation of 1-hexene and cyclohexene in the presence of $[Rh(PPh_3)_3Cl]$ supported on 1% and 4% cross-linked polystyrene were studied, the influence of solvent was found to be dependent on substrate concentration [23, 39]. On changing from benzene to 1 : 1 benzene : ethanol there is a significant enhancement in the hydrogenation rate at low substrate concentrations, whereas at high substrate concentrations there is relatively little change in rate; furthermore for high cyclohexene concentrations the rate decreases as benzene is replaced by ethanol. These complex observations arise from the mutually conflicting effects of an increase in substrate concentration within the resin beads when the polarity of the solvent is increased and the lower resin swelling power of ethanol than benzene. These effects are only significant under conditions when the amount of substrate is rate limiting [23, 39]. Clearly the optimum solvent would be one that combines good swelling ability and high polarity. Tetrahydrofuran meets both requirements and results in enhanced hydrogenation rates for the sterically more demanding cyclohexene than benzene [39, 40].

Selectivity towards substrates of different molecular sizes can also be controlled by altering the loading of the catalyst on to the support. High loadings result in high selectivities (Table I) [29]. This is because the polymer surrounding

TABLE I

Relative rates of reduction of olefins in competition with cyclohexene in the presence of [(polystyrene—⟨ ⟩— PPh_2)Rh(PPh_3)$_2$Cl] in toluene. (From [29]).

Olefin	Relative hydrogenation rates on beads with	
	High Rh[I] loading	Low Rh[I] loading
(cyclopentene)	1.75 ± 0.03	1.80 ± 0.1
(cyclohexene)	1.00 ± 0.05	1.00 ± 0.04
(cycloheptene)	0.805 ± 0.05	0.97 ± 0.06
(cyclooctene)	0.43 ± 0.08	0.64 ± 0.05
(norbornene deriv.)	0.08 ± 0.003	0.35 ± 0.02

the catalyst imposes a diffusion barrier between the bulk solution and the catalytic centre. The resulting concentration gradient across this diffusion barrier will be greatest for the most active catalyst. When comparing olefins of different molar volumes, which give rise to different diffusion rates, the olefin with the largest molar volume should result in the greatest concentration gradient in the polymer. Consequently, the most active catalyst will place the greatest demand on the bulk solution to supply substrate which will result in an enhancement of the differences in the diffusion rates of the two olefins.

[Rh(PPh_3)$_3$Cl] supported on silica gel in the presence of excess triphenyl-phosphine promotes the specific reduction of acetylene and propyne to ethene and propene respectively [111a]. During acetylene reduction the catalyst suffers progressive deactivation, probably due to strong ethylene coordination. This can be reduced, and deactivation minimised by using strongly coordinating solvents and microporous supports; however although these give longer catalyst life they do so with the penalty of reduced activity.

6.9.5.1.d. *Regioselectivity*. Many unsaturated substrates contain more than one double bond, and it is often highly desirable to hydrogenate only one. Polymer supported catalysts should, because of their greater potential selectivity, be valuable in this role, and indeed have been shown to be able to give two to four times greater regioselectivity in hydrogenation of the ester side chain double bond of the steroid **11** than the corresponding homogeneous catalyst [29]. This greater regioselectivity is, of course, achieved at the expense of reduced activity.

$$O-\overset{\overset{\displaystyle O}{\|}}{C}-(CH_2)_n-CH{=}CH-(CH_2)_n-CH_3$$

11

6.9.5.1.e. *Hydrogen/deuterium exchange*. Poly(styrylphosphine) containing rhodium(I) complexes promotes D_2/H_2O and D_2/C_2H_5OH exchange; the most active catalyst is that obtained from $[Rh(C_{10}H_{12})Cl]_2$ [111b].

6.9.5.2. *Other Rhodium(I)-Phosphine Complexes*
Reaction of $[RhPPh_3)_4H]$ with poly(carboxylic acids) such as poly(acrylic acid) and poly(maleic acid) eliminates hydrogen (reaction (11)) and yields active olefin hydrogenation catalysts [71].

$$+ (4-x)(PPh_3) + H_2 \quad (11)$$

$$[Rh(\text{P})\text{-}C_6H_4\text{-}CH_2PPhR)_2(nbd)]^+ ClO_4^- + acacH \quad (12)$$

A detailed study [95, 112–115] of a series of rhodium(I) catalysts prepared by reaction (12) showed that their effectiveness depended upon several factors including:

(i) The steric and electronic properties of the R group played the dominant role in determining catalyst selectivity.

(ii) The nature of the solvent, in particular its ability to swell the polymer, suggested that reaction occurred within the polymer bead.

(iii) The phosphorus : rhodium ratio, which was particularly important in preventing reduction to the metal.

(iv) ESR investigation of the catalysts during and after hydrogenation showed the presence of rhodium(II), which is not observed in homogeneous catalysts. The rhodium(II) is formed during an induction period during which the colour becomes a greenish-brown.

(v) In the absence of solvent hydrogen reduces the rhodium to the free metal.

Supported rhodium(II) hydrogenation catalysts have been prepared by treating polymeric diphenybenzylphosphine with $RhCl_3 \cdot 3H_2O$ [116] and by immobilising $[Rh_2(OAc)_4]$ on phosphinated silica [117]. The latter catalyses the dehydrogenation of isopropanol. Rhodium(II) species were identified by ESR [116].

Cationic rhodium(I) hydrogenation catalyst precursors of the type $Rh(PPh_3)_n^+$ where $n = 2$ or 3 and rhodium(I) complexes containing both PPh_3 and Ph_2-$PCH_2CH_2P^+Ph_2CH_2Ph$, have been introducing into swelling mica-type silicates, such as hectorite (see Section 1.4.b) by exchanging sodium cations for either $[Rh_2(OAc)_{4-x}]^{2+}$ followed by subsequent reaction with triphenylphosphine or $Ph_2PCH_2CH_2P^+Ph_2CH_2Ph$ followed by reaction with triphenylphosphine and $[Rh(cod)Cl]_2$ [118–121]. Both catalysts are more active than their homogeneous analogues for the hydrogenation of 1-hexene in methanol. Their inactivity as benzene hydrogenation catalysts indicated the absence of metallic rhodium. The activities and selectivities of these catalysts can be altered markedly by changing the solvent since solvents can have profoundly different swelling abilities. The initial rates of reduction of small alkynes (1-hexyne or 2-hexyne) in the presence of methanol which swells the interlayer region to an average of 7.7 Å are comparable to those of the corresponding homogeneous catalysts, whereas for large alkynes such as diphenylacetylene the intercalated catalyst can be 100 times less active than its homogeneous analogue. The spatial requirements of the substrate are well illustrated by the decreasing hydrogenation rate of 2-decyne to cis-2-decene in solvents of different swelling abilities (Table II) [119]. $[Rh(nbd)(dppe)]^+ClO_4^-$ intercalated into hectorite catalyses the 1,2- and 1,4-addition of hydrogen to 1,3-dienes at rates that vary between 10^{-5}

TABLE II

The hydrogenation of 2-decyne to *cis*-2-decene by rhodium(I) catalysts intercalated on hectorite (prepared from $[Rh_2(OAc)_{4-x}]^{2+}$ and PPh_3 ($Rh : P = 1 : 6$). (Data from [119]).

Solvent	Rate (ml H_2 min^{-1} mmol Rh^{-1})		R_i/R_h	interlayer swelling (Å)
	intercalated (R_i)	homogeneous (R_h)		
$CH_2Cl_2 + 7\% MeOH$	2800	3300	0.85	10.0
MeOH	1200	2800	0.43	7.7
$Et_2O : MeOH$ (3 : 1v/v)	660	2800	0.24	6.7
$C_6H_6 + 7\% MeOH$	20	1000	0.02	5.7

and 0.83 relative to the corresponding homogeneous catalyst, and with selectivities towards the 1,2-addition products 1.5 to 2.3 times greater than their homogeneous analogues [121a]. When the extent of swelling of the support is comparable to the substrate size, as in methanol, then the spatial requirements of the substrate are particularly important. However when the swelling is much greater, as in acetone, the interlayer region becomes more like a solution and other effects such as differences in solvation between the intercalated and homogeneous states begin to dominate [121a].

Cationic rhodium complexes have been anchored on to polystyrene by long alkyl chains in order to attempt to place the rhodium(I) sites well into the solvent (reaction (13)) [45].

$$\tag{13}$$

When the resulting catalysts were used to hydrogenate 1-octene in acetone it was found that [45]:

(i) The rate of hydrogenation was comparable to or superior to the corresponding homogeneous catalyst based on $PMePh_2$ under comparable conditions.

(ii) The presence of excess free phosphine groups on the support was detrimental, presumably because these were sufficiently free to coordinate to the rhodium(I) so preventing olefin access.

(iii) The supported catalyst normally remained light orange throughout the cycle but if excess rhodium(I) were present or perchloric acid was added then rhodium metal was formed during the hydrogenation.

(iv) The drop in hydrogenation rate due to isomerisation of 1-octene to 2-octene was less for the supported than for the homogeneous catalyst containing PMePh$_2$.

6.9.5.3. *Rhodium(I)-Phosphinite Complexes*

Attempts to use rhodium(I)-polyphosphite complexes as supported catalysts have been unsuccessful [122, 123]. When [Rh(CO)$_2$Cl]$_2$ was added to phosphinated poly(methallyl alcohol) and poly(methyl methacrylate) the products were irreproducible and very dependent on the nature of the polymer. The initial polymers were linear, but rhodium(I) introduced cross-links making the products insoluble. The cross-links also put a strain on the polymer with the result that the system became unstable and spontaneously rearranged probably by dissociating the phosphinite.

6.9.5.4. *Rhodium Carbonyl Complexes*

[RhL(CO)Cl]$_2$, L = CO, PPh$_3$, P(tolyl)$_3$ react with polystyrene bound pyridine, 2,2'-bipyridine and acetylacetone to form supported complexes. However in contrast to the unsupported complexes these are unstable and are readily reduced to rhodium metal [62b, 124, 125]. [Rh(CO)$_2$Cl]$_2$ interacts strongly with γ-alumina to give a product which is more active in the hydrogenation of 1- and 2-pentyne than [Rh(CO)$_2$Cl]$_2$ itself [126, 127]. It also catalyses hydrogenation of the terminal double bond in *trans*-1,3-pentadiene [127a]. In contrast the activity of [Rh$_4$(CO)$_{12}$] is slightly reduced by supporting it on γ-alumina [127]; this catalyst reduces the internal double-bond of *trans*-1,3-pentadiene only once reduction of the terminal double-bond is complete [127b]. Comparative decarbonylation studies of a series of polymer-supported rhodium(I) carbonyl complexes [(Ⓟ₁)Rh(CO)$_2$] and [(Ⓟ₂)RhCl(CO)$_2$], where Ⓟ₁ = polymer functionalised with pentane-2,4-dionate or ethyldithiocarbamate and Ⓟ₂ = polymer functionalised with ethylamine or diphenylphosphine) have shown that their catalytic activities are primarily determined by the ease with which loss of carbon monoxide takes place, and also depends on whether or not the rhodium-polymer bond remains intact under hydrogenation conditions [127c].

[Rh$_4$(CO)$_{12}$] reacts with (Ph$_2$PCH$_2$CH$_2$)$_2$Si(OEt)$_2$ to give [Rh$_4$(CO)$_{10}$ {(Ph$_2$PCH$_2$CH$_2$)$_2$Si(OEt)$_2$}], a cluster that changes its structure upon reaction with silica. Although this product has a higher thermal stability than that formed by absorbing $[Rh_4(CO)_{12}]$ directly onto silica, the slight activity for benzene hydrogenation indicates rhodium metal formation [128]. When Amberlyst A–21 (containing $Me_2NCH_2C_6H_4-$ groups) is treated with $[Rh_4(CO)_{12}]$ in thf at 100 °C for 5 hours under 97 atmospheres pressure of carbon monoxide a red coloured resin is formed which turns green on replacing the carbon monoxide by nitrogen [129]. The red-to-green colour change is accompanied by loss of carbonyl ligands and appears to yield $[Rh_6(CO)_{15}]^{2-}$ bound to the support. This product is a more active catalyst for the selective hydrogenation of α,β-unsaturated carbonyl and nitrile compounds using a mixture of carbon monoxide and water as the source of hydrogen than its homogeneous counterpart probably because the active Rh_6 species dimerise readily to Rh_{12} species in homogeneous solution [129].

$[Rh_6(CO)_{10}]$ bound to silicas bearing phosphine, amine and diamine functional groups retains its Rh_6 structural integrity and is active for the gas phase hydrogenation of olefins and acetylenes below 370 K [130].

6.9.5.5. *Organometallic Rhodium Complexes*
Polystyrene supported $[Rh^{III}CpCl_2]$ prepared by reaction (14) is a good olefin and arene hydrogenation catalyst in the presence of excess triethylamine at 70 °C under 110 psi hydrogen pressure [131, 132].

$$\text{reflux, MeOH, 1 week} \tag{14}$$

The role of the triethylamine is to promote replacement of a chloride ligand by hydride (reaction (15)).

$$\text{...} + H_2 + Et_3N \longrightarrow \text{...} + Et_3NHCl \tag{15}$$

In the absence of triethylamine the product is an effective catalyst for the isomerisation of allylbenzene [132]. The normal lithiated cyclopentadiene route cannot be used when rhodium trichloride is used as the source of rhodium because it would be destroyed by the water in the metal salt. However it can be used when $[Rh(CO)_2Cl]_2$ is the source of rhodium (reaction (16)) and in this case the product was effective in the hydrogenation of olefins, aldehydes and ketones although decomposition occurs to yield rhodium(0), which is probably the active catalyst [133].

$$\text{(16)}$$

The supported complex, 12, is active for the hydrogenation of substituted benzenes PhX to cyclohexenes; the rate decreases in the order X = MeO > Me > H [134]. 12 was prepared by the relatively complex process of copolymerising 13 with styrene and divinylbenzene and then reacting 13 as shown in reaction (17) because there is evidence (reaction (18)) that polystyrene containing cyclopentadiene attached directly to the phenyl ring, prepared by lithiation of polystyrene followed by reaction with cyclopent-2-enone does not have the integrity of structure expected and accordingly behaves anomalously [134].

$$\text{(17)}$$

$$\text{(18)}$$

Silica bound $[\{Si\}-ORh(allyl)H]$, prepared by treating silica with $[Rh(allyl)_3]$ [135], is an effective arene hydrogenation catalyst. With benzene the initial hydrogenation rate of more than 3000 turnovers/hour (22 °C, 500 psi H_2) was maintained indefinitely whereas, with naphthalene under the same conditions, the initial rate of 60 turnovers/hour soon dropped to 7.7 turnovers/hour, which was then maintained [136]. Since propane was evolved during the decreasing rate stage it is believed that the initial η^3-C_3H_5 ligand is replaced by an η^3-$C_{10}H_7$ ligand in the early stages of the reaction. The naphthalene is initially reduced to tetralin which is subsequently hydrogenated to both *cis*- and *trans*-decalin.

6.9.5.6. *Rhodium Carboxylate Complexes*

The main problem with phosphine supports is that rhodium(I) tends to be eluted during the hydrogenation due to rhodium(I)-phosphorus bond cleavage. This process is accelerated if oxygen is adventitiously introduced. A number of workers have attempted to surmount this problem by replacing a phosphine ligand by a carboxylic acid ligand. Treatment of chloromethylated polystyrene with anthranilic acid yields a polymer which, on reaction with rhodium(III) trichloride followed by sodium borohydride, gives rhodium(I) hydrogenation catalysts [7, 8]. These catalysts are exceptionally active hydrogenation catalysts for olefins, aromatics, carbonyl-, nitrile- and nitro-compounds. The catalysts have long term stability and are fairly insensitive to poisons. One of the potential problems of using carboxylate ligands is that they are only modest donor ligands towards rhodium(I). Whilst a modest donor ligand facilitates the formation of coordinatively unsaturated rhodium(I), it may allow facile reduction of the rhodium(I) to rhodium(0). An ESCA study showed that this did not occur with the anthranilic acid supported rhodium(I), probably due to the spatial restrictions imposed by the polymer. In support of this the more active catalysts are those with the greater degrees of cross-linking [7, 8]. Rhodium(III)-anthranilic acid complexes supported on Amberlyst A–27 ion exchange resins catalyse the hydrogenation of 1-decene at room temperature [136a].

Reaction of [Rh(PPh$_3$) (CO)H] with the carboxylic acid functional group of alternating copolymers of maleic acid and different alkyl- and aryl-vinylethers (reaction (19)) yields olefin hydrogenation and isomerisation catalysts [72].

$$L = CO, PPh_3$$

Unlike their ruthenium(II) analogues (see Section 6.12 above), these rhodium(I) catalysts are more effective as hydrogenation than isomerisation catalysts. Their activity increases progressively with use due to the release of carbonyl and triphenylphosphine ligands with the formation of coordinatively unsaturated species. However their activity is always less than their isobutyrate homogeneous analogues [71, 72].

Rhodium trichloride supported on either an acrylic acid-divinylbenzene copolymer, or a crosslinked polystyrene with iminodiacetic acid functional groups, acts as a selective hydrogenation catalyst for olefinic double-bonds with no activity for carbonyl groups or aromatic rings [137, 138]. In the latter case rhodium metal may be the active catalyst, since it is activated by boiling in aqueous methanol, when it goes black; boiling in water alone leaves the polymer yellow and inactive [138].

Dimeric $[Rh(OAc)_2]_2$ in the presence of an activating ligand, triphenylphosphine, catalyses the dehydrogenation of 2-propanol [138a]. When triphenylphosphine is replaced by phosphinated silica, dehydrogenation occurs without an induction period and at a steady rate that shows no deactivation with time, in contrast to the homogeneous catalysts [138b].

6.9.5.7. *Rhodium Amide and Imidazole Complexes*
On boiling rhodium(III) trichloride with polyamides in aqueous solution coordination of rhodium(III) to the amide bond followed by reduction to rhodium(I) occurs. The resulting supported $[(\text{(P)}-CONHR)RhHCl_2]^{2-}$ complexes catalyse the hydrogenation of olefins, aromatics and nitro compounds at room temperature and atmospheric pressure [139]. $[Rh(nbd)L_2]$ complexes prepared by treating $[Rh(nbd)_2]ClO_4$ with a suspension of polyimidazole or polytetramethylimidazole are less active than their homogeneous analogues in the hydrogenation of phenyl acetate, but more active in the hydrogenation of 1-hexene [139a].

6.9.5.8. *Rhodium Thioether Complexes*
Binuclear thioether bridged rhodium(I) complexes supported on chloromethylated polystyrene (reaction (20)) catalyse the hydrogenation of cyclohexene under 3 atmospheres of hydrogen at 40 °C [140].

6.9.5.9. *Organorhodium(III) Complexes*
$[Rh(allyl)_3]$ reacts with silica gel to give the supported rhodium(III) complex $[\{Si\}-ORh(allyl))_2]$ which on hydrogenation under ambient conditions yields $[\{Si\}-ORhH(allyl)]$ [141]. These rhodium complexes, being supported through 'hard' oxide ligands, retain the rhodium in the (+3) oxidation state and,

$$[Rh(\mu\text{-Cl}\{P(OMe)_3\}_2]_2 \quad (20)$$

as a result, promote reactions by acting as Lewis acids. Hence, they promote hydrogen/deuterium exchange (reaction (21) as well as H/D exchange between butane and deuterium gas.

$$(21)$$

The latter reaction may involve hydride abstraction from the butane by rhodium(I) leading to a carbonium ion which is either stabilised or deprotonated by the oxide support. These catalysts are likely to be studied further as carbon-hydrogen bond activation becomes an increasingly popular field of research.

6.9.6. ASYMMETRIC HYDROGENATION

During the past 20 years there has been an increasing demand for optically pure chircal compounds. In part this stems from both a recognition that only one optical isomer of chiral drugs is usually therapeutically active, and the recognition that the 'inactive' isomer may not be merely inactive but may in some instances be positively harmful. In the production of optical isomers, asymmetric synthesis obviates the use of optically active starting materials as well as of reagents for the separation of enantiomers. The high activity of rhodium(I) homogeneous hydrogenation catalysts has led to a search for rhodium(I) complexes that catalyse asymmetric hydrogenation [142]. Considerable success has been achieved with homogeneous catalysts involving a range of optically active ligands including particularly (−)-diop, **14**, which is prepared from the inexpensive, commercially available, natural chiral material L-(+)-tartaric acid.

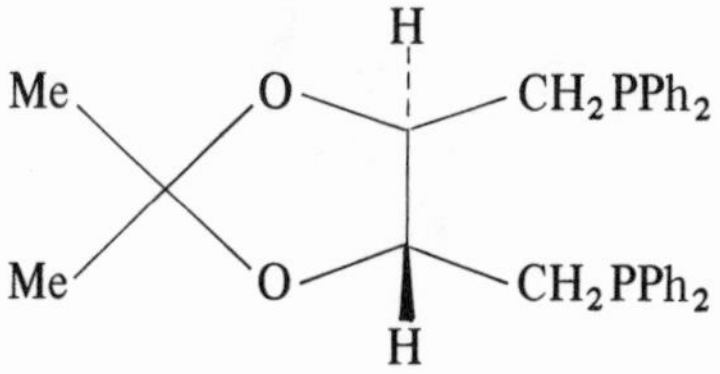

14

Recently a number of groups have attempted to support such optically active rhodium(I) complexes on insoluble supports.

The first attempts to support asymmetric complexes involved rhodium(I)-diop complexes supported on Merrifield's resin (reaction (22)) [143].

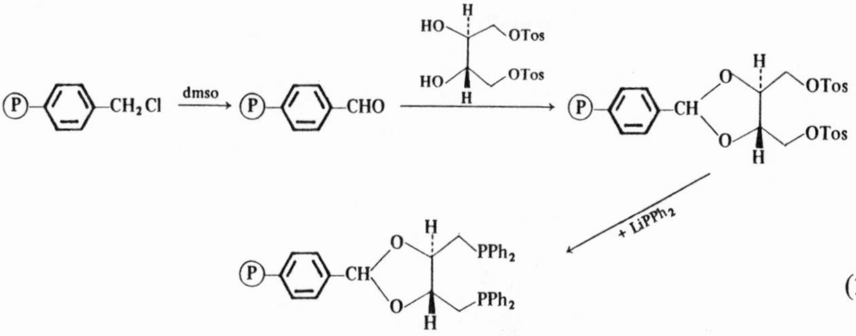

(22)

Whilst such polymers swell in non-polar solvents they shrink in polar solvents, thus preventing access of the substrate to the catalyst site. They are therefore unsuitable for the hydrogenation of substrates such as acylamido acrylic acids that are only soluble in polar solvents. Even with substrates that are soluble in benzene, rhodium(I)-diop catalysts supported in this way give lower optical yields than their homogeneous analogues [143]. In order to obtain polymers capable of swelling in polar, as well as non-polar solvents, copolymerisation of functionalised styrene with hydroxyethylmethacrylate was investigated (reaction (23)) [144, 145]. The resulting polymer, **15**, is swollen by alcohols and other polar solvents.

(23)

Hydrogenation of α-N-acylaminoacrylic acids in ethanol in the presence of **15** gave the same optical yields and produced amino acids having the same absolute configuration as in the presence of the soluble form of the catalyst. However the rates were reduced. The polymeric catalysts could be removed by filtration and reused without loss of optical purity provided that care was taken to exclude oxygen during the work-up [144, 145].

An alternative approach to using a hydrophilic copolymer to increasing the rhodium(I)-diop selectivity has been to investigate the influence of the molecular weight of non-cross-linked polystyrene. The optical selectivity decreases as the molecular weight decreases because of the tendency of low molecular weight polymers to coil up and so depress the effective asymmetric interaction [146–148]. Thus a relatively rigid support that leads to isolated catalytic sites is the ideal. This is confirmed by supporting rhodium(I)-diop complexes on silica gel (reaction (24)) where an asymmetric efficiency comparable to that of the homogeneous catalyst was achieved, although with an order of magnitude reduction of activity [149].

$$SiO_2 + \quad \begin{matrix} Ph_2PCH_2 \\ \\ Ph_2PCH_2 \end{matrix} \overset{H}{\underset{H}{\diagup\diagdown}} \begin{matrix} O \\ \\ O \end{matrix} \diagdown\diagup \begin{matrix} CH_3 \\ \\ (CH_2)_4Si(OEt)_3 \end{matrix} \qquad \overset{Under\ N_2}{\longrightarrow} \qquad Si\!\!-\!\!(CH_2)_4\,diop$$

$$\Big\downarrow [Rh(cyclooctene)_2Cl]_2$$

$$Si\!\!-\!\!(CH_2)_4\,diop\!\!-\!\!Rh^I \qquad (24)$$

This interpretation may be correct or it may be simplistic. The latter is suggested by the fact that when diop is supported on graphite, and then treated with [Rh(cod)Cl]$_2$ (reaction (25)) the resulting catalysts give lower optical yields

$$(S = \text{solvent or olefin})$$

than their homogeneous analogues and more important give products whose absolute configurations are exactly the opposite to those obtained using the corresponding homogeneous catalysts [150]. This could be due to a change of structure in the supported as opposed to homogeneous catalyst, or may be due to participation of surface phenolic groups, formed on the graphite during the initial potassium permanganate oxidation. This latter would be consistent with the observations already mentioned, and others mentioned below, that the local environment of the active site is very important indeed.

In an attempt to significantly alter the local environment of the active site, an optically active alcoholic functional group was introduced adjacent to the rhodium(I)-diop centre (reaction (26)) [35]. By introducing the alcoholic

(26)

functional group initially as a ketone it was possible, by reducing this asymmetrically, to obtain alcoholic groups of either configuration. In a protic solvent such as benzene-ethanol the optical yield was independent of the absolute configuration of the copolymer alcohol group. However in a non-protic solvent the optical yield did depend on the absolute configuration of the alcohol; in tetrahydrofuran, for example, a 40% optical yield was obtained when the copolymer alcohol was of R-configuration whereas when it had an S-configuration the optical yield was only 25%. Thus the copolymer alcohol groups apparently play a role in solvating the transition state in the absence of a protic solvent [35, 151].

Enzymes often have asymmetric active sites. Thus, if an achiral catalyst could be immobilised inside an enzyme, in the vicinity of its active site, the enzyme's chirality might become manifest if the enzyme-metal complex was employed as a catalyst (reaction (27)).

$$+ \; ML_n \longrightarrow \quad \!\!-ML_m \; + \; (n-m)L$$

(27)

| | achiral | enzyme with |
| enzyme with
chiral cavity(*) | homogeneous
catalyst | catalyst in
cavity |

This approach has now been effected by noting that the enzyme avidin, which has a chiral cavity, binds the substrate biotin very firmly. Biotin was converted to a rhodium(I) derivative, **16**, (reaction (28)) which was then bound to avidin. When the resulting avidin complex was used to catalyse the hydrogenation of α-acetamidoacrylic acid at 0 °C under 1.5 atmospheres of hydrogen, enantiomeric excesses of 34–41% were achieved with molar turnovers in excess of 500 [36]. Thus the asymmetric environment of the enzyme cavity had induced asymmetry in an achiral rhodium(I) complex-catalysed hydrogenation. A similar result, though experimentally simpler to achieve, was found when the achiral phosphine Ph_2PCl was reacted with an optical active cellulose and rhodium(I) introduced by equilibration with $[Rh(PPh_3)_3Cl]$. Between 11 and 28% optical purity was obtained in the reduction of N-α-phthalimidoacrylate to methyl-N-phthaloyl-D-alaninate [152].

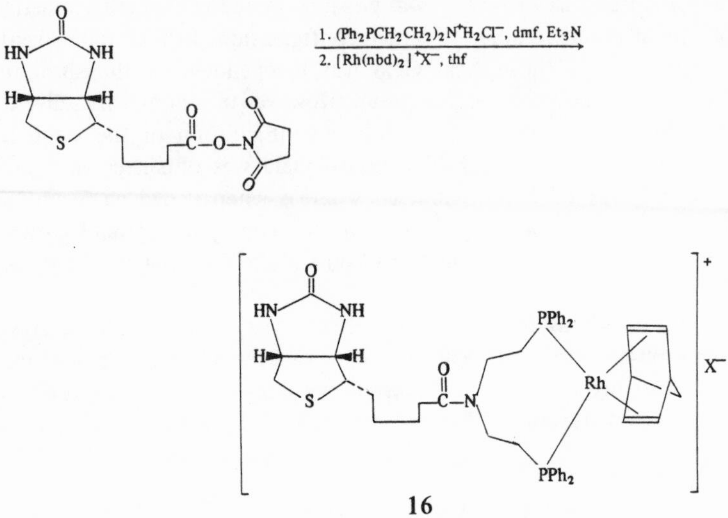

$$1. (Ph_2PCH_2CH_2)_2N^+H_2Cl^-, dmf, Et_3N$$
$$2. [Rh(nbd)_2]^+X^-, thf$$

$$(28)$$

16

Polymer-supported chiral pyrrolidine phosphine ligands, **17a** and **17b**, have been used to support neutral, **18**, or cationic, **19**, rhodium(I) complexes (reaction (29)) [153, 154].

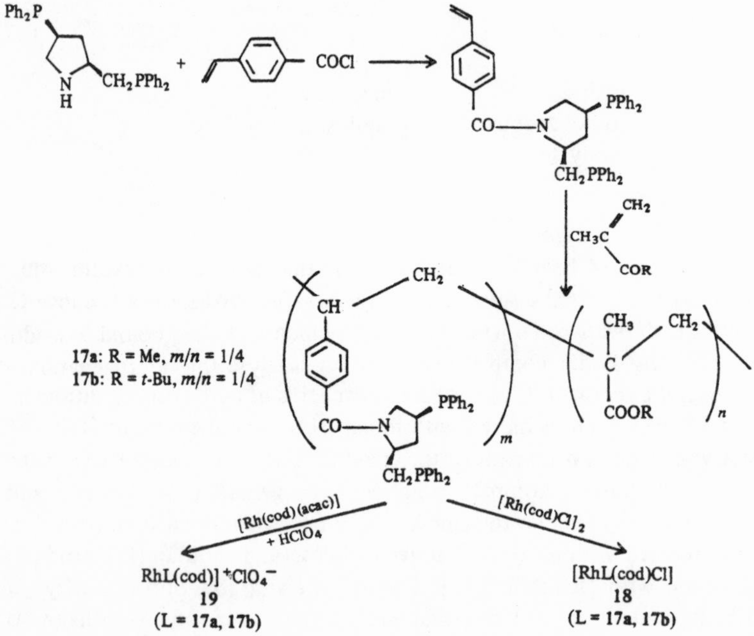

17a: R = Me, m/n = 1/4
17b: R = t-Bu, m/n = 1/4

$$[Rh(cod)(acac)]$$
$$+ HClO_4$$

$$[Rh(cod)Cl]_2$$

RhL(cod)]$^+$ClO$_4^-$
19
(L = **17a, 17b**)

[RhL(cod)Cl]
18
(L = **17a, 17b**)

$$(29)$$

In benzene **18** catalysed the asymmetric hydrogenation of keto-pantolactone to R(−)-pantolactone (reaction (30)) with optical yields of 73.4% (L = **17a**) and 75.7% (L = **17b**) [153].

$$\text{(30)}$$

The cationic supported complex, **19**, was more effective than the neutral supported complex, **18**, in the asymmetric hydrogenation of itaconic acid to methylsuccinic acid and Z-α-acetamidocinnamic acid; **19** was as effective as its homogeneous analogue under optimum conditions, although those optimum conditions needed more careful control with the supported than with the unsupported catalyst [154].

The chiral dimenthylphosphine group has been attached to polystyrene resins to form both **20** and **21** [155, 156].

20   **21**

By supporting rhodium(I) on **20** and **21**, optical yields as high as 58% were achieved in the hydrogenation of Z-α-acetamidocinnamic acid, although the activities of these supported catalysts were low. The optical activity of the product was heavily dependent on the solvent varying from 58% enantiomeric excess in 1 : 1 benzene : ethanol, through 14% ethanol in dioxane to only 8% ethanol in tetrahydrofuran. If oxygen was accidentally introduced during recycling the activity and optical yield decreased remarkably [155, 156]. Only one asymmetric catalyst has been reported using an asymmetric rhodium(I) complex [Rh(cod){(R)-(+)-PhCH$_2$CH(NHAc)COOH}]$^+$ClO$_4^-$ supported on the clay hectorite; this catalyses the asymmetric hydrogenation of Z-α-acetamidocinnamic acid [157].

6.9.7. IRIDIUM

When Vaska's complex is anchored onto phosphinated polystyrene the resulting supported complex is more active than its homogeneous analogue at equivalent P : Ir ratios in promoting the hydrogenation of 4-vinylcyclohexene to 4-ethyl-cyclohexene and 1,5-cod to cyclooctene [46, 158–161]. This suggests that the anchored phosphine moieties are not sufficiently mobile on the time scale of key steps in the mechanism to intercept anchored coordinatively unsaturated iridium intermediates at a rate equivalent to that of dissolved triphenylphosphine. These catalysts showed a very unusual temperature effect in that the rate of hydrogenation increased with decreasing temperature. This is probably due to the greater mobility of the anchored $-PPh_2$ groups at higher temperature, enabling them to compete more readily with the olefin for coordination sites around iridium. The polystyrene supported iridium(I) complexes could be exposed to air without loss of activity, indeed the activity was often enhanced on recycling due in part to the formation of a hydride complex during an induction period in the first cycle. The activity of these catalysts is enhanced in dimethylformamide due to competition between dmf and a phosphine ligand, leading to highly active [(Ⓟ—PPh_2)IrCl(CO) (dmf)] [158].

Iridium(I)-pyridine-phosphine complexes are very active olefin hydrogenation catalysts. However they are rapidly deactivated by a mechanism believed to involve dimerisation. An attempt to stabilise the catalysts by anchoring the iridium(I) on highly (40%) cross-linked phosphinated polystyrene failed [158a], although this was probably due to the use of dichloromethane as the solvent rather than failure of the matrix to prevent dimerisation within the polymer.

In addition to catalysing direct hydrogenation Vaska's complex supported on phosphinated polystyrene catalyses transfer hydrogenation between formic acid and α,β-unsaturated ketones (reaction (31)) [162].

$$RCH{=}CHCOR' + HCOOH$$

$$\xrightarrow[\text{[(Ⓟ—PPh}_2\text{)}_x\text{IrCl(CO) (PPh}_3\text{)}_{2-x}\text{]}]{} RCH_2CH_2COR' + CO_2 \tag{31}$$

As with so many reactions catalysed by supported complexes this reaction is very dependent on the solvent, conversions under otherwise identical conditions increasing in the order toluene (62%) < ethyleneglycol (89%) < decane (98%) < decalin (99%), an effect that is reversible on going back to the original solvent. A comparison of the unsupported and supported iridium(I) catalysts shows the supported catalyst to be more efficient as well as reusable, which the unsupported catalyst is not. Less than 50 ppm of iridium(I) were leached off in the first three runs, and thereafter practically none at all [162].

Tetrairidium clusters supported on phosphinated polystyrene [(Ⓟ—PPh_2)$_x$-$Ir_4(CO)_9(PPh_3)_{3-x}$] are active for the hydrogenation of ethylene at 30 °C under one atmosphere pressure [163, 164]. The catalysts cannot be prepared by direct reaction with $[Ir_4(CO)_{12}]$, but must instead be made indirectly by treating the phosphinated polymer with zinc, carbon monoxide and either $[Ir(CO)_2Cl(p\text{-toluidine})]$ or $[Ir(cod)Cl]_2$ to effect an *in situ* formation of the cluster. Phosphinated silica, $[Si]-(CH_2)_3PPh_2$ reacts directly with $[Ir_4(CO)_{12}]$ to give degradation and/or metal aggregation [165]. However, in the presence of either pyridine or trimethylamine-N-oxide, $[(\{Si\}-(CH_2)_3PPh_3)Ir(CO)_{11-n}\cdot(PPh_3)_n]$ where $n = 0-2$ is formed [166]. These supported clusters are thermally stable below $330\,°C$ and have no activity for ethylene hydrogenation at room temperature. They have some activity at elevated temperatures, although this may be due to degradation products rather than the supported tetrairidium cluster.

6.9.8. NICKEL

Nickel complexes have not been used extensively as homogeneous hydrogenation catalysts. There are accordingly very few reports of supported nickel complex catalysts where the complex survives through the reaction without being reduced to metallic nickel. Phosphinated cross-linked polystyrene treated with $NiCl_2\cdot6H_2O$ gives a more selective catalyst for the reduction of nitrobenzene by sodium borohydride than the corresponding homogeneous catalysts (reaction (32)); in particular polystyrene containing $-CH_2PPh_2$ can result in yields of 60% of **22** if the conditions are right (Table III) [167].

TABLE III

Selectivity of homogeneous and polymer supported nickel(II) complexes in catalysing the reduction of nitrobenzene according to reaction (32). (From [167]).

Catalyst	Yield (%)		
	22	**23**	**24**
$[Ni(PPh_3)_2Cl_2]$	10	40	40
(P)—⟨ ⟩—PPh_2 + $NiCl_2\cdot6H_2O$	55	5	
$[Ni(PhCH_2PPh_2)_2Cl_2]$	50	20	
(P)—⟨ ⟩—CH_2PPh_2 + $NiCl_2\cdot6H_2O$	60	5	

$$PhNO_2 \xrightarrow{\text{NaBH}_4, \text{ cat}} PhNO + PhN=NPh + PhN=NPh +$$
$$\downarrow$$
$$O$$

$$\quad\quad\quad\quad\quad\quad\quad\quad\quad 22 \quad\quad\quad\quad 23 \quad\quad\quad\quad (32)$$

$$+ PhNHNHPh + PhNH_2$$

$$24$$

However the nickel(II) is reduced to a finely dispersed form of metallic nickel during the reaction.

Nickel(II) supported on anthranilic acid anchored to highly cross-linked polystyrene (reaction (33)) promotes the hydrogenation olefins, dienes and

nitrobenzene, although it is not particularly active and has a relatively short lifetime [108]. The reduction of nitrobenzene is not clean, it gives a sticky amorphous solid that contains only a small amount of aniline.

Nickel catalysts prepared by codepositing nickel and an arene at $-196\,^{\circ}C$ onto alumina do not yield supported nickel(0) complexes but free metallic nickel which is active in toluene hydrogenation and hydrodealkylation [169]. Similarly supported nickel catalysts prepared by pyrolysis of $[Cp_2Ni_2(CO)_2]$ and $[Cp_3Ni_3(CO)_2]$ dispersed on silica gel and Vycor glass are active for ethylene and benzene hydrogenation [170]. Triethylaluminium has been used to activate nickel catalysts prepared by absorbing nickel(II) stearate on to γ-alumina [171] and nickel(II) naphthenehydroxamate on to polyacetylene [172, 173]; the former catalyses polyolefin hydrogenation and the latter phenylacetylene hydrogenation.

6.9.9. PALLADIUM AND PLATINUM

The first published work on supported metal complex catalysts involved the use of cationic $[Pd(NH_3)_4]^{2+}$ and $[Pt(NH_3)_4]^{2+}$ supported on sulphonated polystyrene. However to be of use as olefin hydrogenation catalysts the metal(II) salts must be reduced with hydrazine to give supported finely dispersed metal. Palladium catalysts prepared in this way have been used to catalyst olefin hydrogenation [174, 175]. Palladium catalysts prepared similarly by calcining NaY molecular sieves treated with $[Pd(NH_3)_4]^{2+}$ have been used to catalyst the oxidative dehydrogenation of cyclohexane [176]. These reactions are typical of many catalysed by supported palladium and platinum complexes in that the role of the complex is to be reduced to an extremely finely dispersed, and hence very active, metallic species.

A number of authors have reported investigations on the use of $[MCl_4]^{2-}$, (M = Pd, Pt) supported on basic ion-exchange resins. The large variations in activity and selectivity of these catalysts are principally due to different degrees of cross-linking, the greater activity and selectivity being observed with lower degrees of cross-linking. These catalysts have been used in the hydrogenation of olefins, acetylenes, allyl alcohol, propargyl alcohol cinnamaldehyde, cyclo-pentadiene and nitrobenzene [30–32, 177–183]. In several cases there is doubt as to whether the active catalyst is the anionic metal(II) complex or the free metal, although in others the presence of the free metal has been confirmed. Inevitably, therefore, these catalysts have been compared to such heterogeneous catalysts as palladium on charcoal. The greater dispersion of palladium on the anion exchange resin than on charcoal accounts for the greater activity of many of these resin-supported catalysts [32]. However, steric hindrance plays an important role in the resin-supported system since $[PdCl_4]^{2-}$ on an ion-exchange resin is five times as effective as palladium on charcoal for the hydrogenation of cyclohexene (80 °C; 25 atmospheres) whereas the latter is 33 times as effective as the resin supported catalyst for 1,2-dimethylcyclohexene reduction and 11 times as effective for 2,3-dimethylcyclohexene [30]. Hydrogenation of allyl alcohol over $[PdCl_4]^{2-}$ on Amberlyst-27 is much more selective for the formation of propanol than using palladium on charcoal [32]. In the hydrogena-tion of styrene over $[PdCl_4]^{2-}$ supported on Amberlyst-27 the rate decreases as the molecular weight of the alcohol solvent increases: MeOH > EtOH > i-PrOH [183].

A comparison of the effectiveness of a series of catalysts in the hydrogenation of several terpenes showed that palladium on charcoal was more active than $[PdCl_4]^{2-}$ on either Amberlyst-27 or Amberlyst-15 which in turn were more active than $[PdCl_4]^{2-}$ on the polysaccharide basic resin Sephadex QAE 50. However the ion-exchange supported catalysts had greater regio- and stereo-selectivity than palladium on charcoal; this selectivity depended on the solvent and was greater in acetic acid than ethanol (Table IV) [184].

TABLE IV

Comparison of selectivities in the hydrogenation of α-pinene over supported palladium complexes. (From [184]).

| | | | | | 25 | | | 26 |

Catalyst	Solvent	Selectivity	
		% 25	% 26
$[PdCl_4]^{2-}$/Amberlyst-27	EtOH	72	28
	HOAC	82	18
$[PdCl_4]^{2-}$/Sephadex QAE 50	EtOH	76.5	23.5
	HOAC	85.5	14.5
Palladium/charcoal	EtOH	68.5	31.5
	HOAC	72	28

Palladium(II) and platinum(II) bis-phosphine dichloro complexes have been found to be very effective catalysts for the selective reduction of polyenes to monoenes [185–188]. However the real interest in this reduction lies in the selective hydrogenation of vegetable oils to enable them to be used as components of margarine and other hard fats, and for this application it is essential to remove all the catalyst at the end of the reaction. This is difficult with homogeneous catalysts and accordingly a number of groups have attempted to support $PdCl_2$ and $PtCl_2$ on polymeric phosphine ligands [189–193]. When

PtCl$_2$ is supported on phosphinated polystyrene it behaves in essentially the same way as its unsupported analogues, requiring tin(II) chloride activation and having an activity that increases as the amount of platinum on the polymer increases [189, 190]. For the palladium(II) system, however, marked differences are observed between the polymer-supported catalysts and their homogeneous analogues (Table V) [189–193]. A detailed investigation of these differences

TABLE V

Differences in the behaviour of homogeneous and supported palladium(II) complexes in the hydrogenation of polyolefins.

Characteristic	[Pd(PPh$_3$)$_2$Cl$_2$]	PdCl$_2$ supported on phosphinated polystyrene
Hydrogen pressure necessary for activity (atm).	35	1
Induction period	None	Induction period depends on temperature
Colour change during hydrogenation	None, remains yellow	Changes from light yellow to grey-green
Effect of addition of SnCl$_2$	Essential for activity	Makes the catalyst inactive at 1 atm. H$_2$
Effect of addition of PPh$_3$	Not reported	Makes catalyst inactive at 1 atm H$_2$

suggests that the higher activity of the supported palladium complexes is probably due to having fewer than two phosphine ligands per palladium in the supported system. This reduces the stability of palladium(II) with probable formation of very active palladium(0) finely dispersed within the support [193]. PdCl$_2$ supported on phosphinated polystyrene may also be used to catalyse the hydrogenation of monoolefins as well as acetylenes. Its high selectivity allows semi-hydrogenation of isolated triple bonds to be accomplished readily [194].

Binuclear palladium(II) complexes supported on phosphinated silica gel can selectively semihydrogenate, as well as fully hydrogenate, cyclopentadiene [196]. Monophosphine coordination is essential for activity; bidentate and tridentate phosphines yield catalytically inactive products [197]. A comparison of the activities of the supported palladium(0) complex, [({Si}—CH$_2$CH$_2$PPh$_2$)Pd(dba)], and the palladium(II) complex, [({Si}—CH$_2$CH$_2$PPh$_2$)PdCl$_2$], shows the former to be three times more active in cyclopentadiene hydrogenation [197]. Monodentate phosphine coordination is essential if palladium(II) acetate supported on phosphinated polystyrene is to be active as a phenylacetylene or styrene hydrogenation catalyst [198]. Unusually these catalysts are more active in hydrogenating the olefin than the acetylene. Palladium(II) anchored on to phosphinated silica containing $-(CH_2)_3 PPh_2$, $-p\text{-}C_6H_4CH_2PPh_2$, $-p\text{-}C_6H_4CH_2OCH(CH_2PPh_2)_2$ and $-(CH_2)_3OCH(CH_2PPh_2)_2$ has been used to hydrogenate a series of olefins [198a–198d].

The bimetallic cluster complexes $[Pt_2Co_2(CO)_8(\textcircled{P}-PPh_2)_2]$, $[Pt_2Ru(CO)_5 (\textcircled{P}-PPh_2)_3]$, $[PtRu_2(CO)_{8-x}(dppe)(\textcircled{P}-PPh_2)_x]$ and $[PtFe_2(CO)_8(\textcircled{P}-PPh_2)_2]$ supported on phosphinated polystyrene are active ethylene hydrogenation catalysts at temperatures below $100\,^{\circ}C$ and at atmospheric pressure [68a, 199, 200].

A number of groups have reported palladium and platinum hydrogenation catalysts supported by nitrogen donor ligands. A catalyst of particular interest is that of $PdCl_2$ supported on poly(vinylpyridine) where workers in the Institute of Chemistry in Shanghai have shown that the activity and selectivity of these catalysts in olefin hydrogenation depend on both the N : Pd molar ratio and the coordination ability of the solvent [201]. In contrast to the phosphine supported catalysts mentioned above and the bipyridine supported species described below, these are reported not to be readily reduced to palladium metal by hydrogen [201]. Other Chinese groups have used acrylonitrile-divinylbenzene and 2-N-vinylpyrrolidone-divinylbenzene copolymers as supports [201a].

Palladium and platium complexes supported on bipyridine attached to polystyrene (reaction (34)) are active olefin hydrogenation catalysts at ambient temperature and pressure [125, 202, 203].

$$[(\textcircled{P}-bipy)PdX_2] \qquad (34)$$

27

M = Pd,   X = OAc
M = Pd, Pt,   X = Cl

The palladium(II) acetate derivative of 27 is a useful hydrogenation catalyst for olefins and acetylenes. More substituted or sterically demanding olefins are less readily reduced than less hindered olefins, whereas there are only minor differences between a variety of acetylenes. Acetylenes, however, are reduced preferentially to olefins and the olefins can be obtained in good yield without significant isomerisation if the reaction is stopped after the consumption of one equivalent of hydrogen [203]. The palladium in the product at the end of the hydrogenation has been reduced to palladium(0) although the exact structure of the palladium is unclear. Certainly the palladium(0) product is an active catalyst with a similar selectivity to the palladium(II) material. A similar result was obtained with the $PdCl_2$ derivative of 27 whereas the $PtCl_2$ derivative was not reduced to platinum metal; it was, however, an active hydrogenation catalyst under 3 atmospheres of hydrogen at 70 °C [125].

PdCl$_2$ bound to γ-aminopropyl groups linked to silica gel is a useful selective catalyst for the semihydrogenation of methylcyclopentadiene and $HC \equiv CCMe_2$-OH [204]. However its ability to catalyse the reduction of nitrobenzene cleanly to aniline suggests that the true catalyst may well be palladium metal [1]. This is also probably true of the silica-supported polyacrylonitrile palladium complex reported by the Chinese to be active in the clean reduction of nitrobenzene to aniline [206], as well as the silica anchored nitrile palladium olefin hydrogenation catalysts, $\{Si\}—CH_2CH_2CN/Pd$, reported by Russian workers [207].

Both palladium(II) [208] and platinum(II) [209] have been supported on polyamides to yield very active catalysts for olefins. The activity and selectivity of the palladium complexes are both reactant and solvent dependent, which is in contrast to conventional palladium catalysts. This, coupled with the different product distributions to those usually found was used to support the argument that palladium(II) may be the active catalyst [208]; however the platinum catalysts also cleanly reduced nitrobenzene to aniline [209], a reaction that is considered characteristic of platinum metal [1].

Palladium(II) bound to anthranilic acid anchored to chloromethylated polystyrene is less active than palladium on alumina in the hydrogenation of soybean esters and methylsorbate; however it is as selective with the soybean esters and more selective with methyl sorbate [9, 9a]. The anthranilic acid supported catalyst is very stable with a lifetime of at least 10 000 cycles per palladium atom. The major deactivation mechanism is not elution of palladium(II), nor reduction (XPS confirmed the presence of palladium(II)), but changes in the structures of the polystyrene beads [7, 10]. Although less selective than the Lindlar catalyst in the hydrogenation of alkynes, where it gives cis-hydrogenation, 28 is stable to air and storage [11].

28

Palladium catalysts in which the palladium is bound directly to polystyrene are only active hydrogenation catalysts after reduction to the free metal (reaction (35)) [210].

$$(35)$$

6.9.10. ACTINIDES

The organoactinides $[M(\eta^5\text{-}C_5Me_5)_2Me_2]$, M = U or Th, react with alumina to eliminate methane and form the supported complexes $[(\{Si\}\text{—}O\text{—})_nM(\eta^5\text{-}C_5Me_5)_2Me_{2-n}]$, $n = 1, 2$ [211]. These products are very active catalysts for the hydrogenation of propene in spite of the fact that only about 12.5% of the adsorbed actinide species are catalytically active. In particular the activity of the catalytic sites is considerably greater than that of their homogeneous analogues.

6.10. Reduction of Inorganic Molecules

Cobalt tetraphenylporphyrin complexes catalyse the reduction of nitric oxide by both hydrogen and carbon monoxide to yield nitrogen, nitrous oxide and ammonia. The activities of the cobalt complexes are remarkably enhanced by supporting them either on titanium dioxide or on polymers carrying imidazole functional groups [212–215]. The enhancement on titanium dioxide was 3000 times that of the unsupported complex, indicating a very specific role for the support, since silica — by enhancing the effective surface area of the cobalt tetraphenylporphyrin — gave only a ten-fold enhancement relative to the homogeneous system [213, 215].

A review of the properties of immobilised transition metal complex catalysts and their application to nitrogen fixation has been published by a group from Jilin University at Changchun in China [216]. The same group has described the use of an iron-molybdenum cluster complex supported on polyvinylpyridine for nitrogen fixation and acetylene reduction [217].

6.11. Michael Addition

Michael addition involves nucleophiles adding in a 1,2- or 1,4-fashion across an activated double-bond system [218]. Supported nickel(II) acetylacetonate, prepared as in reaction (36), catalyses the Michael addition of β-diketones to β-nitrostyrenes (reaction (37)).

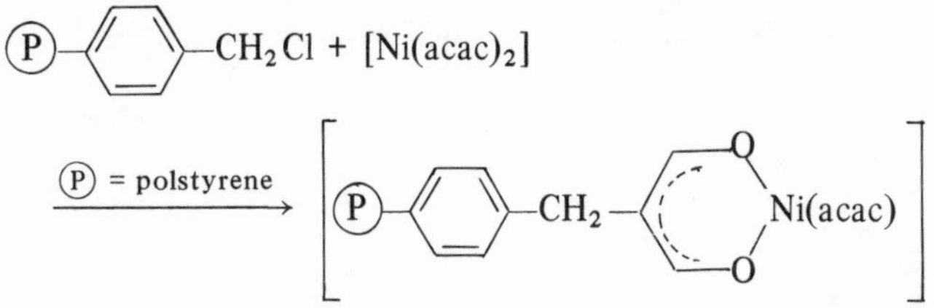

$$RCOCH_2COR' + ArCH{=}CHNO_2 \xrightarrow{\ Ni^{(II)}\ cat.\ } \begin{array}{l} RCOCHCOR' \\ \ \ \ |\\ \ \ \ CHAr \\ \ \ \ | \\ \ \ \ CH_2NO_2 \end{array} \qquad (37)$$

The activity of the supported catalyst is comparable to that of its homogeneous analogue, [Ni(acac)$_2$], when the amounts of nickel(II) are the same [219]. However when the β-diketone is ethylacetoacetate (R$'$ = OEt) the polymer-supported catalyst is better. This can be understood in terms of Scheme 1.

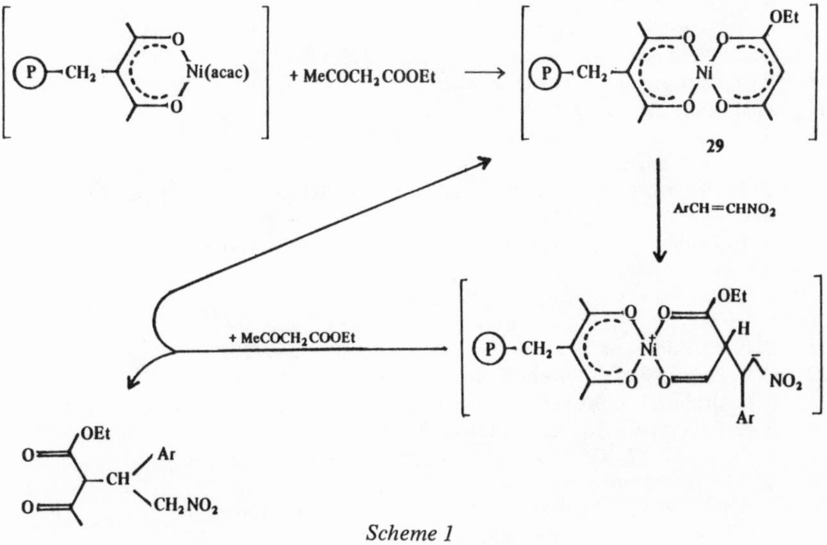

Scheme 1

Formation of the mixed ligand complex, **29**, leads to activation of ethylaceto-acetate only with the supported catalyst, whereas when [Ni(acac)$_2$] is the catalyst acetylacetone becomes activated as well and the product is not clean, alghough this can be obviated by using [Ni(ethylacetoacetate)$_2$].

References

1.   P. N. Rylander: *Catalytic Hydrogenation over Platinum Metals*, Academic Press, New York (1967).
2.   F. H. Jardine, J. A. Osborn, G. Wilkinson and J. F. Young: *Chem. Ind. (London)*, 560 (1965).
3.   B. R. James: *Homogeneous Hydrogenation*, Wiley, New York (1973).
4.   B. R. James in: *Comprehensive Organometallic Chemistry* (Ed. by G. Wilkinson, F. G. A. Stone and E. W. Abel) vol. 8, Chapter 51 (1982).
5.   J. P. Collman and L. S. Hegedus: *Principles and Applications of Organotransition Metal Chemistry*, University Science Books, Mill Valley, California, Chapter 6 (1980).
6.   G. Webb: 'Catalysis', *Chem. Soc. Spec. Per. Rep.* **2**, 145 (1978).
7.   N. L. Holy in: *Fundamental Research in Homogeneous Catalysis* (Ed. by M. Tsutsui) Plenum Press, New York, **3**, 691 (1979).
8.   N. L. Holy: *J. Org. Chem.* **44**, 239 (1979).
9.   E. N. Frankel, J. P. Friedrich, T. R. Bessler, W. F. Kwolek and N. L. Holy: *J. Amer. Oil Chem. Soc.* **57**, 349 (1980).
9a. N. L. Holy, W. A. Logan and K. D. Stein: *US Patent* 4,313,018 (1982); *Chem. Abs.* **96**, 141870 (1982).
10. N. L. Holy: *J. Org. Chem.* **43**, 4686 (1978).
11. N. L. Holy and S. R. Shelton: *Tetrahedron* **37**, 25 (1981).
12. N. L. Holy: *ChemTech.* **10**, 366 (1980).
13. N. L. Holy: *JCS Chem. Commun.* 1074 (1978).
14. M. Terasawa, K. Kaneda, T. Imanaka and S. Teranishi: *J. Catal.* **5**, 406 (1978).
15. British Petroleum Co. Ltd.: *British Patent* 1,295,675 (1972).
16. R. H. Grubbs and L. C. Kroll: *J. Amer. Chem. Soc.* **93**, 3062 (1971).
17. Z. M. Michalska and D. E. Webster: *Platinum Metals Review* **18**, 65 (1974).
18. R. H. Grubbs, C. Gibbons, L. C. Kroll, W. D. Bonds and C. H. Brubaker: *J. Amer. Chem. Soc.* **95**, 2373 (1973).
19. C. H. Brubaker, W. D. Bonds and S. Chandrasekaran: *166th Amer. Chem. Soc. Meeting*, Chicago (1973) *Ind. and Eng. Chem. Div.*, paper 31.
19a. R. H. Grubbs, C. P. Lau, R. Cukier and C. Brubaker: *J. Amer. Chem. Soc.* **99**, 4517 (1977).
19b. E. S. Chandrasekaran: *Diss. Abs.* **B36**, 6143 (1976).
20. L. Verdet and J. K. Stille: *Organometallics* **1**, 380 (1982).
21. C. P. Nicolaides and N. J. Coville: *J. Organometal. Chem.* **222**, 285 (1981).
22. C. U. Pittman and G. Wilemon: *Ann. N.Y. Acad. Sci.* **333**, 67 (1980).
23. R. H. Grubbs, L. C. Kroll and E. M. Sweet: *J. Macromol. Sci. Chem.* **A7**, 1047 (1973).
24. E. M. Sweet: *Diss. Abs.* **B38**, 4821 (1978).
25. R. H. Grubbs, L. C. Kroll and E. M. Sweet: *Polymer Preprints, Amer. Chem. Soc. Division of Polymer Chem.* **13**, 828 (1972).
26. R. H. Grubbs, L. C. Kroll and E. M. Sweet: *166th Amer. Chem. Soc. Meeting*, Chicago (1973) *Ind. Eng. Chem. Div.*, paper 30.

27. R. H. Grubbs and L. C. Kroll: *162nd Amer. Chem. Soc. Meeting*, Washington (1971) *Polymer Section*, paper 68.

28. J. M. Moreto, J. Albaiges and F. Camps in: *Catalysis, Heterogeneous and Homogeneous* (Ed. by B. Delmon and G. Jannes) Elsevier, Amsterdam, p. 339 (1975).

29. R. H. Grubbs, E. M. Sweet and S. Phisabut in: *Catalysis in Organic Syntheses* (Ed. by P. N. Rylander and H. Greenfield) Academic Press, N.Y., p. 153 (1976).

30. J. Sabadie and J. E. Germain: *Bull. Soc. Chim. France*, 1133 (1974).

31. H. Dupin, I. Schifter, A. Perrard, J. E. Germain, A. Guyot and H. Jacobelli: *Bull. Soc. Chim. France*, 619 (1977).

32. R. L. Lazcano, M. P. Pedrosa, J. Sabadie and J. E. Germain: *Bull. Soc. Chim. France*, 1129 (1974).

33. T. Uematsu, F. Saito, M. Muira and H. Hashimoto: *Chem. Lett.* 113 (1977).

34. Z. C. Brzezinska, W. R. Cullen and G. Strukul: *Can. J. Chem.* 58, 750 (1980).

35. T. Matsuda and J. K. Stille: *J. Amer. Chem. Soc.* 100, 268 (1978).

36. M. E. Wilson and G. M. Whitesides: *J. Amer. Chem. Soc.* 100, 306 (1978).

37. R. H. Grubbs, L. C. Kroll and E. M. Sweet: *J. Macromol. Sci.* A7, 1047 (1973).

38. C. U. Pittman, L. R. Smith and R. M. Hanes: *J. Amer. Chem. Soc.* 97, 1742 (1975).

38a. C. Zhao, Y. Zhou, L. Wang and S. Chin: *Lanzhou Daxue Xuebao, Ziran Kexueban* 18, 121 (1982); *Chem. Abs.* 97, 164906 (1982).

39. G. Innorta, A. Modelli, F. Scagnolari and A. Foffani: *J. Organometal. Chem.* 185, 403 (1980).

40. A. J. Naaktgeboren, R. J. M. Nolte and W. Drenth: *J. Mol. Catal.* 11, 343 (1981).

41. M. Capka, P. Svoboda, M. Cerny and J. Hetflejs: *Tetrahedron Lett.* 4787 (1971).

42. Y. Tajima and E. Kunioka: *J. Org. Chem.* 33, 1689 (1968).

43. J. E. Bercaw, R. H. Marvich, L. G. Bell and H. H. Brintzinger: *J. Amer. Chem. Soc.* 94, 1219 (1972).

44. K. Kochloefl, W. Liebelt and H. Knoezinger: *JCS Chem. Commun.* 510 (1977).

45. J. M. Brown and H. Molinari: *Tetrahedron Lett.* 2933 (1979).

46. S. Jacobson, W. Clements, H. Hiramoto and C. U. Pittman: *J. Mol. Catal.* 1, 73 (1975).

47. C. Andersson and R. Larsson: *J. Amer. Oil. Chem. Soc.* 58, 675 (1981).

47a. B. M. Trost: *Chem. in Brit.* 20, 315 (1984).

48. I. V. Howell, R. D. Hancock, R. C. Pitkethly and P. J. Robinson in: *Catalysis, Heterogeneous and Homogeneous* (Ed, by B. Delmon and B. Jannes) Elsevier, Amsterdam, p. 349 (1975).

49. K. G. Allum, R. D. Hancock, I. V. Howell, T. E. Lester, S. McKenzie, R. C. Pitkethly and P. J. Robinson: *J. Organometal. Chem.* 107, 393 (1976).

50. K. G. Allum, R. D. Hancock, I. V. Howell, T. E. Lester, S. McKenzie, R. C. Pitkethly and P. J. Robinson: *J. Catal.* 43, 331 (1976).

51. C.-P. Lau: *Diss Abs.* B38, 4804 (1978).

52. E. S. Chandrasekaran, R. H. Grubbs and C. H. Brubaker: *J. Organometal. Chem.* 120, 49 (1976).

53. C. H. Brubaker in: *Catalysis in Organic Synthesis* (Ed. by G. V. Smith) Academic Press, New York, p. 25 (1977).

54. W. D. Bonds, C. H. Brubaker, E. S. Chandrasekaran, C. Gibbons, R. H. Grubbs and L. C. Kroll: *J. Amer. Chem. Soc.* 97, 2128 (1975).

55. R. Jackson, J. Ruddlesden, D. J. Thompson and R. Whelan: *J. Organometal. Chem.* 125, 57 (1977).

56. C. U. Pittman, B. T. Kim and W. M. Douglas: *J. Org. Chem.* 40, 590 (1975).

57. R. A. Awl, E. N. Frankel, J. P. Friedrich and E. H. Pryde: *J. Amer. Oil Chem. Soc.* **55**, 577 (1978).

58. H. B. Gray and C. C. Frazier: *US Patent* 4,228,035 (1980).

59. Yu. I. Yermakov. A. N. Startsev, V. A. Burmistrov and B. N. Kuznetsov: *React. Kinet. Catal. Lett.* **14**, 155 (1980).

60. Yu. I. Yermakov, B. N. Kuznetsov, A. N. Startsev, P. A. Zhdan, A. P. Shepelin, V. I. Zaikovskii, L. M. Plyasova and V. A. Burmistrov: *J. Mol. Catal.* **11**, 205 (1981).

61. T. Imanaka, Y. Katoh, Y. Okamoto and S. Teranishi: *Chem. Lett.* 1173 (1981).

61a. N. Oguni, S. Shimazu, Y. Iwamoto and A. Nakamura: *Polymer J. (Tokyo)* **13**, 849 (1981).

62. S. J. Hardwick: *Diss. Abs.* **B42**, 624 (1981).

62a. J.-B. N'G. Effa, B. Djebailli, J. Lieto and J.-P. Aune: *JCS Chem. Commun.* 408 (1983).

62b. S. Bhaduri, H. Khwaga and V. Khanwalkar: *JCS Dalton Trans.* 445 (1982).

63. I. Duran, J. Monfort, E. Rodriguez and R. A. Sanchez-Delgado: *Simp. Iberoam. Catal.* **149** (1980); *Chem. Abs.* **94**, 208282 (1981).

64. R. A. Sanchez-Delgado, I. Duran, J. Monfort and E. Rodriguez: *J. Mol. Catal.* **11**, 193 (1981).

65. Z. Otero-Schipper, J. Liels, J. J. Rafalko and B. C. Gates: *Stud. Surf. Sci. Catal.* **4**, 535 (1980).

65a. Y. Doi and K. Yano: *Inorg. Chim. Acta* **76**, L71 (1983).

66. S. C. Brown and J. Evans: *JCS Chem. Commun.* 1063 (1978).

67. S. C. Brown and J. Evans: *J. Mol. Catal.* **11**, 143 (1981).

68. J.-B. N'G. Effa, J. Lieto and J.-P. Aune: *J. Mol. Catal.* **15**, 367 (1982).

68a. K. J. McQuade, R. Pierantozzi, M. B. Freeman and B. C. Gates: *Preprints Amer. Chem. Soc. Div. Petr. Chem.* **25**, 75 (1980).

68b. B. Besson, A. Chôplin, L. D'Ornelas and J. M. Basset: *JCS Chem. Commun.* 843 (1982).

69. P. Pertici, G. Vitulli, C. Carlini and F. Ciardelli: *J. Mol. Catal.* **11**, 353 (1981).

70. G. Braca, C. Carlini, F. Ciardelli and G. Sbrana: *6th Int. Conf. Catal.*, London, 528 (1976).

71. G. Sbrana, G. Braca, G. Valentini, G. Pazienza and A. Altomare: *J. Mol. Catal.* **3**, 111 (1977).

72. G. Valentini, G. Sbrana and G. Braca: *J. Mol. Catal.* **11**, 383 (1981).

73. N. J. Coville and C. P. Nicolaides: *J. Organometal. Chem.* **219**, 371 (1981).

74. F. R. Hartley, S. G. Murray and P. N. Nicholson: *J. Mol. Catal.* **16**, 363 (1982).

75. W. Strohmeier, B. Grasser, R. Marcec and K. Holke: *J. Mol. Catal.* **11**, 257 (1981).

76. H. Hirai and T. Furuta: *J. Poly. Sci., Poly. Chem. Lett.* **9**, 459 (1971).

77. W. K. Rybak and J. J. Ziolkowski: *J. Mol. Catal.* **11**, 365 (1981).

78. M. V. McCabe and M. Orchin: *Fuel* **55**, 266 (1976).

79. V. L. Kuznetsov, B. N. Kuznetsov and Yu. I. Yermakov: *Kinet. Katal.* **19**, 346 (1978).

80. N. F. Noskova and N. I. Marusich: *Tr. Inst. Org. Katal. Elektrokhim., Akad. Nauk. Kaz. SSR* **4**, 48 (1973); *Chem. Abs.* **80**, 82259 (1974).

81. F. H. Jardine: *Prog. Inorg. Chem.* **28**, 63 (1981).

82. D. E. Bergbreiter and M. S. Bursten in: *Advances in Organometallic and Inorganic Polymer Science* (Ed. by C. E. Carraher, J. E. Sheats and C. U. Pittman) Marcel Dekker, New York, p. 369 (1982).

83. D. E. Bergbreiter, M. S. Bursten, K. Cook and G. L. Parsons: *Amer. Chem. Soc., Symp, Ser.* **192**, 31 (1982).

84. E. Bayer and V. Schurig: *Angew. Chem. Int. Ed.* 14, 493 (1975).
85. G. Bernard, Y. Chauvin and D. Commereuc: *Bull. Soc. Chim. France* 7–8, 1163 and 1168 (1976).
86. T. H. Kim: *Diss. Abs.* B37, 356 (1976).
87. J. Kiji, S. Kadoi and J. Furukawa: *Angew. Makromol. Chem.* 46, 163 (1975).
88. K. G. Allum, R. D. Hancock and R. C. Pitkethly: *UK Patent* 1,295,675 (1972).
88a. T. Okano, K. Tsukiyama, H. Konishi and J. Kiji: *Chem. Lett.* 603 (1982).
89. J. Halpern: *Colloqu. Inst. CNRS* 281, 27 (1977); *Chem. Abs.* 91, 174549 (1979).
90. M. Czakova and M. Capka: *J. Mol. Catal.* 11, 313 (1981).
91. J. Conan, M. Bartholin and A. Guyot: *J. Mol. Catal.* 1, 375 (1976).
92. M. Bartholin, J. Conan and A. Guyot: *J. Mol. Catal.* 2, 307 (1977).
92a. K. Hahn and D. Zerpner: *German Patent* 3,042,410 (1982); *Chem. Abs.* 97, 116128 (1982).
93. M. Bartholin, C. Graillat, A. Guyot, G. Coudurier, J. Bandiera and C. Naccache: *J. Mol. Catal.* 3, 17 (1977).
94. G. Strukul, P. D'Olimpio, M. Bonivento, F. Pina and M. Graziani: *J. Mol. Catal.* 2, 179 (1977).
95. G. Strukul, M. Bonivento, M. Graziani, E. Cernia and N. Palladino: *Inorg. Chim. Acta* 12, 15 (1975).
96. J. Reed, P. Eisenberger, B.-K. Teo and B. M. Kincaid: *J. Amer. Chem. Soc.* 99, 5217 (1977).
97. J. Reed, P. Eisenberger, B.-K. Teo and B. M. Kincaid: *J. Amer. Chem. Soc.* 100, 2375 (1978).
98. J. Manassen and Y. Dror: *J. Mol. Catal.* 3, 227 (1978).
99. K. G. Allum, R. D. Hancock, I. V. Howell, R. C. Pitkethly and P. J. Robinson: *J. Catal.* 43, 322 (1976).
100. M. H. J. M. De Croon and J. W. E. Coenen: *J. Mol. Catal.* 11, 301 (1981).
101. S. Lamalle, A. Mortreux, M. Evrard, F. Petit, J. Grimblot and J. P. Bonnelle: *J. Mol. Catal.* 6, 11 (1979).
102. M. K. Neuberg: *Diss. Abs.* B39, 5929 (1979).
103. D. W. Meek, J. A. Tiethof and D. L. DuBois: *7th Int. Conf. Organometal. Chem.*, Venice, 128 (1975).
104. B. A. Sosinsky, W. C. Kalb, R. A. Grey, V. A. Uski and M. F. Hawthorne: *J. Amer. Chem. Soc.* 99, 6768 (1977).
105. E. S. Chandrasekaran, D. A. Thompson and R. W. Rudolph: *Inorg. Chem.* 17, 760 (1978).
106. A. J. Birch and K. A. M. Walker: *Tetrahedron Lett.* 1935 (1967).
107. H. Singer and G. Wilkinson: *J. Chem. Soc. (A)* 2516 (1968).
108. A. Guyot, C. Graillat and M. Bartholin: *J. Mol. Catal.* 3, 39 (1977).
109. R. H. Grubbs and E. M. Sweet: *J. Mol. Catal.* 3, 259 (1978).
110. M. Bartholin, C. Graillat and A. Guyot: *J. Mol. Catal.* 10, 361 (1981).
111. G. Bernard, Y. Chauvin and D. Commereuc: *Colloq. Int. CNRS* 281, 165 (1977); *Chem. Abs.* 91, 210624 (1979).
111a. J. M. G. Figueroa and J. M. Winterbottom: *J. Chem. Technol. Biotechnol.* 32, 857 (1982).
111b. G. Strathdee and R. Given: *Can. J. Chem.* 52, 3000 (1974).
112. M. Graziani, G. Strukul, M. Bonivento, F. Pinna, E. Cernia and N. Palladino in: *Catalysis, Heterogeneous and Homogeneous* (Ed. by B. Delmon and G. Jannes) Elsevier, Amsterdam, p. 331 (1975).

113. F. Pinna, M. Bonivento, G. Strukul, M. Graziani, E. Cernia and N. Palladino: *7th Int. Conf. Organometal. Chem.*, Venice, 195 (1975).
114. F. Pinna, M. Bonivento, G. Strukul, M. Graziani, E. Cernia and N. Palladino: *J. Mol. Catal.* 1, 309 (1976).
115. F. Pinna, C. Candilera, G. Strukul, M. Bonivento and M. Graziani: *J. Organometal. Chem.* 159, 91 (1978).
116. T. Imanaka, K. Kaneda, S. Teranishi and M. Terasawa: *6th Int. Congr. Catal.*, London, 509 (1976).
117. S. Shinoda, T. Kojima and Y. Saito: *Stud. Surf. Sci. Catal.* 7b, 1504 (1981); *Chem. Abs.* 95, 219566 (1981).
118. T. J. Pinnavaia: *J. Amer. Chem. Soc.* 97, 3819 (1975).
119. T. J. Pinnavaia, R. Raythatha, J. G. S. Lee, L. J. Halloran and J. F. Hoffman: *J. Amer. Chem. Soc.* 101, 6891 (1979).
120. W. H. Quayle and T. J. Pinnavaia: *Inorg. Chem.* 18, 2840 (1979).
121. T. J. Pinnavaia: *Amer. Chem. Soc., Symp. Ser.* 192, 241 (1982).
121a. R. Raythatha and T. J. Pinnavaia: *J. Organometal. Chem.* 218, 115 (1981).
122. W. R. Cullen, D. J. Patmore, A. J. Chapman and A. D. Jenkins: *J. Organometal. Chem.* 102, C12 (1975).
123. W. R. Cullen, B. R. James, A. D. Jenkins, G. Strukul and Y. Sugi: *Inorg. Nucl. Chem. Lett.* 13, 577 (1977).
124. R. S. Drago, E. D. Nyberg and A. G. El A'mma: *ONR-TR-5 Report* (1981); *Chem. Abs.* 95, 192884 (1981).
125. R. S. Drago, E. D. Nyberg and A. G. El A'mma: *Inorg. Chem.* 20, 2461 (1981).
126. P. M. Lausarot, G. A. Vaglio and M. Valle: *13th Congr. Naz. Chim. Inorg.* 37 (1980); *Chem. Abs.* 95, 24175 (1981).
127. P. M. Lausarot, G. A. Vaglio and M. Valle: *J. Organometal. Chem.* 204, 249 (1981).
127a. P. M. Lausarot, G. A. Vaglio and M. Valle: *J. Organometal. Chem.* 215, 111 (1981).
127b. P. M. Lausarot, G. A. Vaglio and M. Valle: *J. Organometal. Chem.* 240, 441 (1982).
127c. S. Bhaduri and H. Khwaja: *JCS Dalton*, 419 (1983).
128. A. K. Smith, W. Abboud, J. M. Basset, W. Reimann, G. L. Rempel, J. L. Bilham, V. Bilhou-Bougnol, W. F. Graydon, J. Dunogues, N. Ardon and N. Duffaut: *Fund. Res. Homog. Catal.* (Ed. by M. Tsutsui) Plenum Press, New York, 3, p. 621 (1979).
129. T. Kitamura, T. Joh and N. Hagihara: *Chem. Lett.* 203 (1975).
130. H. Knoezinger and E. Rumpf: *Inorg. Chim. Acta* 30, 51 (1978).
131. H. S. Tung: *Diss. Abs.* B42, 627 (1981).
132. H. S. Tung and C. H. Brubaker: *J. Organometal. Chem.* 216, 129 (1981).
133. B. H. Chang, R. H. Grubbs and C. H. Brubaker: *J. Organometal. Chem.* 172, 81 (1979).
134. A. Sekiya and J. K. Stille: *J. Amer. Chem. Soc.* 103, 5096 (1981).
135. M. D. Ward and J. Schwartz: *J. Mol. Catal.* 11, 397 (1981).
136. M. D. Ward and J. Schwartz: *J. Amer. Chem. Soc.* 103, 5253 (1981).
136a. I. Rajca: *Polish Patent* 144,594 (1982); *Chem. Abs.* 98, 16393 (1983).
137. Y. Nakamura and H. Hirai: *Noguchi Kenkyusho Jiho* 20, 2 (1977); *Chem. Abs.* 87, 207139 (1977).
138. Y. Nakamura and H. Hirai: *Chem. Lett.* 823 (1975).
138a. S. Shinoda, T. Kojima and Y. Saito: *J. Mol. Catal.* 18, 99 (1982).
138b. S. Shinoda, Y. Tokushige, T. Kojima and Y. Saito: *J. Mol. Catal.* 17, 81 (1982).
139. E. N. Rasadkina, I. D. Rozhdestvenskaya and I. V. Kalechits: *Kinet. Catal.*, (*Eng. Trans.*) 17, 799 (1976).

139a. R. Sariego and L. A. Oro: *Bol. Soc. Chil. Quim.* 27, 62 (1982); *Chem. Abs.* 97, 45017 (1982).

140. P. Kalck, R. Poilblanc, A. Gaset, A. Rovera and R. P. Martin: *Tetrahedron Lett.* 21, 459 (1980).

141. M. D. Ward and J. Schwartz: *Organometallics* 1, 1030 (1982).

142. V. Caplar, G. Comisso and V. Sunjic: *Synthesis*, 85 (1981).

143. W. Dumont, J. C. Poulon, T. P. Dang and H. B. Kagan: *J. Amer. Chem. Soc.* 95, 8295 (1973).

144. N. Takaishi, H. Imai, C. A. Bertelo and J. K. Stille: *J. Amer. Chem. Soc.* 98, 5400 (1976).

145. N. Takaishi, H. Imai, C. A. Bertelo and J. K. Stille: *J. Amer. Chem. Soc.* 100, 264 (1978).

146. K. Ohkubo, K. Fujimori and K. Yoshinaga: *Inorg. Nucl. Chem. Lett.* 15, 231 (1979).

147. K. Ohkubo, M. Haga, K. Yoshinaga and Y. Motozato: *Inorg. Nucl. Chem. Lett.* 16, 155 (1980).

148. K. Ohkubo, M. Haga, K. Yoshinago and Y. Motozato: *Inorg. Nucl. Chem. Lett.* 17, 215 (1981).

149. I. Kolb, M. Cerny and J. Hetflejs: *React. Kinet. Catal. Lett.* 7, 199 (1977).

150. H. B. Kagan, T. Yamagishi, J. C. Motte and R. Setton: *Isr. J. Chem.* 17, 274 (1979).

151. G. L. Baker: *Diss. Abs.* **B41**, 4115 (1981).

152. H. Pracejus and M. Bursian: *East German Patent* 92031 (1972); *Chem. Abs.* 78, 72591 (1973).

153. K. Achiwa: *Heterocycles* 9, 1539 (1978).

154. K. Achiwa: *Chem. Lett.* 905 (1978).

155. H. W. Krause: *React. Kinet. Catal Lett.* 10, 243 (1979).

156. H. Krause and H. Mix: *East German Patent* 133,199 (1978); *Chem. Abs.* 91, 63336 (1979).

157. M. Mazzei, M. Riocci and W. Marconi: *German Patent*, 2,845,216 (1979); *Chem. Abs.* 91, 56329 (1979).

158. S. E. Jacobson and C. U. Pittman: *JCS Chem. Commun.* 187 (1975).

158a. M. Bartholin, C. Graillat and A. Guyot: *J. Mol. Catal.* 10, 377 (1981).

159. C. U. Pittman, S. E. Jacobson and H. Hiramoto: *J. Amer. Chem. Soc.* 97, 4774 (1975).

160. C. U. Pittman, S. E. Jacobson, L. R. Smith, W. Clements and H. Hiramoto: *Catalysis in Organic Synthesis* (Ed. by P. N. Rylander and H. Greenfield) Academic Press, New York, p. 161 (1976).

161. C. U. Pittman, A. Hirao, J. J. Y. Q. Ng, R. Hares and C. C. Lin: *Preprints Div. Petrol. Chem. Amer. Chem. Soc.* 22, 1196 (1977).

162. J. Azran, O. Buchman and J. Blum: *Tetrahedron Lett.* 22, 1925 (1981).

163. J. J. Rafalko, J. Lieto, B. C. Gates and G. L. Schrader: *JCS Chem. Commun.* 540 (1978).

164. J. Leito, J. J. Rafalko and B. C. Gates: *J. Catal.* 62, 149 (1980).

165. T. Castrillo, H. Knoezinger, J. Lieto and M. Wolf: *Inorg. Chim. Acta* 44, 4239 (1980).

166. T. Castrillo, H. Knoezinger, M. Wolf and B. Tesche: *J. Mol. Catal.* 11, 151 (1981).

167. B. Loubinoux, J. J. Chanot and P. Caubere: *J. Organometal. Chem.* 88, C4 (1975).

168. N. L. Holy and R. Shalvoy: *J. Org. Chem.* 45, 1418 (1980).

169. D. H. Ralston: *Diss. Abs.* **B41**, 4128 (1981).

170. M. Ichikawa: *JCS Chem. Commun.* 26 (1976).

171. T. A. Petrova and N. F. Noskova: *Izv. Akad. Nauk. Kaz. SSR* 27 (1980); *Chem. Abs.* 94, 45082 (1981).

172. N. F. Noskova and D. V. Sokol'skij: *Zh. Fiz. Khim.* 49, 2668 (1975).

173. N. F. Noskova, N. I. Marusich, M. P. Grechko and D. V. Sokol'skij: *Kinet. Katal.* 18, 1436 (1977).

174. W. O. Haag and D. D. Whitehurst: *German Patent* 1,800,380 (1969); *Chem. Abs.* 71, 33823 (1969).

175. W. O. Haag and D. D. Whitehurst: *US Patent* 3,578,609 (1971).

176. I. Mochida, T. Jitsumatsu, A. Kato and T. Seiyama: *Bull. Chem. Soc. Japan* 44, 2595 (1971).

177. J. Sabadie, G. Descotes and J. E. Germain: *Bull. Soc. Chim. Fr.* 1855 (1975).

178. J. Sabadie and G. Descotes: *Bull. Soc. Chim. Fr.* 515 (1977).

179. J. Sabadie, I. Schifter and J. E. Germain: *Bull. Soc. Chim. Fr,* 616 (1977).

180. V. Z. Sharf, V. D. Kopylova, L. P. Karapetyan, E. L. Frumkina, L. Kh. Friedlin, K. M. Saldadze and V. N. Krutii: *Izv. Akad. Nauk SSSR, Ser. Khim.* 2746 (1977).

181. Mobil: *Belgian Patent* 721,686 (1969).

182. M. Paez-Pedrosa, I. Schifter and J. E. Germain: *Bull. Soc. Chim. Fr.* 1977 (1974).

182a. M. V. Klyuev: *Izv. Vyssh. Uchebn. Zaved.* 25, 751 (1982); *Chem. Abs.* 97, 99077 (1982).

183. R. Linarte-Lazcano, H. D. Perez-Villagomez and M. E. Ruiz-Vizcaya: *Rev. Port. Quim.* 19, 273 (1977); *Chem. Abs.* 93, 25600 (1980).

184. G. Allandrieu, G. Descotes, J. P. Praly and J. Sabadie: *Bull. Soc. Chim. Fr.* 519 (1977).

185. J. C. Bailar and H. Itatani: *J. Amer. Chem. Soc.* 89, 1592 (1967).

186. H. Itatani and J. C. Bailar: *J. Amer. Chem. Soc.* 89, 1600 (1967).

187. J. C. Bailar: *J. Amer. Oil Chem. Soc.* 47, 475 (1970).

188. J. C. Bailar: *Platinum Met. Rev.* 15, 2 (1971).

189. H. S. Bruner and J. C. Bailar: *J. Amer. Oil Chem. Soc.* 49, 533 (1972).

190. H. S. Bruner and J. C. Bailar: *Inorg. Chem.* 12, 1465 (1973).

191. D. J. Baker and J. C. Bailar in: *6th Conference on Catalysis in Organic Synthesis* (Ed. by G. V. Smith) p. 1 (1977).

191a. C. Pan, P. Feng and B. Liang: *Cuihua Xuebao* 2, 152 (1981); *Chem. Abs.* 96, 6106 (1982).

192. C. Andersson and R. Larsson: *Chem. Scr.* 15, 45 (1980).

193. C. Andersson and R. Larsson: *J. Amer. Oil Chem. Soc.* 58, 675 (1981).

194. K. Kaneda, M. Terasawa, T. Imanaka and S. Teranishi: *Chem. Lett.* 1005 (1975).

195. K. Kaneda, M. Terasawa, T. Imanaka and S. Teranishi in: *Fundamental Research in Homogeneous Catalysis* (Ed. by M. Tsutsui) Plenum Press, N. Y. 3, p. 671 (1979).

196. V. A. Semikolenov, D. Kh. Mikhailova, Ya. V. Sobchak, V. A. Likholobov and Yu. I. Yermakov: *Kinet. Katal.* 21, 526 (1980).

197. V. A. Semikolenov, D. Mikhailova and J. W. Sobczak: *React. Kinet. Catal. Lett.* 10, 105 (1979).

198. V. A. Semikolenov, V. A. Likholobov, G. Valentini, G. Braca and F. Ciardelli: *React. Kinet. Catal. Lett.* 15, 383 (1980).

198a. Z. Feng and H. Liu: *Gaofenzi Tongxun* 261 (1982); *Chem. Abs.* 98, 71427 (1983).

198b. Y. Chen, J. Liu, Y. Lin, C. Xiao and C. Mai: *Wuhan Daxue Xuebao, Ziran Kexueban,* 29 (1982); *Chem. Abs.* 98, 82741 (1983).

198c. Y. Chen, J. Liu, Y. Lin, J. Ni, C. Xiao and Y. Wan: *Wuhan Daxue Xuebao, Ziran Kexueban* 41 (1982); *Chem. Abs.* 97 61687 (1982).

198d. Y. Lin, J. Liu, J. Ni, Y. Chen, C. Xiao and Y. Wang: *Cuihua Xuebao* 3, 220 (1982); *Chem. Abs.* 97, 215076 (1982).
199. R. Pierantozzi, K. J. McQuade and B. C. Gates: *Stud. Surf. Sci. Catal.* 7B, 941 (1981).
200. R. Bender, P. Braunstein, J. Fischer, L. Ricard and A. Mitschler: *Nouv. J. Chim.* 5, 81 (1981).
201. B.-S. Chen, Z.-M. Feng and R.-Y. Chen: *Ts'ui Hua Hsueh Pao* 1, 213 (1980); *Chem. Abs.* 94, 120773 (1981).
201a. H. Liu: *Cuihua Xuebao* 3, 315 (1982); *Chem. Abs.* 98, 125157 (1983).
202. R. J. Card and D. C. Neckers: *Inorg. Chem.* 17, 2345 (1978).
203. R. J. Card, C. E. Leisner and D. C. Neckers: *J. Org. Chem.* 44, 1095 (1979).
204. V. Z. Sharf, A. S. Gurovets, L. P. Finn, I. B. Slinyakova, V. N. Krutii and L. Kh. Friedlin: *Izv. Akad. Nauk SSSR, Ser. Khim.* 104 (1979).
205. V. Z. Sharf, A. S. Gurovets, V. N. Krutii, I. B. Slinyakova, L. P. Fini and S. I. Shcherbakova: *Izv. Akad. Nauk SSSR, Ser. Khim.* 2533 (1979).
206. Y.-J. Li and Y. Y. Jiang: *Ts'ui Hua Hsueh Pao* 2, 42 (1981); *Chem. Abs.* 95, 96618 (1981).
207. V. A. Semikolenov, V. A. Likholobov and Yu. I. Yermakov: *Kinet. Katal.* 18, 1294 (1977).
208. R. P. MacDonald and J. M. Winterbottom: *J. Catal.* 57, 195 (1979).
209. E. N. Rasadkina, T. S. Kukhareva, I. D. Rozhdestvenskaya and I. V. Kalechits: *Kinetics and Catalysis (Eng. Trans).* 16, 1273 (1975).
210. D. E. Bergbreiter, J. M. Killough and G. L. Parsons in: *Fundamental Research in Homogeneous Catalysis* (Ed. by M. Tsutsui) Plenum Press, N.Y., 3, 651 (1979).
211. R. G. Bowman, R. Nakamura, P. J. Fagan, R. L. Burwell and T. J. Marks: *JCS Chem. Commun.* 257 (1981).
212. I. Mochida, H. Fujitsu, K. Takeshita and K. Tsuji: *Stud. Surf. Sci. Catal.* B7, 1516 (1981).
213. I. Mochida, K. Suetsugu, H. Fujitsu and K. Takeshita: *JCS Chem. Commun.* 166 (1982).
214. K. Tsuhi, M. Imaizumi, A. Oyoshi, I. Mochida, H. Fujitsu and K. Takeshita: *Inorg. Chem.* 21, 721 (1982).
215. I. Mochida, K. Suetsugu, H. Fujitsu and K. Takeshita: *J. Phys. Chem.* 87, 1524 (1983).
216. C. Sun and X. Liu: *Cuihua Xuebao* 3, 154 (1982); *Chem. Abs.* 97, 61679 (1982).
217. C. Sun, S. Li, Q. Huang and S. Niu: *Gaodeng Xuexiao Huaxue Xuebao* 3, 398 (1982); *Chem. Abs.* 97, 169696 (1982).
218. R. Brettle in: *Comprehensive Organic Chemistry* (Ed. by D. Barton and W. D. Ollis) Pergamon, Oxford, vol. 1, p. 1043 (1979).
219. C. P. Frei and T. H. Chan: *Synthesis* 467 (1982).

CHAPTER 7

HYDROSILYLATION

Hydrosilylation is formally the addition of the units of R_3Si and H (R may be alkyl, alkoxy or halide) to an unsaturated group such as an olefin, acetylene or ketone (reactions (1) and (2)) [1–4].

$$R_3SiH + {>}C{=}C{<} \longrightarrow H-\overset{|}{\underset{|}{C}}-\overset{|}{\underset{|}{C}}-SiR_3 \qquad (1)$$

$$R_3SiH + {>}C{=}O \longrightarrow -\overset{|}{\underset{H}{C}}-OSiR_3 \qquad (2)$$

The reaction is formally and mechanistically rather similar to hydrogenation. The principle catalysts that have been used in the homogeneously catalysed reaction have been chloroplatinic acid (which is reduced to platinum(II)), rhodium(I) complexes such as Wilkinson's complex [Rh(PPh$_3$)$_3$Cl] and transition metal carbonyls. Their mode of action probably involves oxidative addition of H–SiR$_3$ to give a platinum(IV) [5] or rhodium(III) [6] intermediate which then transfers the H and SiR$_3$ groups to the originally unsaturated bond (reaction (3)).

A similar mechanism almost certainly takes place on polymer-supported complexes, indeed some evidence has been put forward for the formation of some platinum(II) species when a styrene/divinylbenzene copolymer bearing dimethylamino groups is treated with chloroplatinic acid [7].

Metal complex hydrosilylation catalysts have been supported on a wide range of supports (Table I). Whilst most of these are phosphinated supports, particularly polystyrene and silica, other supports have also been used particularly anion exchange resins with quaternary ammonium groups and polymers with amino- and nitrile functional groups. Although in most cases the donor groups have been introduced on to an already existing support some workers have polymerised organosilicon precursors (reaction (4)) [14].

$$(EtO)_3 SiHC = CH_2 + Me_3 COOCMe_3 + Ph_2 PH \longrightarrow Ph_2 PCH_2 CH_2 Si(OEt)_3$$

$$\Bigg\downarrow \begin{array}{l} \text{reflux in acetic acid} \\ \text{with } Si(OEt)_4 \text{ and a} \\ \text{trace of conc. HCl} \end{array}$$

Nonlinear polymer
containing 11%
phosphorus   (4)

There are in principle three ways in which the supported catalyst could act:

(1) The reaction could take place at the metal site which itself remains bonded to the surface throughout the reaction.

(2) The catalytically active species or its precursor may be abstracted from the support into the solution by a reversible process so that the catalysis is effectively a homogeneous process.

(3) The third possibility is essentially the same as the second except that the abstraction is irreversible. The fact that most of the catalysts can be reused several times, and indeed that their catalytic activity increases slightly after the first use, effectively eliminates the third possibility. It perhaps should remain a possibility, although an unlikely one, in the supported palladium(II) catalysis for the addition of trichlorosilane to butadiene. These polymer-supported catalysts are more active than their homogeneous analogues (suggesting that reaction occurs on the surface), but the palladium is pulled off the surface and left in the liqud as palladium metal so that the catalyst cannot be reused [9].

A distinction between mechanisms (1) and (2) is not easy, and indeed both probably occur. In support of mechanism (1) it has been found that in some cases the overall yield following reuse of polymer-supported catalysts is greater

TABLE I.

Hydrosilylations catalysed by supported metal catalysts

Metal Complex	Support	Substrates	Reference
H_2PtCl_6	Cross-linked polystyrene substituted with cyanomethyl groups	Acetylene with $HSiCl_3$	[7, 8]
$PdCl_2$ $RhCl_3$ H_2PtCl_6	γ-alumina, silica, molecular sieves or glass with one of the following groups:- $-PPh_2$, $\equiv Si(CH_2)_2PPh_2$, $\equiv Si(Me)(CH_2)_3CN$, $\equiv Si(CH_2)_3NMe_2$, $\equiv Si(CH_2)_3CN$ or $\equiv Si(CH_2)_3C_5H_4N$.	Butadiene, 1-heptene, 1-hexene, 1-decene, vinylethylether or styrene with $HSiCl_3$, $HSiEt_3$ or $HSi(OEt)_3$.	[9]
H_2PtCl_6	Anion-exchange resin	1-Heptene with $HSiMe_2Ph$	[10, 11]
H_2PtCl_6	$H_2NCH_2CH_2NHCH_2CH_2CH_2Si(OMe)_3$/ octamethylcyclotetrasiloxane copolymer	1-Hexene with $HSiMeCl_2$	[12]
H_2PtCl_6	Silica treated with $MeCl_2Si(CH_2)_3PPh_3^+Br^-$	Not specified in abstract	[13]
Na_2PdCl_4	Phosphinated silica prepared by reaction (4)	Allyl chloride and butadiene with $HSiCl_3$	[14]
$[Pd(PPh_3)_3]$ $[Pd(PPh_3)_2Cl_2]$ $[Pd(PhCN)_2Cl_2]$ $[Pd(\eta^3\text{-}C_3H_5)Cl]_2$ $PdCl_2$	Cross-linked polystyrene containing $-CH_2PPh_2$ or $-CH_2CN$ Amberlyst A21	Butadiene with Me_3SiOH	[15]

(continued)

Metal Complex	Support	Substrates	Reference
$[Pd(\eta^3\text{-}C_3H_5)Cl]_2$ $[(PhCN)_2PdCl_2]$ $[(PPh_3)_2PdCl_2]$ $[(PPh_3)_2PdBr_2]$ $[\{o\text{-}C_6H_4(CH_2CN)_2\}PdCl_2]$	Cross-linked polystyrene with $-CH_2PPh_2$ or $-CH_2CN$ groups	Butadiene, isoprene, piperylene, chloroprene or 2,3-dimethyl-1,3-butadiene with $HSiCl_3$, $HSiMeCl_2$, $HSiEtCl_2$ or $HSi(OEt)_3$.	[16]
$[Rh(PPh_3)_3Cl]$	poly(p-diphenylphosphinostyrene)	$CH_2{=}C(Me)COOMe$, $CH_2{=}C{=}CHMe$, $CH_2{=}CHCMe_3$, $CH_2{=}CHOEt$ or $CH_2{=}CH(CH_2)_4Me$ with $HSi(OEt)_3$	[17]
$[Rh(PPh_3)_3Cl]$ $[Rh(PPh_3)_2(CO)Cl]$ $[Rh(PPh_3)_3(CO)H]$ $RhCl_3$ $[Pt(PPh_3)_4]$	Silica phosphinated with $Ph_2P(CH_2)_2Si(OEt)_3$	1-Hexene with $HSiMe_2Ph$	[18]
$[Rh(PPh_3)_3Cl]$	poly-p-(diphenylphosphino)styrene	1-Hexene with $HSiPr_3$ and $RCH{=}CH_2$ (R = Bu, Me_3C, EtO, NC, AcO) with $HSi(OEt)_3$	[19]
$[Rh(PPh_3)_3Cl]$ H_2PtCl_4	Silica treated with $Cl_3Si(CH_2)_3Cl$ and functionalised with R_2NH, R_2 = piperidine, pyrrolidine, morphine, cyclo-C_5H_{10}, cyclo-C_4H_8, Ph_2, Et_2	Allylchloride with $HSiCl_3$; allylchloracetate, 1-hexene, 1-heptene, 1-octene with $HSi(OEt)_3$	[20]
$[Rh(PPh_3)_3Cl]$	Phosphinated silica prepared by reaction (4)	1-Hexene with $HSi(OEt)_3$ and $HSiEt_3$	[21]

(continued)

Metal Complex	Support	Substrates	Reference
$[Rh(PPh_3)_3Cl]$	Phosphinated chrysotile asbestos (a mineral silicate)	1-Hexene with $HSi(OEt)_3$	[21a]
$[Rh(PPh_3)_3Cl]$	Silica phosphinated with $Ph_2P(CH_2)_2Si(OEt)_3$	R_3SiOH with $HSiR_3$, H_2SiR_2 and H_3SiR; $H(OSiMe_2)_nH$ with H_2SiR_2 or $HMe_2Si(OSiMe_2)_nH$	[22]
$[Rh(CO)_2Cl_2]$	Dimethylaminomethylated cross-linked polystyrene	1-Heptene or styrene with $HSiCl_3$. Styrene, butadiene, 1-heptene/3,3-dimethyl-1-butene or 1-heptene/4,4-dimethyl-1-pentene with $HSiEt_3$.	[23]
$[Rh(CO)_2Cl]_2$	poly(p-chloromethylstyrene)	2-Hexene	[24]
$[Rh(C_2H_4)_2Cl]_2$	phosphinated acetal of a cross-linked polystyrene	Acetophenone with H_2SiPh_2; acetophenone and isobutyrophenone with various mono- and di-hydrosilanes	[25, 26]
$[Rh(C_2H_4)_2Cl]_2$	Silica treated with $Ph_2P(CH_2)_nSi(OEt)_3$, $n = 1$–6	1-Hexene with $HSiMe_2Ph$	[27]
$[Rh(C_2H_4)_2Cl]_2$	poly-(p-R_2Pstyrene) cross-linked with divinylbenzene; R = Ph, menthyl	Not specified in abstract	[28]
$[Rh(C_8H_{14})_2Cl]_2$	silica treated with $(EtO)_3Si(CH_2)_3PPh(menthyl)$	$PhCOR$ (R = Me, Et) with H_2SiPh_2	[28]

(continued)

Metal Complex	Support	Substrates	Reference
$RhCl_3$	Cross-linked polystyrene substituted with $—CH_2PPh_2$, $—CH_2NMe_2$ or $—CH_2CN$. Polymethacrylate substituted with $—O(CH_2)_2CN$, $—OC_6H_4PPh_2$, $—O(CH_2)_3PPh_2$ or $—O(CH_2)_2NMe_2$. Amberlyst A21 Allylchloride-divinylbenzene copolymer substituted with $—CH_2PPh_2$.	1-Hexene or 1-heptene with $HSi(OEt)_3$, $HSiEt_3$ or $HSiCl_3$	[30]
$RhCl_3$ $RhCl_3$ followed by C_2H_4	Phosphinated 20% cross-linked polystyrene	1-Hexene, vinylethylether, acrylonitrile or *trans*-2-heptene with $HSi(OEt)_3$	[31]
$RhCl_3 \cdot H_2O$	Phosphinated cross-linked polystyrene	Vinyl acetate with $HSi(OEt)_3$	[29]
$[(\text{Ⓟ}—PPh_2)_{5-n}Fe(CO)_n]$ $(n = 3, 4)$	$\text{Ⓟ}—$ = phosphinated polystyrene-1%-divinylbenzene	1-Pentene with $HSiEt_3$	[30]
$[(\text{Ⓟ}—PhP(CH_2)_2PPh_2)_x \{Fe(CO)_m\}_y]$ $(x = 1, 2; n = 3, 4; y = 1, 2)$			

than for the corresponding homogeneous catalyst [9]. The greater activity of a number of supported catalysts as compared to their homogeneous analogues provides further support for mechanism (1). Thus the activities of $[Pt(PPh_3)_4]$, $[Rh(PPh_3)_2(CO)Cl]$ and $[Rh(PPh_3)_3(CO)H]$ on phosphinated silica were all greater than their homogeneous analogues in promoting the reaction of 1-hexene with dimethyl(phenyl)silane [18]. $[Pt(PPh_3)_4]$, in particular, was 22 times as active when supported, which was ascribed to equilibrium (5) lying further to the right due to the rigidity of the support [18].

$$[(\textcircled{P}-PPh_2)_n Pt(PPh_3)_{4-n}]$$
$$\rightleftharpoons \quad [(\textcircled{P}-PPh_2)_{n-1} Pt(PPh_3)_{4-n}] + \textcircled{P}-PPh_2 \qquad (5)$$

When rhodium(I) was supported on silica phosphinated using $(EtO)_3 Si(CH_2)_n$-PPh_2 the catalyst prepared with $n = 1$ was 10 times as active as those prepared with $n = 2-6$ [27]. This was ascribed to the different rhodium species formed on the surface (reaction (6)), and provides strong support for the true catalyst being an immobilised rhodium complex.

$$[Rh(C_2H_4)_2 Cl]_2 + \{Si\}-(CH_2)_n PPh_2$$

$$\xrightarrow{n=1} \quad [(\{Si\}-CH_2 PPh_2)Rh(C_2H_4)_2 Cl]$$

$$\xrightarrow{n=2-4} \quad trans\text{-}[(\{Si\}-(CH_2)_n PPh_2)Rh(C_2H_4)(\mu\text{-}Cl)]_2 \qquad (6)$$

$$\xrightarrow{n=5,6} \quad [Rh\{\{Si\}-(CH_2)_n PPh_2\}(\mu\text{-}Cl)]_2$$

It is possible that the very high activity of the material obtained with $n = 1$ is due in part to the proximity of the supported rhodium to surface oxide ions [27]. A kinetic study of rhodium(I) complexes prepared by reaction of Wilkinson's complex with phosphinated silica prepared by reaction (4) showed that supporting the metal had essentially no effect on the kinetic parameters and mechanism of the hydrosilylation of 1-hexene with triethylsilane [21].

Further support for mechanism (1) arises from the greater selectivity of supported rather than homogeneous catalysts [9]. This is particularly well illustrated by the greater optical yield obtained when a supported catalyst is used in the hydrosilylation of isobutyrophenone as compared to the corresponding homogeneous catalyst (see below) [26]. In support of mechanism (2) some authors have observed that part of the metal complex is abstracted out of the support into solution during the reaction [23]. The fact that only a proportion of the metal suffers this fate supports the reversibility of the attachment of the metal complex to the polymer.

An investigation of the influence of surface area on the activity of the supported catalysts has shown that the activity increases with increase in surface area, whilst the selectivity is virtually independent of surface area [23]. This result is consistent with both mechanisms (1) and (2). Thus in mechanism (1) the reaction takes place in the pores of the catalyst, which are sufficiently large that they impose no steric demands on the reactants, so that the activity of the catalyst is dependent on the ease with which this species is abstracted from the polymer support; clearly this will increase with increasing surface area.

The specificities with respect to the silane of the rhodium- and platinum-based catalysts are complementary since the order of activity of the rhodium catalysts is $HSi(OEt)_3 > HSiEt_3 > HSiCl_3$ whereas for the platinum catalysts it is $HSiCl_3 > HSi(OEt)_3 > HSiEt_3$ [9]. These specificities are the same as found for the corresponding homogeneous catalysts.

A very important example of substrate specificity was observed in the asymmetric hydrosilylation of ketones using a catalyst based on polystyrene containing the optically active diphosphine, 1. Thus whereas the optical yields

1

obtained on treating acetophenone with dihydrosilanes were very similar to those obtained with the corresponding homogeneous catalyst, with isobuty-rophenone it was found that the supported catalyst gave a significantly higher optical yield that the homogeneous catalyst [26]. This strongly suggests that the support takes a specific role in the catalysis, which would of course be analogous to the role of the groups surrounding the active site in many metallo-enzymes. Clearly such specificity imparted to the catalyst by the substrate suggests that supported catalysts may become a very important class of highly specific catalysts in the future. Although it is not certain what specific role the support has in the catalyst it is suggested that some of the phenyl groups may help to protect the catalytic rhodium species by occupying or blocking vacant sites around the rhodium thus preserving a coordinatively unsaturated species.

Chiral rhodium(I) complexes, prepared as in Scheme 1, are less active than their homogeneous counterparts in promoting the enantioselective hydrosilyla-tion of ketones [29]. Their asymmetric efficiency depends to some extent

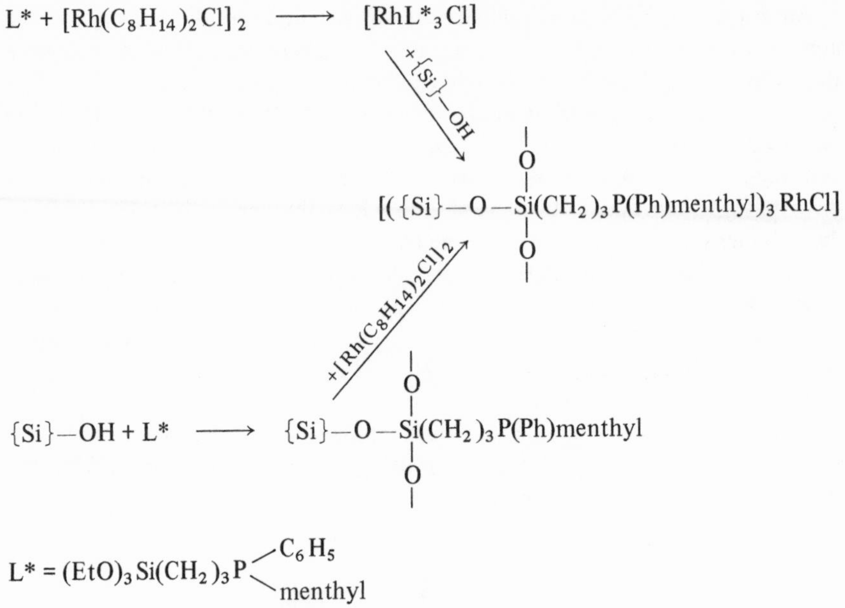

L* + [Rh(C$_8$H$_{14}$)$_2$Cl]$_2$ $\longrightarrow$ [RhL*$_3$Cl]

$[(\{Si\}-O-Si(CH_2)_3P(Ph)menthyl)_3\,RhCl]$

$\{Si\}-OH + L^* \longrightarrow \{Si\}-O-Si(CH_2)_3\,P(Ph)menthyl$

$$L^* = (EtO)_3\,Si(CH_2)_3P \overset{\displaystyle -C_6H_5}{\underset{\displaystyle \text{menthyl}}{\Big\langle}}$$

Scheme 1

on their detailed method of preparation. Further support for the suggestion that they operate by mechanism (1) comes from the observation that in several cases these anchored catalysts give alcohols of the opposite configuration to their homogeneous counterparts, whereas on reuse the product has the same configuration as that obtained homogeneously. This arises from P—Rh bond cleavage during the catalysis, leáding to mechanism (2).

Hydrosilylation yields obtained with supported catalysts are usually high, but lower than with the corresponding homogeneous catalysts. However several supported catalysts have been prepared which have no homogeneous equivalent [30] (in which cases the supported catalysts could be said to be infinitely better). For many of the catalysts the yield on the second time of use is slightly greater than on the first and thereafter alters very little.

An interesting hydrosilylation reaction catalysed by supported palladium complexes occurs between trimethylsilanol and butadiene (reaction (7)).

$2\,CH_2{=}CH{-}CH{=}CH_2 + Me_3\,SiOH$

$$\xrightarrow[\text{Pd}^0 \text{ or Pd}^{II}]{\text{supported}} Me_2\,SiO{-}CH_2{-}CH{=}CH{-}(CH_2)_3{-}CH{=}CH_2 \qquad (7)$$

The hydrosilylation is accompanied by butadiene dimerisation. Although the catalysts were effective, it was found that, during the course of the reaction, the metal was pulled off the support and thus the major advantage of the supported catalyst — namely ease of separation by filtration at the end of the reaction — was lost [15]. It was surmised that this occurred because, during the course of the reaction, an intermediate, such as 2, was involved in which the butadiene dimer occupied sufficient coordination sites to displace the palladium from the support. If this explanation is correct it suggests that polymer-supported catalysts may be of little use whenever the reaction components occupy several coordination sites around the catalyst, as in butadiene cyclooligomerisation. However the successful use of supported nickel complexes as butadiene cyclooligomerisation catalysts suggested that such a pessimistic prediction is unwarranted [34].

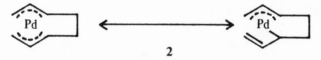

2

Iron(0) carbonyl catalysts supported on polystyrene cross-linked with 1% divinylbenzene (reactions (8) and (9)) are active catalysts for the hydrosilylation of 1-pentene when the reaction is carried out under UV-irradiation (reaction (10)) to promote the formation of coordinatively unsaturated iron(0) complexes [33].

$$\text{(P)}-PPh_2 + [Fe_3(CO)_{12}] \xrightarrow[n = 3, 4]{thf} [(\text{(P)}-PPh_2)_{5-n}Fe(CO)_n] \qquad (8)$$

$$3$$

$$\text{(P)}-PhPCH_2CH_2PPh_2 + [Fe_3(CO)_{12}]$$

$$\xrightarrow[\substack{x = 1, 2; n = 3, 4 \\ y = 1, 2}]{thf} [(\text{(P)}-PhPCH_2CH_2PPh_2)_x \{Fe(CO)_n\}_y] \qquad (9)$$

$$4$$

$$\text{1-pentene} + HSiEt_3 \xrightarrow[+ 3 \text{ or } 4]{h\nu} \text{pentyl-SiEt}_3 + \text{pentenyl-SiEt}_3 +$$

$$+ \text{pentyl-SiEt}_3 + \text{pentenyl-SiEt}_3 + \text{pentene} \qquad (10)$$

In addition to promoting the hydrosilylation of unsaturated compounds, supported rhodium(I) complexes have been used to promote the reaction of organosilicon hydrides with organosilanols (reaction (11)) [22].

$$R_3SiOH + HSiR'_3 \xrightarrow{\text{cat}} R_3SiOSiR'_3 + H_2 \qquad (11)$$

Although very active, the supported catalysts are less active than their homogeneous analogues, due to the restricted accessibility of some of the supported sites.

A series of silica supported iridium(I) complexes has been evaluated as catalysts for the silylation of alcohols (reaction (12)) [35].

$$R_3SiH + R'OH \longrightarrow R_3Si(OR') + H_2 \qquad (12)$$

The silica supported catalyst $[IrCl(CO)(\{Si\}-CH_2CH_2PPh_2)_2]$, like its homogeneous analogue, is very active initially, although it suffers rapid deactivation. By contrast the catalyst prepared by treating phosphinated silica, $\{Si\}-CH_2CH_2PPh_2$, with $[IrCl(cod)_2]_2$ has a slightly lower initial activity but much greater longevity [35].

References

1. J. P. Collman and L. S. Hegedus: *Principles and Applications of Organotransition Metal Chemistry*, University Science Books, Mill Valley, California, p. 384 (1980).
2. D. A. Armitage in: *Comprehensive Organometallic Chemistry* (Ed. by G. Wilkinson, F. G. A. Stone and E. W. Abel) Pergamon, Oxford, Vol. 2, p. 117 (1982).
3. J. L. Speier: *Adv. Organometal. Chem.* 17, 407 (1979).
4. E. Lukevics, Z. V. Belyakova, M. G. Pomerantseva and M. G. Voronkov: *J. Organometal. Chem. Library* 5, 1 (1977).
5. F. R. Hartley: *Chem. Rev.* 69, 799 (1969).
6. I. Ojima, M. Nihonyanagi and Y. Nagai: *JCS Chem. Commun.* 938 (1972).
7. M. Mejstrikova, R. Rericha and M. Kraus: *Coll. Czech. Chem. Commun.* 39, 135 (1974).
8. M. Kraus: *Coll. Czech. Chem. Commun.* 39, 1318 (1974).
9. M. Capka and J. Hetflejs: *Coll. Czech. Chem. Commun.* 39, 154 (1974).
10. T. N. Zaslavskaya, V. O. Reikhsfel'd and N. A. Filippov: *Zh. Obshch. Khim.* 50, 2286 (1980).
11. V. O. Reikhsfel'd, N. I. Flerova, N. A. Filippov and T. N. Zaslavskaya: *Zh. Obshch. Khim.* 50, 2017 (1980).
12. E. R. Martin: *US Patent* 3,795,656 (1974); *Chem. Abs.* 81, 92240 (1974).
13. N. K. Skvortsov, N. A. Filippov, L. L. Brokhina, N. P. Pron, V. S. Brovko, A. V. Nikitin, T. N. Zaslavskaya and V. O. Reikhsfel'd: *USSR Patent* 743717 (1980); *Chem. Abs.* 94, 8195 (1981).
14. F. G. Young: *German Patent* 2,330,308 (1974); *Chem. Abs.* 80, 121737 (1974).
15. M. Capka, P. Svoboda and J. Hetflejs: *Coll. Czech. Chem. Comm.* 38, 1242 (1973).
16. P. Svoboda, V. Vaisarova, M. Capka, J. Hetflejs, M. Kraus and V. Bazant: *German Patent* 2,260,260 (1973); *Chem. Abs.* 79, 78953 (1973).
17. V. Bazant, M. Capka, H. Jahr, J. Hetflejs, V. Chvalovsky, H. Pracejus and P. Svoboda: *East German Patent* 100,267 (1973); *Chem. Abs.* 80, 83225 (1974).

18. Z. M. Michalska: *J. Mol. Catal.* **3**, 125 (1977).
19. P. Svoboda, M. Capka, V. Chvalovsky, V. Bazant, J. Hetflejs, H. Jahr and H. Pracejus: *Z. Chem.* **12**, 153 (1972).
20. B. Marciniec, Z. W. Kornetka and W. Urbaniak: *J. Mol. Catal.* **12**, 221 (1981).
21. B. Marciniec, W. Urbaniak and P. Pawlak: *J. Mol. Catal.* **14**, 323 (1982).
21a. B. Marciniec and W. Urbaniak: *J. Mol. Catal.* **18**, 49 (1983).
22. Z. M. Michalska: *Trans. Met. Chem.* **5**, 125 (1980).
23. I. Dietzmann, D. Tomanova and J. Hetflejs: *Coll. Czech. Chem. Commun.* **39**, 123 (1974).
24. V. Bazant, M. Capka, I. Dietzmann, H. Fuhrmann, J. Hetflejs and H. Pracejus: *East German Patent* 103,903 (1974); *Chem. Abs.* **81**, 49815 (1974).
25. J.-C. Poulin, W. Dumont, T.-P. Dang and H. B. Kagan: *Compt. Rend.* **277C**, 41 (1977).
26. W. Dumont, J.-C. Poulin, T.-P. Dang and H. B. Kagan: *J. Amer. Chem. Soc.* **95**, 8295 (1973).
27. Z. M. Michalska, M. Capka and J. Stock: *J. Mol. Catal.* **11**, 323 (1981).
28. H. Krause and H. Mix: *German Patent* 133,199 (1978).
29. M. Capka: *Coll. Czech. Chem. Commun.* **42**, 3410 (1977).
30. M. Capka, P. Svoboda, M. Kraus and J. Hetflejs: *Chem. Ind. (London)*, 650 (1972).
31. M. Capka, P. Svoboda, M. Cerny and J. Hetflejs: *Tetrahedron Lett.* 4787 (1971).
32. V. Bazant, M. Capka, M. Tscherny, J. Hetflejs, M. Kraus and P. Svoboda: *German Patent* 2,245,187 (1973); *Chem. Abs.* **79**, 67025 (1973).
33. C. U. Pittman, W. D. Honnick, M. S. Wrighton, R. D. Sanner and R. G. Austin in: *Fundamental Research in Homogeneous Catalysis* (Ed. by M. Tsutsui) Plenum Press, N. Y. **3**, p. 603 (1979).
34. C. U. Pittman, L. R. Smith and R. M. Hanes: *J. Amer. Chem. Soc.* **97**, 1742 (1975).
35. J. Dwyer, H. S. Hibal and R. V. Parish: *J. Organometal. Chem.* **228**, 191 (1982).

CHAPTER 8

REACTIONS INVOLVING CARBON MONOXIDE

8.1. Introduction

Carbon monoxide is a very important industrial raw material, whose importance is increasing steadily as conventional organic hydrocarbon feedstocks become scarce [1÷2b]. Carbon monoxide is available from a wide variety of sources. Thus the major sources of synthesis gas, a mixture of carbon monoxide and hydrogen, are steam reforming of natural gas (methane); naphtha and heavier hydrocarbon fractions could be used as well (reactions (1) and (2)).

$$C_nH_{2n+2} + n\,H_2O \xrightleftharpoons{700-900°C} n\,CO + (2n+1)\,H_2 \tag{1}$$

$$C_nH_{2n+2} + 2n\,H_2O \xrightleftharpoons{700-900°C} n\,CO_2 + (3n+1)\,H_2 \tag{2}$$

The three products of these reactions are linked through the water-gas shift reaction (3), as is discussed in more detail in Section 8.5.

$$CO + H_2O \rightleftharpoons CO_2 + H_2 \tag{3}$$

The technology is also available to permit coal to be used as a source of carbon monoxide and hydrogen by treating the coal at 1000°C with alternate bursts of air, which effects the two exothermic processes shown in reactions (4) and (5), and steam, which effects the endothermic production of carbon monoxide and hydrogen (reaction (6)).

$$C + O_2 \longrightarrow CO_2 \tag{4}$$

$$2\,C + O_2 \longrightarrow 2\,CO \tag{5}$$

$$C + H_2O \longrightarrow CO + H_2 \tag{6}$$

By the year 2000 coal-based synthesis gas will probably be a major source of chemicals. Accordingly there is great interest in this field and catalysts of all types — heterogeneous, homogeneous and polymer-supported — are under extensive investigation at the present [3–4c].

Carbon monoxide and hydrogen can be obtained separately from synthesis gas. The methods of separation include:

(i) *Cryogenic Fractionation (Linde Process)*. This involves partial condensation of the carbon monoxide and residual methane at about −180°C and 40 atmospheres followed by fractional distillation. Any carbon dioxide and water present must first be removed since they will otherwise solidify and cause blockages.

(ii) *Diffusion (Monsanto Prism Process)*. The hydrogen is removed by preferential diffusion through a specially prepared, supported palladium membrane.

(iii) *Selective Adsorption*. Carbon monoxide is either selectively adsorbed on to zeolites or into copper(I) salt solutions such as a solution of copper(I) chloride and aluminium trichloride in toluene used in the Tenneco 'Cosorb' process.

The relationship between the various fuels and the chemicals described in this chapter is shown schematically in Figure 1. Those arrows which are not qualified refer to reactions that lie outside the scope of the present book. However they do enable future routes to chemicals to be envisaged.

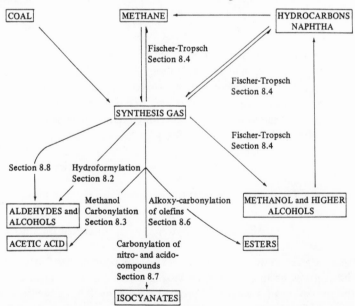

Fig 1. Relationship between sources of carbon monoxide and chemicals described in the present chapter.

8.2. Hydroformylation

Hydroformylation literally involves the addition of a hydrogen atom to one end of an olefinic double bond and a formyl group to the other [4—6a]. In practice molecular hydrogen and carbon monoxide are used as the sources of these two groups. The process is big business with about 3 million tons of aldehydes and their derivatives being produced annually by this route. Of these the biggest single product is the manufacture of butyraldehyde from propene and synthesis gas (reaction (7)).

$$CH_3CH{=}CH_2 + H_2 + CO \longrightarrow CH_3CH_2CH_2CHO \qquad (7)$$

Some of this is hydrogenated to n-butanol but the greater part is self-condensed to form 2-ethylhexanal derivatives [7] (reaction (8)).

Much of the 2-ethylhexanol is converted to phthalate esters for use as plasticizers. Butanol and other short-chain alcohols are used extensively as solvents.

The overall reaction between olefins, hydrogen and carbon monoxide can be complex, since not only are both linear and branched aldehydes (and hence alcohols) formed, but the same catalysts promote olefin isomerisation. Reaction (9) summarises the products formed.

Clearly, then, a given olefin can give rise to a considerable range of products and one of the features of interest will be how selective a given catalyst is at promoting the formation of a particular product. In fact in all but a few specialised cases the linear, or normal, isomer is the preferred commerical product and the normal : branched ratio is an important parameter in industrial hydroformylation processes; in general the higher that ratio is the better.

Three major types of transition metal complexes have been used in commercial hydroformylation processes. In order of historic development these are:

(i) simple cobalt carbonyl, or hydrido cobalt carbonyl complexes;
(ii) hydrido cobalt carbonyl complexes having tertiary phosphine ligands;
(iii) tertiary phosphine hydrido rhodium carbonyl complexes.

Their characteristics are summarised in Table I [61a]. The optimum catalyst depends on the desired end-product. If this is the alcohol, then the cobalt carbonyl plus tertiary phosphine catalyst will usually be preferred because it gives a reasonable yield of linear product at a catalyst cost about 3500 times less than the rhodium(I) system. However the rhodium(I) system is usually preferred when the aldehyde itself is the desired product. The milder operating conditions of the rhodium(I) system are very important in days of high energy costs.

TABLE I

The characteristics of cobalt and rhodium catalysed hydroformylation processes (from [6a]).

Parameter	Cobalt Carbonyl	Cobalt Carbonyl + tertiary phosphine	Rhodium(I)
Temperature °C	140–180	160–200	80–120
Pressure, atm	250–350	50–100	15–25
Metal concentration, % metal/olefin	0.1–1.0	0.5–1.0	10^{-2}–10^{-3}
Normal : *iso* ratio[a]	3–4 : 1	6–8 : 1	10–14 : 1
Aldehydes, %	*ca.* 80	–	ca. 96
Alcohols, %	*ca.* 10	ca. 80	–
Alkanes, %	*ca.* 1	ca. 15	ca. 2
Other products, %	*ca.* 9	ca. 5	ca. 2

[a] With terminal olefins as feed.

Both the rhodium and the cobalt complexes catalyse olefin isomerisation as well as olefin hydroformylation. In the case of the rhodium(I) catalysts the amount of isomerisation decreases as the ligands are altered in the order $CO >$ $NR_3 > S > PR_3$; when homogeneous and supported amine-rhodium complexes were compared it was found that they both gave similar amounts of isomerisation whereas with the tertiary phosphine complexes the supported catalysts gave rather less olefin isomerisation than their homogeneous counterparts [8–10].

The formation of alcohols as an end-product is promoted by using more vigorous conditions, both of temperature and pressure, for both the cobalt and rhodium catalysts. In the case of cobalt carbonyls the presence of tertiary phosphine ligands also promotes alcohol formation as well as olefin hydrogenation to the paraffin [11]. For rhodium(I) complexes, both homogeneous and polymer-supported, the presence of amine ligands promotes the formation of alcohols; for a given amine there is an optimum concentration, the optimum decreasing as the basicity of the amine increases [12]. This is consistent with the role of the amine being that of promoting the formation of a hydrido-complex (reaction (10)), which is thought to be the active intermediate in hydrogenation of aldehydes over cobalt and rhodium compelx catalysts [13, 14].

$$HNR_3 + [Rh_x(CO)_y]^- \; \rightleftharpoons \; [R_3N-\overset{\overset{\displaystyle H}{|}}{Rh}_x(CO)_y]$$

$$\rightleftharpoons \; R_3N + [HRh_x(CO)_y] \tag{10}$$

The present chapter does not include the use of supported liquid- and gas-phase hydroformylation catalysts. These are described in Section 1.8.2.

8.2.1. COBALT HYDROFORMYLATION CATALYSTS

Relatively few hydroformylations using supported cobalt complexes have been reported. Moffatt showed that poly-2-vinylpyridine reversibly reacted with both $[Co_2(CO)_8]$ and $[HCo(CO)_4]$, the cobalt carbonyl being displaced by excess carbon monoxide [15, 16]. This enabled the polymer to pick up the cobalt carbonyl at the end of the reaction, so enabling it to be separated from the products by filtration. The polymer was believed to act as a 'catalyst reservoir' by rapidly releasing the cobalt carbonyl into solution in the presence of further carbon monoxide, so that the actual catalysis was a homogeneous process. However more recent work by us has cast doubt upon this explanation, and suggests that the cobalt is bound to pyridine throughout the reaction [17]. Thus we carried out a series of four experiments (Table II) in which 1-hexene was hydroformylated in the presence of cobalt carbonyl alone (run A), 4-vinyl-pyridine γ-radiation grafted polypropylene plus cobalt carbonyl (run D), and two systems midway between these extremes, namely: cobalt carbonyl plus pyridine (run B), and cobalt carbonyl plus polypropylene (run C). It is apparent that run D, in which the hydroformylation was catalysed by cobalt carbonyl in the presence of 4-vinylpyridine supported on polypropylene not only showed total conversion of 1-hexene with no other hexene isomers present at all, but also more than doubled the normal : branched selectivity of the catalyst compared to cobalt carbonyl alone (run A). This strongly suggests that some catalytically active sites involve cobalt bound to the supported pyridine residues. The greater selectivity of the pyridine-supported cobalt carbonyl in run D as compared to the unsupported cobalt carbonyl in run A could arise from at least 3 effects:

TABLE II

Hydroformylation of 1-hexene (0.25mol) under 100 atmospheres of H_2 + CO (1 : 1) at 185°C for 15 minutes in the presence of $[Co_2(CO)_8]$ (3.2 mmol) in benzene (94 ml) (from [17]).

Run	A	B	C	D
Additive	None	Pyridine (20 mmol)	Polypropylene (7.04g)	4-vinylpyridine γ-radiation grafted on to polypropylene (7.04g; equivalent to 20 mmol of pyridine)
Analysis of Products (%)				
1-Hexene	0	8	4	0
n-Heptanal	30	36	37	55
2-Methylhexanal	7	26	22	21
2-Ethylpentanal	13	13	12	9
Hexane	10	12	8	5
Alcohols and Aldehyde Condensation products	20	5	17	10
Ratio of normal/ branched aldehydes	0.75	0.92	1.09	1.83

(i) The presence of pyridine in the coordination sphere of the active cobalt catalyst: run B where the same amount of pyridine was present as in run D, but in the form of free pyridine indicates that the presence of pyridine bound to cobalt gives only a small enhancement by a factor 1.23 in the normal: branched selectivity of the catalyst.

(ii) The proximity of the polypropylene groups to the active site: in run C polypropylene itself was added to the cobalt carbonyl and, even in the absence of specific binding sites for cobalt on the polymer, gave a 1.45 fold enhancement of the normal: branched selectivity.

(iii) A combination of both pyridine coordination to the cobalt and the total environment of the active site would account for the enhancement of selectivity by a factor of 2.5 in run D in the presence of 4-yinylpyridine γ-radiation grafted on to polypropylene as compared to run A. This arises because γ-radiation grafting yields a product in which several 4-vinylpyridine sites are grafted to a single site on polypropylene (reaction (11)).

$$\underset{\substack{| \\ -CH_2-C- \\ |}}{\overset{CH_3}{}} + \overset{}{\underset{N}{\bigcirc}} \longrightarrow \underset{\substack{CH_2 CH-CH_2-CH \\ }}{\overset{CH_3}{\left(CH_2-C\right)}} \tag{11}$$

As a result the cobalt carbonyl is bound within a 3-dimensional site within the polymer, **1**, in which the pyridine nuclei and the polypropylene backbone

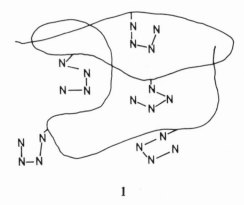

1

contribute to the total geometry of the site. Such a model gives a supported catalyst that has additional properties to its homogeneous analogue and, in particular, is approaching an enzyme in the sense that it has a 3-dimensional active site [17].

$[Co_4(CO)_{12}]$ does not catalyse the hydroformylation of olefins, either alone or in the presence of poly-4-vinylpyridine at room temperature and 10 atmospheres pressure of a 4 : 1 mixture of H_2 and CO. However, photolysis of a solution of $[Co_4(CO)_{12}]$ in the presence of cross-linked poly-4-vinylpyridine microspheres (5μ diameter) yields a supported cobalt complex which does catalyse the hydroformylation of 1-pentene under these conditions [18]. A series of amino-phosphines such as $P\text{-}(C_6H_4NMe_2\text{-}p)_3$ have been linked to sulphonated polystyrene by ion-exchange and then used to support a number of metal complexes. $[Co_2(CO)_8]$, which binds to the nitrogen donors, hydroformylates 1-hexene at $120°C$ to yield a ratio of normal : branched aldehydes of about 2.5 : 1 [19].

Polystyrene and poly-1,4-phenylene oxide supported cyclopentadienylcobalt dicarbonyl catalyse the hydroformylation of 1-pentene to a mixture of linear and branched aldehydes at 130°C. The initial linear : branched ratio of 3 : 1 on the polystyrene supported catalyst falls during the reaction due to isomerisation of the 1-pentene. This isomerisation can be suppressed by addition of ditertiary phosphines such as $Ph_2PCH_2CH_2PPh_2$ [22] so that there has been interest in supporting cobalt complexes on phosphinated silica [23–24a] and polystyrene [25].

After heating $[Co_4(CO)_{12}]$ adsorbed on to metal oxides above 90°C to release some carbon monoxide, active hydroformylation catalysts are obtained [26, 27]. Basic oxides such as ZnO, MgO, TiO_2 and La_2O_3 give higher hydroformylation activity than acidic oxides such as Al_2O_3, $SiO_2–Al_2O_3$ and V_2O_5, which showed negligible activity. In comparison with their rhodium analogues cobalt catalysts are less active but give greater normal : branched selectivity; thus, on zinc oxide, this selectivity decreases in the order $[Co_4(CO)_{12}]$ 9 : 1 > $[RhCo_3-(CO)_{12}]$ 8 : 2 > $[Rh_2Co_2(CO)_{12}]$ 7 : 3 > $[Rh_4(CO)_{12}]$ 58 : 42 [27]. A series of patents describe hydroformylation catalysts based on $[Rh_nCo_m(CO)_{12}]$, n = 1–3, m = 4, supported on aminated cross-linked polystyrene [23–30]. $[Co_2(CO)_6(PBu_3)_2]$ adsorbed on to diatomaceous earth and activated carbon has been used as a supported liquid-phase and solid hydroformylation catalyst in a gas chromatograph type system (See section 1.8.2) [31]. Heteronuclear cobalt-palladium carbonyl phosphine complexes anchored on phosphinated silica have much greater activity in the gas-phase hydroformylation of propene than either palladium(0) or cobalt(0) homonuclear complexes [31a].

8.2.2. RHODIUM(I) HYDROFORMYLATION CATALYSTS

The replacement of carbonyl ligands in rhodium(I) complexes such as [Rh(acac)-$(CO)_2$] by tertiary phosphines, whether in solution or polymer supported, increases the ratio of normal to branched aldehydes [32, 33]. The great advantage of supporting the rhodium(I) is that it is not necessary to swamp the rhodium catalyst with excess tertiary phosphine in order to obtain a high normal to branched ratio of product [33]. This is well illustrated in Table III where, for 1-hexene under identical conditions, the homogeneous catalyst [Rh(p-styrylPPh_2)(acac)(CO)] requires a phosphorus : rhodium ratio of 250 : 1 to yield a normal : branched aldehyde ratio of 5 : 1, whereas the corresponding supported catalyst in which all the phosphine ligands are bound to the polymer only requires a phosphorus : rhodium ratio of 2 : 1 to yield the same selectivity. Since large excesses of tertiary phosphines inhibit olefin coordination and hence suppress catalytic activity [35] it is possible to obtain the highest normal : branched aldehyde ratios at higher activities using supported rather than homogeneous catalysts. The enhancement of normal : branched selectivity in the

TABLE III

Effect of phosphorus : rhodium ratio on the normal : branched selectivity of the hydro-formylation of 1-hexene in benzene under 10 atmospheres H_2 + CO (1 : 1) at 65°C (data from [33]).

1. *Supported Catalyst* [Rh(ⓟ—PPh₂)(CO)(acac)], 300 minutes[a]

P : Rh ratio[b]	8.36	5.86	4.05	2.17	1.32
Selectivity[c]	16.0	14.1	13.2	5.2	2.5

2. *Homogeneous Catalyst* [Rh(*p*-styrylPPh₂)(CO)(acac)], 100 minutes

P : Rh ratio[b]	250	80	8.0	2.0	1.0
Selectivity[c]	5.0	3.7	2.6	2.2	2.7

[a] ⓟ—PPh₂ = *p*-styrylPPh₂, γ-radiation grafted on to polypropylene (prepared as in [34]).
[b] gm atom/gm atom
[c] ratio of normal : branched aldehydes.

supported catalysts arises in part from the different positions of the equilibria for replacement of phosphine by carbon monoxide in the rhodium coordination sphere in the two media. In part these different positions will arise because the concentration of carbon monoxide within the polymer will probably be very different from that in the bulk solution. As would be expected from this, although supported catalysts normally have higher selectivities than their un-supported counterparts, this is not always so, particularly at low phosphorus loadings [36].

When functionalised olefins are hydroformylated the sensitivity of the ratio of normal : branched products on the phosphorus . rhodium ratio depends on the functional group present. Thus with allyl alcohol the ratio of normal to branched products formed at 60°C under 7 atmospheres in the presence of phosphinated polystyrene supported [RhH(CO)(PPh₃)₃] only varies from 1.93 : 1 at P : Rh = 3 : 1 to 2.33 : 1 at P : Rh = 30 : 1, although it can be increased further to 4 : 1 by increasing the ratio of H_2 : CO in the gas mixture [37]. In the case of methyl methacrylate the ratio of normal : branched aldehyde actually decreased as the phosphorus : rhodium ratio increased [38].

Many supported hydroformylation catalysts have comparable activities to their homogeneous analogues, so long as the comparisons are made at temperatures above room temperature (e.g. 60°C or above). This is not always true at low temperatures because their activities become limited by the rates of diffusion to and from the catalytic site [8, 9, 35, 39]. Thus at 40 °C and 17 atmospheres [(ⓟ—PPh₂)RhH(CO)(PPh₃)] where ⓟ— = polystyrene-1%-divinylbenzene was 0.22 times as active as its homogeneous analogue in 1-

pentene hydroformylation, whereas at 60° and 53 atmospheres the supported catalyst was 1.08 times as active as the homogeneous material [35, 39].

In many cases the selectivity of supported catalysts decreases as the hydroformylation proceeds. There are two reasons for this:

(i) Many rhodium(I) hydroformylation catalysts also promote olefin isomerisation leading to formation of internal olefins which inevitably yield branched aldehydes.

(ii) Many commerical sources of mixtures of hydrogen and carbon monoxide contain traces of oxygen. This oxidises the phosphine support to phosphine oxide, which does not coordinate to rhodium. Consequently there is a steady decrease in the phosphorus: rhodium ration during the reaction, which inevitably leads to a reduction in the ratio of normal: branched aldehydes (see Table III).

Rhodium(I) catalysed olefin hydroformylation occurs by a dissociative mechanism (Scheme 1) in which the first stage involves replacement of a molecule

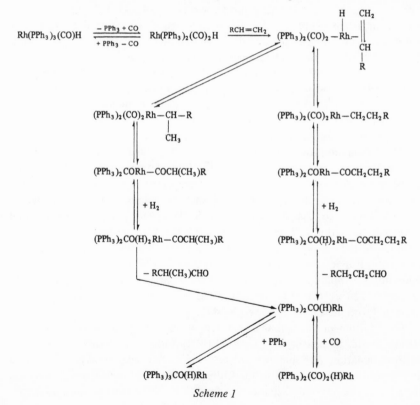

Scheme 1

of triphenylphosphine from the coordination sphere around rhodium by gaseous carbon monoxide [4, 6, 40]. Clearly, in a supported catalyst if the phosphine group lost in this stage were that binding the complex to the support, then leaching of the rhodium from the support would be observed. This is often observed, although on one occasion the rhodium content of a supported catalyst, far from decreasing on use, actually increased [35]! The increase in rhodium content probably arises from the release of triphenylphosphine, since immobilisation of $[RhH(CO)_2(PPh_3)_2]$ on phosphinated polymers gives both square-planar and trigonal-bipyramidal complexes of rhodium(I) depending on the number of anchor sites. The actual structure depends on the separation of the anchor sites, the flexibility of the side chain linking the anchor site to the polymer and the degree of rhodium loading [41, 42]. Rhodium leaching from phosphinated polymers has been found to decrease when the temperature and hydrogen pressure are increased and when the carbon monoxide pressure is decreased as expected from the equilibria suggested in Scheme 1 [43, 44]. Leaching can be greatly reduced, if not eliminated altogether, if instead of displacing a tertiary phosphine ligand from the rhodium(I), a carbonyl ligand is displaced [32, 33, 44]. Thus, when $[Rh(acac)(CO)_2]$ is used as the source of rhodium(I), carbon monoxide is displaced and this readily migrates out of the polymer so preventing reformation of the initial rhodium(I) complex which would facilitate leaching. The nature of the solvent can profoundly influence the degree of leaching. This is particularly true with aminated polymers where elution was particularly promoted by the aldehydes formed. Aldehydes had a much greater effect than alcohols [43, 44].

We noted above (Table III and associated text) that increasing the phosphorus : rhodium ratio enhances the normal : branched selectivity. This effect can be temperature dependent due to the increasing mobility of the polymer supported $-PPh_2$ groups with increasing temperature. Thus at high phosphorus : rhodium ratios, $[RhH(CO)(PPh_3)_3]$ anchored on phosphinated polystyrene showed normal to branched selectivity in 1-pentene hydroformylation that increased on warming from $40°(2.4)$ to $60°C(6.9)$ and decreased to 3.2 on warming further to $80°C$, in contrast, at low phosphorus : rhodium ratios the selectivity dropped steadily from $40°C(3.0)$ to $80°C(2.5)$, whereas at very high phosphorus : rhodium ratios the selectivity increased on heating from $100°C-(12.1)$ to $120°C(16.1)$ [36, 46, 47]. The selectivity of these latter catalysts decreased with increasing pressure [36, 47].

The addition of an equimolar amount of an α, ω-bis(diphenyl phosphine)-alkane to the homogeneous $[RhH(CO)(PPh_3)_3$ not only promotes the ready hydroformylation of functionalised terminal olefins such as allyl alcohol and vinyl acetate [48], but it also markedly improves the thermal stability of the rhodium(I) catalyst and decreases the degree of accompanying olefin isomerisation [49–51]. Polymer-supported phosphines may behave in a similar way, since they often act as bidentate phosphines. There has also been interest in supporting bidentate phosphines themselves on to polymers. Thus [RhH(CO)(PPh$_3$)$_3$] exchanged on to the polystyrene supported bidentate phosphine (P)—PPhCH$_2$-CH$_2$PPh$_2$ gave higher normal to branched selectivity than [RhH(CO)(Ph$_2$-PCH$_2$CH$_2$PPh$_2$)] [52] and that selectivity increased as the temperature increased in the range 60–120°C, and as the pressure decreased [53]. No rhodium leaching could be detected [53]. These catalysts isomerise 1-pentene readily, which is a disadvantage [36].

[((P)—PPh$_2$)RhH(CO)(PPh$_3$)], where (P)—PPh$_2$ is a soluble non-crosslinked phosphinated polystyrene, has been used to hydroformylate 1-pentene with a normal : branched selectivity of 77 : 32; the catalyst can be separated at the end of the reaction by membrane filtration or by precipitation with n-hexane [54, 55]. A series of polyorganosiloxane polymers functionalised with —PPh$_2$ groups to form both soluble and insoluble $trans$-[RhCl(CO)({Si}—PPh$_2$)$_2$] gave low normal : branched selectivities of between 0.9 and 2.4 in the hydroformylation of 1-hexene at 100°C under 66.6 atmospheres of hydrogen and carbon monoxide; both phosphorus and rhodium were lost from the polymers even when oxygen was excluded [56].

Rhodium complexes chemically bound to silica and alumina through phosphorus [57–58b], nitrogen [57] and sulphur [57, 59] donor ligands have been used to catalyse the liquid-phase hydroformylation of 1-hexene. The phosphorus supported catalysts gave only aldehyde products whereas the nitrogen supported catalysts gave some alcohols as well [57, 58a, 60]. Reducing the number of residual silanol groups on the silica surface reduced the amount of olefin isomerisation [57]. The hydroformylation activity of $trans$-[RhCl(CO){Ph$_2$PCH$_2$-CH$_2$Si(OEt)$_3$}$_2$] bound on to γ-alumina increased dramatically in the presence of added triphenylphosphine [58]. This is believed to block the deactivating Lewis acid sites on γ-alumina. The addition of triphenylphosphine to silica supported $trans$-[{Si}—OCH$_2$CH$_2$PPh$_2$)$_2$RhCl(CO)] did not greatly affect either the activity or selectivity of this catalyst [58].

A series of amino-phosphines such as P(C_6H$_4$NMe$_2$-p)$_3$ linked to sulphonated polystyrene by ion-exchange has been used to support rhodium(I) by reaction with [Rh(CO)$_2$Cl]$_2$. These catalysts hydroformylated 1,5-cyclooctadiene with few side-products other than 1,3-cyclooctadiene and very little leaching (0.4–0.6 ppm Rh/hour) [19]. A catalyst formed by mixing a solution of [Rh(nbd)(Ph$_2$PCH$_2$CH$_2$NMe$_3$)$_2$]$^{3+}$ (NO$_3^-$)$_3$ with a macroreticular exchange resin was less active in the hydroformylation of 1-hexene than the cationic rhodium complex used in an aqueous/organic two-phase solvent system; however the supported catalyst gave a better ($\sim$7 : 1) ratio of linear : branched heptanal [60a].

Resins with iminodiacetic acid moieties have been used to bind rhodium. When rhodium trichloride is the source of rhodium the normal to branched aldehyde selectivity was 2.3 and when $[Rh(acac)(CO)_2]$ was used the selectivity was even worse at 0.7 [61].

Polystyrene and poly(1,4-phenylene oxide) supported cyclopentadienyl rhodium dicarbonyl are more active as hydroformylation catalysts than their unsupported analogues due to the tendency of the latter to form inactive poly-nuclear species in solution [21, 21a, 62]. However the linear : branched selec-tivities are very poor, being in the range 0.6–1.5 for the poly(phenylene oxide) supported catalyst [21a].

8.2.3. ASYMMETRIC HYDROFORMYLATION

Monodentate asymmetric tertiary phosphine rhodium(I) complexes supported on polystyrene give only low optical yields in the hydroformylation of prochiral olefins [63, 64]. Similar results are obtained with supported rhodium(I)–(–)-diop catalysts although the optical yields were as high as 20% and did depend somewhat on the density of (–)diop sites in the polymer and the phosphorus : rhodium ratio used [55, 65–68].

Somewhat disappointing optical yields were obtained from the $PtCl_2 - SnCl_2$ complexes of (–)diop where slightly lower optical yields were obtained with the polymer supported ligand (25–30%) than with its homogeneous analogue [69].

8.2.4. OTHER TRANSITION METAL HYDROFORMYLATION CATALYSTS

Polymer supported $[TiCp_2(CO)_2]$ has been reported to be an effective hydro-formylation catalyst although no details have been given [70].

$[Ru(CO)_3(PPh_3)_2]$ supported on phosphinated cross-linked polystyrene catalyses the hydroformylation of 1-pentene without any accompanying olefin isomerisation [71]. Normal : branched selectivities (3.5–3.8) are higher than for the corresponding homogeneous catalysts except when the latter are used in the presence of very large amounts of added triphenylphosphine. The formation of large amounts of the normal product depends on equilibrium (12) lying well to the left, which in turn depends on the phosphorus : ruthenium ratio and the degree of ligand mobility within the resin. Swelling drives equilibrium (12) to the right by pushing the phosphine groups further apart [71]. $[Ru(NH_3)_6]^{2+}$ exchanged on the faujasite-type zeolite catalyses the hydroformylation of ethylene to a 1 : 3 : 1 mixture of C_2H_5CHO, C_3H_7OH and $CH_3CH(CHO)C_2H_5$ [72].

$$[((P)-PPh_2)_2 RuH_2(olefin)(CO)] \rightleftharpoons (P)-PPh_2 +$$
$$+ [((P)-PPh_2)RuH_2(olefin)(CO)] \quad (12)$$

When $[Pt(PPh_3)_2Cl_2]$ supported on phosphinated polystyrene was used to hydroformylate 1-tetradecene and 1-decene in the presence of tin(II) chloride the initial ratio of normal : branched aldehydes rose from 8 : 1 in the homogeneous case to between 80 : 1 and 190 : 1 with the supported catalysts [73]. This remarkable enhancement was thought to be steric in origin. Unfortunately in a long run the selectivity drops back to the homogeneous level due to elution of platinum(II) off the polymer. When H_2PtCl_6 was supported on nitrogen donor supports, such as cross-linked poly(4-vinylpyridine) or silica treated with $H_2NCH_2CH_2NH(CH_2)_3Si(OMe)_3$, hydroformylation of α-olefins gave a 54 : 46 ratio of normal : branched aldehydes [74]. Both these platinum systems require very high pressures of between 100 and 200 atmospheres. $[M(PhCN)_2Cl_2]$ where M = Pd and Pt supported on poly(diphenylstyrylphosphine) have been investigated as olefin hydroformylation catalysts [75] as have $[PtH(SnCl_3)-(PPh_3)_2]$, $[PtCl_2(PBu_3)_2]$ and $[PtCl(CO)(PPh_3)_2]ClO_4$ supported on sulphonated polystyrene to which amino-phosphines such as $P-(C_6H_4NMe_2\text{-}p)_3$ have been linked by ion-exchange [19].

8.3. Carbonylation of Methanol

As the costs and availability of raw materials have changed over the last 100 years the preferred industrial routes to acetic acid have also changed significantly (Tahle IV) [3, 76]. The best current process is the rhodium catalysed carbonylation of methanol developed by Monsanto [77] who started up their first plant in 1970. This is more selective and operates under milder conditions than the earlier cobalt catalysed process developed by BASF [78] (Table V). The success of the homogeneous iodide promoted rhodium catalysed process has led to a number of attempts to develop supported catalysts for this process.

The homogeneous rhodium catalysed carbonylation of methanol is believed to take place as shown in Scheme 2. The by-products (<1%) are dimethyl ether and acetaldehyde, the latter being evident if the reaction is carried out in methanol as solvent; acetic acid is sometimes esterified to methyl acetate. Rhodium catalysts have been supported on polystyrene, carbon, silica, alumina, magnesium oxide and type-X-zeolite molecular sieves [19].

$[RhCl(CO)(PPh_3)_2]$ exchanged on to phosphinated polystyrene has been used in the form of a membrane with vapour phase reactants and in solution [80] Infrared spectroscopy showed that $[((P)-PPh_2)_2RhCl(CO)]$ was the

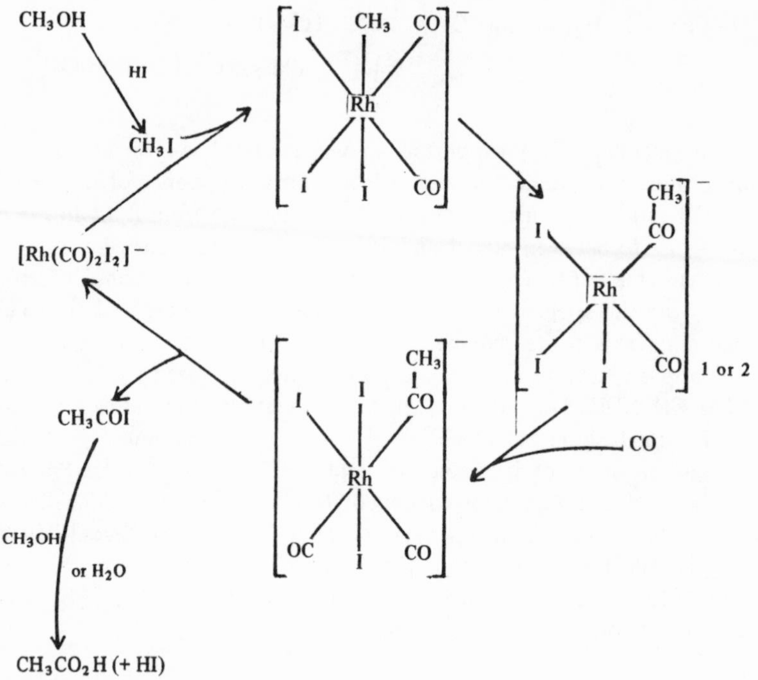

Scheme 2

main rhodium species present on the polymer. At one atmosphere pressure the gas phase reaction yielded methyl acetate as the main product with dimethyl ether as the side product. Over tens of hours in the vapour phase reaction the catalyst was slowly deactivated by conversion to a coordinatively saturated rhodium(III) species; in solution leaching resulted in the loss of activity. The kinetic results show a second-order dependence on rhodium(I) which is in contrast to the homogeneous system, suggesting that the slow step involves methyl iodide addition to the two rhodium(I) centres in the flexible polymer network, one acting as a nucleophile and attacking the carbon whilst the other plays a solvent-like role stabilising the transition state (reaction (13)).

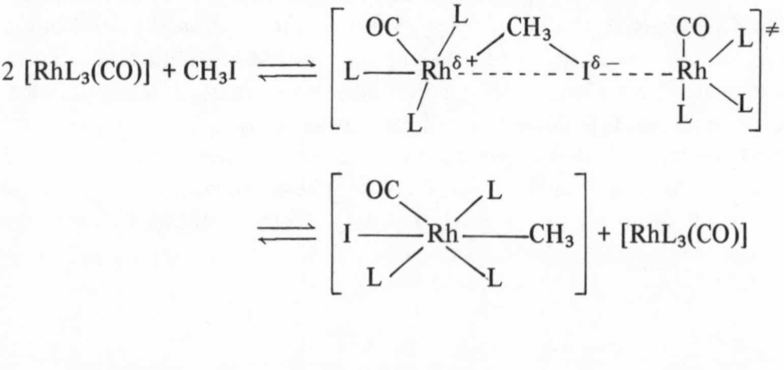

$$2\,[RhL_3(CO)] + CH_3I \;\rightleftharpoons\; \left[\, L{-}Rh^{\delta+}{-}{-}{-}{-}{-}I^{\delta-}{-}{-}{-}{-}Rh\, \right]^{\neq}$$

$$\rightleftharpoons \left[\, I{-}{-}{-}Rh{-}{-}{-}CH_3 \,\right] + [RhL_3(CO)] \qquad (13)$$

TABLE IV

Industrial routes to acetic acid over the last 100 years

Chemical route	Comments
Carbohydrate $\xrightarrow{\text{enzyme}}$ CH_3CO_2H aq	Fermentation route (19th century)
$HC{\equiv}CH$ $\xrightarrow[\text{Hg(II)}]{H_2O}$ CH_3CHO $\longrightarrow$ CH_3CO_2H	First synthetic route
Butane $\xrightarrow[\text{Mn/Co}]{O_2}$ CH_3CO_2H + by-products	Celanese, BP (ca. 1955)
$CH_2{=}CH_2$ $\xrightarrow[\text{Pd(II)/Cu(II)}]{O_2}$ CH_3CHO $\xrightarrow{O_2}$ CH_3CO_2H	Wacker process (1962)
CH_3OH $\xrightarrow[\text{Co(I)/500 atm}]{CO}$ CH_3CO_2H	BASF (1966)
CH_3OH $\xrightarrow[\text{Rh(I)/20 atm}]{CO}$ CH_3CO_2H	Monsanto (1970)
CO/H_2 $\xrightarrow[\text{40 atm}]{\text{Rh(I)}}$ CH_3CO_2H	Union Carbide (under development)

TABLE V

Comparison of the cobalt- and rhodium-catalyzed methanol carbonylation reactions

Parameter	Cobalt process	Rhodium process
Metal concentration	$\sim10^{-1}$M	$\sim10^{-3}$M
Reaction temperature	$\sim230°$C	$\sim180°$C
Reaction pressure	500–700 atm	30–40 atm
Selectivity (for methanol)	90%	>99%
Hydrogen effect	CH_4, CH_3CHO, C_2H_5OH formed as by-products	No effect

Thereafter the mechanism occurs in a similar manner to the homogeneous process. The lower activity of the polymer supported catalyst is probably partly due to the replacement of some iodide ligands by phosphines [81] and partly due to accessibility of the catalytic sites.

Rhodium(I) methanol carbonylation catalysts require a co-catalyst which in the homogeneous system is methyl iodide. It is difficult to build methyl iodide into a support. However a bifunctional catalyst has been prepared by supporting a pseudo alkyl halide, 2 [82, 83]. The support is prepared as in reaction (14) and when treated with rhodium trichloride and exposed to carbon monoxide yields 3, which undergoes oxidation addition with methylthioether groups on the polymer formed by reaction of the thiol groups with methyl iodide (reaction (15)). This catalyst is an active methanol carbonylation catalyst,

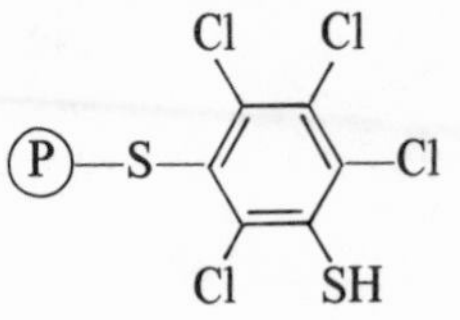

2 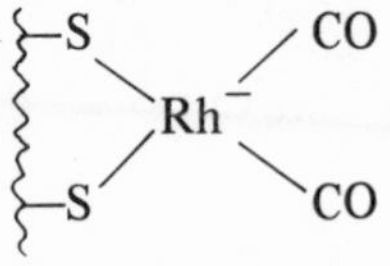   3

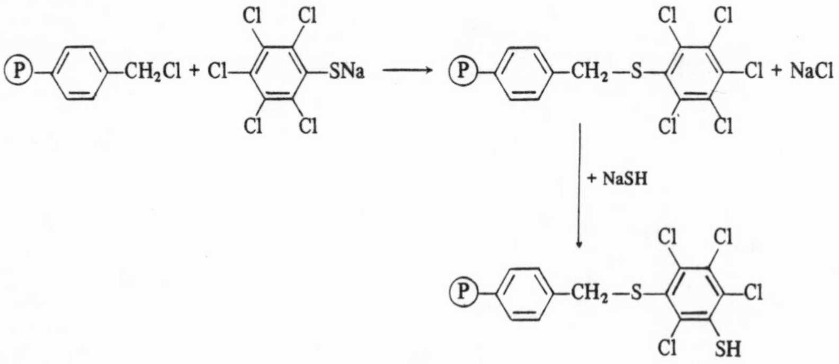

(14)

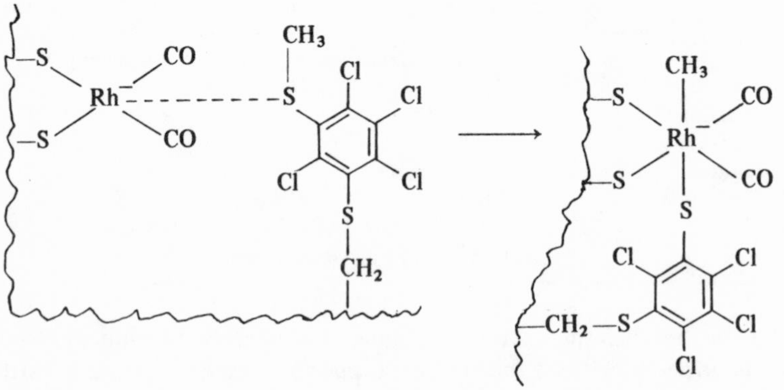

(15)

although a lot less active than its homogeneous counterpart [84]. On continuous use the activity drops due to loss of both rhodium and sulphur. This occurs because although reaction (15) shows the desirable sulphur-rhodium coordination for activity, there is an alternative sulphur that can coordinate through oxidative-addition (reaction (16)). Indeed this latter has a weaker sulphur-carbon bond and, were it not for steric effects, reaction (16), which leads to loss of rhodium from the support, would completely dominate reaction (15), which leads to catalysis [82].

(16)

Anionic rhodium(I) complexes, $[Rh(CO)_2I_2]^-$, have been supported ionically on a polystyrene-poly(4-vinylpyridine) copolymer quaternised with methyl iodide. The resulting catalyst was as active as its homogeneous counterpart. Leaching could be reduced by lowering the dielectric constant of the solvent. Thus substituting a 60 : 40 benzene/methanol mixture for acetic acid/methanol/water (60/145 : 60/145 : 25/145) reduced the leaching from 25% to 7%. A reduction in the percentage of rhodium on the support also reduced the leaching as did reducing the volume of solvent used [85].

Carbon has been found to be a useful support [86–89]; although rhodium trichloride may be used [88], rhodium nitrate is preferred [86, 87] selectivities of up to 99% have been reported [86]. The nitrate based catalysts are up to ten times more active than those prepared from $[Rh(acac)_3]$, $[RhCl(CO)(PPh_3)_2]$ and $[RhCl(PPh_3)_3]$ [87]. A particularly active material is produced by impregnation of the carbon with a solution of rhodium nitrate followed by decomposition at 300°C or above. Higher temperatures give greater activity, probably due to conversion of the nitrate to the oxide. Decomposition in hydrogen gives an inactive catalyst suggesting that rhodium metal is inactive, although hydrogen treatment has a favourable effect on the rhodium trichloride catalysts.

A number of rhodium(I) complexes have been examined for carbonylation activity when supported on metal oxides [90–94]. Catalysts supported on γ-alumina predried at 650°C were active at 200°C and one atmosphere for methanol conversion, but the selectivity was low (<50%) due to dimethyl ether formation; this was particularly true with $[RhCl(PPh_3)_3]$, $[Rh(cod)Cl]_2$ and $[Rh(cod)(OMe)]_2$, whereas $[RhCl(CO)(PPh_3)_2]$ had about the same activity but a selectivity approaching 99% [90]. The formation of ether may be due to γ-alumina catalysed dehydration of methanol although the reason for its limited formation in the presence of $[RhCl(PPh_3)_3]$ is not clear. Zeolite supported rhodium complexes give selectivities approaching 90%, although the activity/rhodium increases as the rhodium loading decreases, the selectivity for acetate formation decreases a little [91–93]. The nature of the support in these systems is clearly very important [94, 94a]. Thus treatment of rhodium trichloride on silica gel with carbon monoxide yields rhodium(I) dicarbonyl species which on further treatment with methyl iodide gives limited rhodium(III)-acyl formation [94]. By contrast, when rhodium trichloride supported on magnesium oxide is treated with carbon monoxide, rhodium(I) mono- and di-carbonyl species are formed, the former particularly if methyl iodide is present, as well as a rhodium-(III)-carbonyl species, the amount of which can be increased by treating the surface with oxygen. No formation of any rhodium-acyl species is observed on adding methyl iodide to the magnesium oxide supported rhodium carbonyls [94]. Carbon monoxide reacts with rhodium trichloride on TiO_2 to yield the rhodium(I) dicarbonyl species [94b].

A comparison of the activities of supported rhodium(I) methanol carbonylation catalysts with their homogeneous counterparts shows that although their selectivities can be very high the activities of the supported catalysts are typically two orders of magnitude less per rhodium atom [79]. Since the rhodium catalyst can be recovered and recycled fairly easily in the homogeneous reaction, this is likely to be preferred commercially for some time to come [3].

8.4. Fischer-Tropsch Reaction

Although the Fischer-Tropsch reaction can be traced back to 1902, when it was reported that certain heterogeneous nickel systems catalysed the reduction of carbon monoxide to methane [95], the modern reaction dates to the observation, in the early 1920s, that alkalised iron turnings promote the conversion of mixtures of carbon monoxide and hydrogen to hydrocarbons and alcohols [96, 97]. The reaction had great potential for the production of synthetic liquid fuels and from the outset generated considerable interest, especially in Germany. During the 1930s and 1940s this was stimulated by the high price of oil (ten times that of coal) and the need for self sufficiency in liquid hydrocarbon fuels. By the end of World War II nine plants were operational in Germany. After the war the greater availability of oil eroded the cost advantage of coal, and the multi-step nature of the process and its relatively low thermal efficiency led to a loss in interest. Only in South Africa was interest maintained by the need to be self-sufficient in liquid hydrocarbons in a country devoid of large deposits of oil. The currently operated SASOL-II plant is shown schematically in Figure 2 [3, 97a].

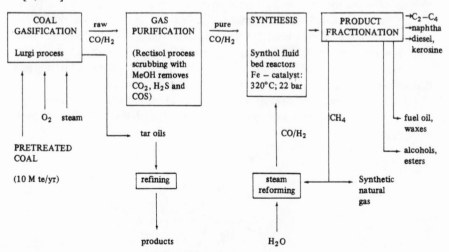

Fig. 2. The SASOL-II Fischer-Tropsch Process

In the present section we shall restrict our attention to supported metal complex Fischer-Tropsch catalysts, and consider them in three sections; (a) catalysts yielding mainly paraffins, (b) catalysts yielding mainly olefins and (c) catalysts yielding mainly alcohols. Readers requiring information on heterogeneous [3, 98–101a] or homogeneous [4, 6b, 102] Fischer-Tropsch catalysts are referred elsewhere.

8.4.1. FISCHER-TROPSCH FORMATION OF PARAFFINS

When [$Mo(CO)_6$] adsorbed onto dehydroxylated γ-alumina is heated to between 300 and 500°C a product is generated (see Section 3.6) in which the oxidation number of molybdenum is about +0.3 [103]. Such materials convert a mixture of carbon monoxide and hydrogen (1 : 5) to methane together with smaller amounts of ethane and propane with turnover frequency of about 0.04 sec^{-1} per surface atom of molybdenum at 300°C [104].

[$Fe_3(CO)_{12}$], [$Ru_3(CO)_{12}$], [$Os_3(CO)_{12}$] and [$Co_2(CO)_8$] impregnated on to metal oxides all form catalysts active for the conversion of mixtures of carbon monoxide and hydrogen into hydrocarbons [105–114]. The length of the paraffin chain can be tailored by careful selection of both the metal complex and the oxide support. Thus [$Fe_3(CO)_{12}$] and [$FeCp(CO)_2$]$_2$ on NaY faujasite promote the formation of C_1 up to at least C_{11} hydrocarbons whereas on silica gel or alumina [$Fe_3(CO)_{12}$] is less active and only forms C_1–C_4 hydrocarbons. [$Co_2(CO)_8$] on NaY faujasite is more active than [$Fe_3(CO)_{12}$], and gives hydrocarbons up to C_9, the exact amounts of hydrocarbons formed depend in both cases on the ratio of hydrogen to carbon monoxide [105]. When [Co_2-$(CO)_8$] is impregnated on to calcined alumina the molecular weight of the paraffin formed at low cobalt loading depended on the pore diameter; with pore diameters of 65 Å and less C_3–C_{10} paraffins were formed, whereas with the largest pores C_{14}–C_{21} paraffins could be obtained [108]. This suggests that diffusion effects can be important in determining selectivity.

[$Os_3(CO)_{12}$] reacts on heating with magnesium oxide to form [$HOs_3(CO)_{10}$-$(OMg \leqslant)$]. On oxidation a mononuclear osmium(II) carbonyl complex, believed to be 4, is formed [109].

4

These are held so tightly on the surface that the three original osmium atoms essentially retain their original positions and so can easily be reunited into Os_3 clusters upon introduction of carbon monoxide. Accordingly, both the original clusters and the oxidised complexes after reduction to the free metal are active for the conversion of mixtures of hydrogen and carbon monoxide (4 : 1) to mixtures of C_1-C_4 paraffins at 300°C and 31.8 atmospheres. Essentially the same series of reactions occurs when $[Os_3(CO)_{12}]$ is absorbed on to silica and alumina; the osmium(II) species has been shown to be an equilibrium mixture of a tricarbonyl and a dicarbonyl (reaction (16)) [114].

$$[Os^{II}(CO)_3(OM)_2]_2 \underset{+CO}{\overset{-CO}{\rightleftharpoons}} [Os^{II}(CO)_2(OM)_2]_n \qquad (16)$$

(M = Si, Al)

These complexes are active for the selective formation of methane from carbon monoxide and hydrogen; small amounts of C_2-C_4 hydrocarbons are formed at lower temperatures, but above 300°C methane is virtually the only hydrocarbon formed.

Polystyrene supported cyclopentadienylcobalt dicarbonyl, 5, (see Section 3.3) converted a mixture of carbon monoxide and hydrogen predominantly to

5

6

methane although other paraffins up to C_{20} were also formed, especially at lower hydrogen : carbon monoxide ratios [115–117]. The closely related material, 6, was two orders of magnitude less active than 5. The activity of 5 (0.011 mmol CO/mmol catalyst-Co/hour) was enhanced by thermolysis to yield the monocarbonyl complex (activity 0.130 mmol CO/mmol catalyst-Co/hour). All attempts to compare these activities with those of the homogeneous counterparts failed because of the instability of the homogeneous analogues, thus demonstrating the special stability exhibited by the polymer-bound species.

8.4.2. FISCHER-TROPSCH FORMATION OF OLEFINS

The interaction of $[Fe_3(CO)_{12}]$ and $[Fe(CO)_5]$ with alumina or magnesium oxide or functionalised silica such as $\{Si\}-(CH_2)_3NH_2$, results in the formation of the anionic hydrido cluster carbonyl $[HFe_3(CO)_{11}]^-M^+$, where $M = Al(O-)_y^+$, $Mg(O-)_y^+$, or $\{Si\}-(CH_2)_3NH_3^+$ together with carbon dioxide [118, 118a].

After thermally decomposing these clusters under 10^{-4} torr at 130°C for 16 hours the resulting catalysts promoted the conversion of a 2 : 1 molar ratio of carbon monoxide and hydrogen at atmospheric pressure to a mixture of hydrocarbons [118–121]. The hydrocarbons formed in a typical experiment at 176°C on the metal oxides were propene (32.0%), methane (26.1%), ethylene (9.2%), but-1-ene (7.3%), cis-but-2-ene (3.6%), trans-but-2-ene (5.5%) and C_5 (7.6%). All the paraffins, except methane were present in much smaller amounts than the olefins. The high selectivity for propene, which can under suitable conditions be as high as 45%, and the low selectivity for ethylene, which can be as low as 5%, suggested that ethylene could be a primary product which undergoes a secondary reaction to propene. This was indeed shown to be the case, suggesting a carbene-metallocyclobutane mechanism (reactions (17–19)) [119, 120].

$$C_2H_4 \longrightarrow 2\underset{Fe}{\overset{CH_2}{\|}} \overset{2H}{\underset{\searrow}{\nearrow}} \begin{array}{c} CH_4 \\ \underset{Fe}{\overset{CH}{\mathrel{\vert\vert\vert}}} + \underset{Fe}{\overset{H}{\vert}} \end{array} . \tag{17}$$

$$\underset{Fe}{\overset{CH_2}{\|}} + C_2H_4 \longrightarrow \overset{\diamondsuit}{Fe} \longrightarrow CH_3CH=CH_2 \tag{18}$$

$$\underset{Fe}{\overset{CH_2}{\|}} + CH_3CH=CH_2 \longrightarrow \overset{\diamondsuit-CH_3}{Fe} \longrightarrow C_2H_5CH=CH_2 \tag{19}$$

When $\{Si\}-(CH_2)_3NH_3^+[FeCo_3(CO)_{12}]^-$ was activated and decarbonylated at atmospheric pressure and 200°C, hydrocarbons in the range C_1 to C_5 were formed. Exposure of the product to synthesis gas at 240°C gave a relatively narrow product range [118a]. At atmospheric pressure there was a relatively low selectivity for methane, a peak at C_6 and >95% of the C_3-C_5 product was olefin (Figure 3). As the synthesis gas pressure was increased, so selectivity for methane formation increased and that for olefin formation decreased, reflecting enhanced hydrogenation activity at elevated pressures. In addition the hydrocarbon peak moved from C_6 (atmospheric pressure) to C_5 (20 atmospheres) to C_3 (40 atmospheres) (Figure 3). This can be explained in terms of a high degree of incorporation of the primary product, ethylene, into the growing chains on the catalyst surface. An increase in the synthesis gas pressure leads to a lower partial pressure of ethylene favouring chain growth by insertion of one carbon units as in reaction (18) [118a].

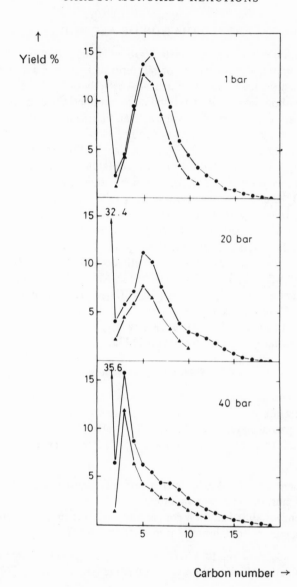

Fig. 3. Product distribution in Fischer-Tropsch synthesis using $\{Si\}-(CH_2)_3-NH_3^+$ [FeCo$_3$(CO)$_{12}$]$^-$ as the catalyst precursor at 1, 20 and 40 atmospheres of CO + H$_2$ (1 : 2): • total hydrocarbons, ▲ olefins. (Reproduced with permission from [118a]).

8.4.3. FISCHER-TROPSCH FORMATION OF ALCOHOLS

Rhodium, platinum and iridium cluster carbonyls supported on MgO, La_2O_3 and ZrO_2 promoted the conversion of mixtures of hydrogen and carbon monoxide to methanol with between 65 and 98% selectivity [122, 122a]. Methane, ethanol and carbon dioxide were formed as by-products. The selectivity of the rhodium clusters arose from the lower activation energy for the formation of methanol (58.5 ± 4 kJ. mol^{-1}) as opposed to 113 ± 8 kJ. mol^{-1} for the formation of methane, ethanol and C_2-C_4 hydrocarbons [122]. When Rh_4 to Rh_{13} cluster carbonyls were adsorbed on to silica containing either ZrO_2 or TiO_2 the resulting catalysts promoted the formation of ethanol as the major product. ZrO_2 and TiO_2 decreased the activation energy for ethanol formation to about 16 kcal/mole. It is suggested that their role is through their basic sites, firstly to prevent higher aggregation of the supported rhodium species and secondly to promote both carbon-oxygen bond cleavage and CO insertion into a rhodium-carbon bond to form the acyl presursor to ethanol [123]. Silica, ZrO_2 and NaY zeolites treated with M^+naphthalide$^-$ (M^+ = Li^+, Na^+ or K^+), to eliminate surface acidity, function as novel supports for $[Rh_4(CO)_{12}]$. The resulting catalysts give greater than 90% selectivity in the conversion of mixtures of carbon monoxide and hydrogen to methanol, whereas $[Rh_4(CO)_{12}]$ supported on basic oxides such as MgO promotes methane formation [123a].

8.5. Water Gas Shift Reaction

In most commercial processes for the production of hydrogen from hydrocarbons, a mixture of carbon monoxide and hydrogen is produced by steam reforming or partial oxidation. The so-called water gas shift reaction (reaction (3)) is then performed to maximise the yield of hydrogen. Typically, this is done in two stages:

(i) Treatment at $350-450°C$ over a chromium-based catalyst gives rapid conversion of most of the carbon monoxide.

(ii) A second catalyst based on copper and zinc is then used at $200-300°C$, at which temperature equilibrium 3 is more favourable for complete conversion.

$[Rh_6(CO)_{16}]$ adsorbed on to silica, alumina or magnesium oxide undergoes reaction with surface hydroxyl groups on heating (reaction (20)) [124–126a].

$$[Rh_6(CO)_{16}] + 6\,M—OH \longrightarrow 3 \begin{bmatrix} OC \diagdown \quad \diagup CO \; OC \diagdown \quad \diagup CO \\ \quad Rh \qquad \qquad Rh \\ O \diagup \quad \diagdown O \diagup \quad \diagdown O \\ | \qquad \quad | \qquad \quad | \\ M \qquad \quad M \qquad \quad M \end{bmatrix} \qquad (20)$$

(M = Si, Al, Mg)

On treatment with either hydrogen at 100°C or excess water a smooth intramolecular rearrangement of the carbonyl groups occurs to yield **7**, which on further treatment with oxygen yields **8**. With carbon monoxide **8** initially forms **9** which at 200°C reforms **7** and carbon dioxide (reaction (22)).

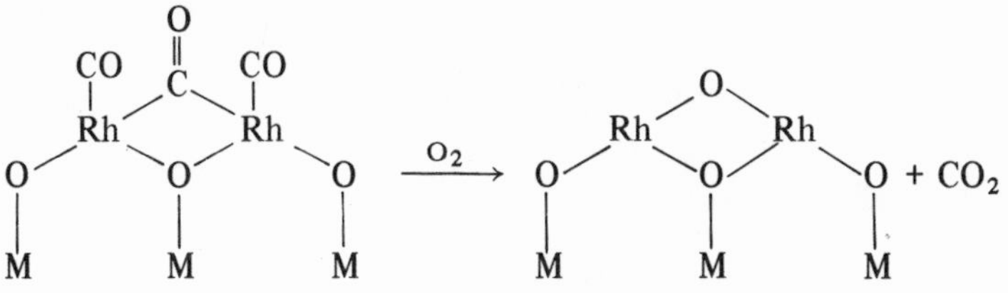

$$\text{(21)}$$

7   **8**

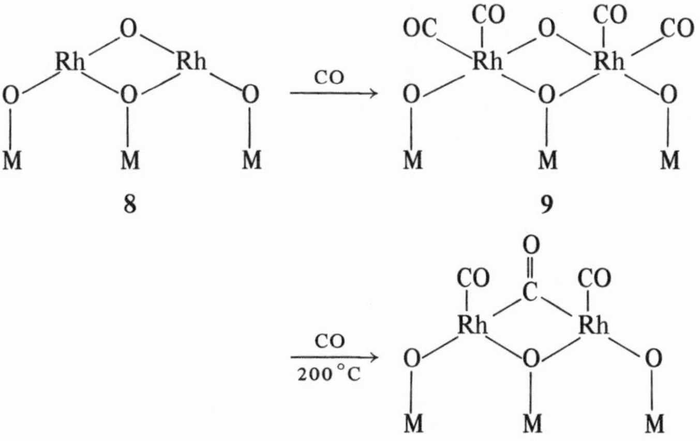

8   **9**

$$\text{(22)}$$

7

The net result of these reactions is a water gas shift reaction catalyst, shown schematically in Scheme 3. Both cycles in Scheme 3 occur readily on alumina and magnesium oxide [124]; the lower cycle occurs less readily on silica [125]. The transition from upper to lower cycle corresponds in fact to a transition from a highly dispersed rhodium(I) carbonyl species to a particle of metallic rhodium covered with linear and bridging carbonyl groups. Once the metallic particle is formed, the lower cycle can be repeated several times without any detectable change of nuclearity.

At room temperature $[Os_3(CO)_{12}]$, $[H_2Os_3(CO)_{10}]$ and $[Os_6(CO)_{18}]$ are weakly physisorbed on silica and alumina [113, 114, 126–130]. On heating to 150°C oxidative addition of M—OH (M = Si, Al) groups to the Os—Os bonds occurs forming $[Os_3H(CO)_{10}(OM)]$ which in turn reacts with carbon monoxide

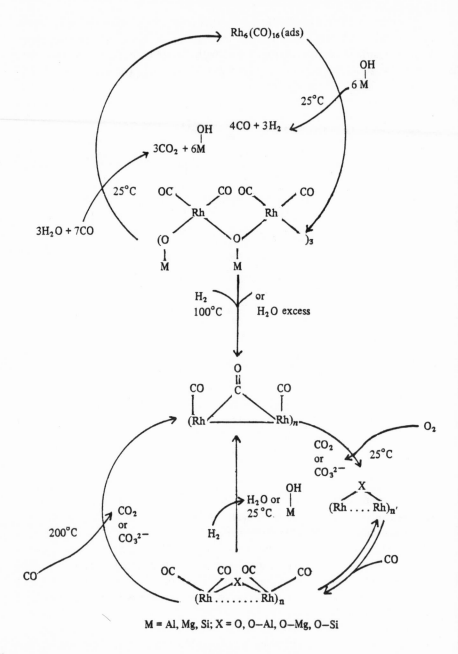

M = Al, Mg, Si; X = O, O–Al, O–Mg, O–Si

Scheme 3

and water to generate $[H_2Os_3(CO)_{10}]$, surface M—OH groups and carbon dioxide, thus effecting the water gas shift reaction. If hydrogen is present during this reaction then Fischer-Tropsch synthesis of hydrocarbons occurs rather than the water gas shift reaction. Osmium and ruthenium cluster carbonyls bound to polystyrene through metal-carbon bonds have been studied as water gas shift reaction catalysts [131], as have $[Ir(NH_3)_6]^{3+}$ and $[Ir(en)_2(NH_3)_2]^{3+}$ supported on 13—X faujasite [131a]. $[Ru(NH_3)_6]^{3+}$ in zeolite Y, activated in a mixture of water and carbon monoxide, is transformed into an active low-temperature water-gas shift catalyst [131b, 131c]. Below 400 K the main ruthenium species formed is $[Ru(NH_3)_5(CO)]^{2+}$, at higher temperatures in carbon monoxide a triscarbonylruthenium(I) complex is formed while in the presence of water a biscarbonylruthenium(I) complex dominates [131d].

$[Rh(P^iPr_3)_2H_2(O_2COH)]$ exchanged on to phosphinated silica containing $\{Si\}—(CH_2)_3PPh_2$ groups is an active catalyst for the hydrogenation of methyl cinnamate using a mixture of carbon monoxide and water as the source of hydrogen [131e]. The catalyst can be repeatedly used with little loss of activity whereas a polystyrene supported analogue is not recoverable due to loss of rhodium.

8.6. Alkoxycarbonylation of Olefins

$[Pd(PPh_3)_2Cl_2]$ supported on phosphinated polymers catalyses the ethoxycarbonylation of 1-pentene in 1 : 1 ethanol/thf under 26.6 atmospheres of carbon monoxide (reaction (23)). The selectivity for ethyl hexanoate formation of these supported catalysts is greater than that of the corresponding homogeneous catalyst [132].

$$C_3H_7CH{=}CH_2 + CO + EtOH \xrightarrow{\text{cat}} C_4H_9CH_2COOEt \qquad (23)$$

8.7. Isocyanates Formed by Carbonylation of Nitro Compounds and Azides

Aromatic nitro compounds react with carbon monoxide in the presence of palladium catalysts to form the corresponding isocyanates (reaction (24)).

$$ArNO_2 + 3\,CO \longrightarrow ArNCO + 2\,CO_2 \qquad (24)$$

This unusual reaction is of interest because diisocyanates are used to prepare polyurethane foams and elastomers. $[Pd_2Mo_2Cp_2(CO)_6(PPh_3)_2]$ supported on alumina has been compared with the well-known $PdCl_2$/pyridine/MoO_3 catalyst [133] which is difficult to recover because, after the reaction, the metal complexes are all in solution. The supported bimetallic complex is readily recovered and overcomes the usual inverse relation between activity and selectivity (selectivity is defined as the proportion of nitrobenzene consumed that is converted to phenylisocyanate) [134]. Catalysts that more closely resemble the homogeneous $PdCl_2$/pyridine/MoO_3 system but in which the pyridine functional group has been incorporated into a polymer have been described [135, 135a].

[RhCl(CO){$Ph_2PCH_2CH_2Si(OEt)_3$}$_2$] bonded to glass catalysed the carbonylation of aromatic azides to isocyanate (reaction (25)) [136].

$$EtOC-\langle\_\rangle-N_3 + CO \xrightarrow{Rh^I \ cat} EtOC-\langle\_\rangle-NCO + N_2 \qquad (25)$$
with O below the $EtOC$ carbonyls

8.8. Syntheses of Aldehydes and Ketones

For convenience a number of syntheses of aldehydes and ketones are collected together in this Section although not all involve carbon monoxide.

Amberlyst A-26 treated with $K^+[FeH(CO)_4]^-$, prepared from potassium hydroxide and iron pentacarbonyl (reaction (26)) converts alkyl halides to aldehydes (reaction (27)) [137]. This reaction involves both carbonylation and dehalogenation, and the same catalyst has been shown to be effective for dehalogenation alone as in reaction (28).

$$\text{P}-\langle\_\rangle-CH_2N^+Me_3Cl^- + K^+[FeH(CO)_4]^-$$

$$\longrightarrow (\text{P}-\langle\_\rangle-CH_2NMe_3)^+[Fe(CO)_4]^- + KCl \qquad (26)$$

$$RX \xrightarrow[\text{thf, reflux}]{(\text{P}-\langle\_\rangle-CH_2NMe_3)^+[FeH(CO)_4]^-} RCHO \qquad (27)$$

$$PhCOCH_2Br \xrightarrow[\text{thf, room temperature, 5 hrs}]{(\text{P}-\langle\_\rangle-CH_2NMe_3)^+[FeH(CO)_4]^-} PhCOCH_3 \qquad (28)$$

Phosphinated, cross-linked, polystyrene-supported rhodium(I), [(P—PPh_2)$_2$-RhCl(CO)] promotes the selective addition of alkyl lithium to acyl chlorides to form ketones in the presence of cyano, aldehyde and ester functional groups [138]. The polymeric rhodium complex is first treated with a lithium reagent to generate a supported alkyl-rhodium(I) complex at low temperature. On addition of the acid chloride oxidative addition occurs, so that, on warming, the ketone can be formed by reductive elimination with regeneration of the original supported complex (Scheme (4)).

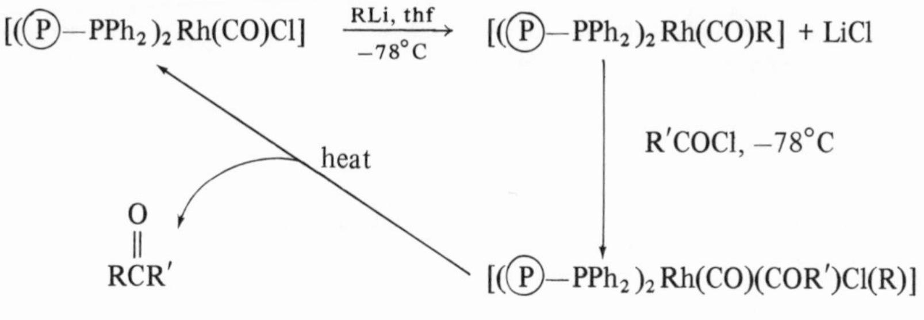

$$[(\textcircled{P}-PPh_2)_2 Rh(CO)Cl] \xrightarrow[-78°C]{RLi, thf} [(\textcircled{P}-PPh_2)_2 Rh(CO)R] + LiCl$$

$$R'COCl, -78°C$$

$$\begin{matrix} O \\ \parallel \\ RCR' \end{matrix}$$

heat

$$[(\textcircled{P}-PPh_2)_2 Rh(CO)(COR')Cl(R)]$$

Scheme 4

The reaction of aluminium trichloride with sulphonated polystyrene yields $[\textcircled{P}-SO_2OAlCl_2]$, an 'organometallic superacid' which may find use as an insoluble Friedel-Crafts catalyst; it is also active as a cracking and isomerisation catalyst for normal alkanes [139]. A study of polystyrene-supported molybdenum tricarbonyl, prepared as in reaction (29), in the Friedel-Crafts alkylation and acylation of benzene derivatives showed that, whilst it was less active than its homogeneous analogue $[Mo(PhCH_3)(CO)_3]$, its selectivity towards *para*-substitution was high and it could be reused several times. No leaching was detected [140].

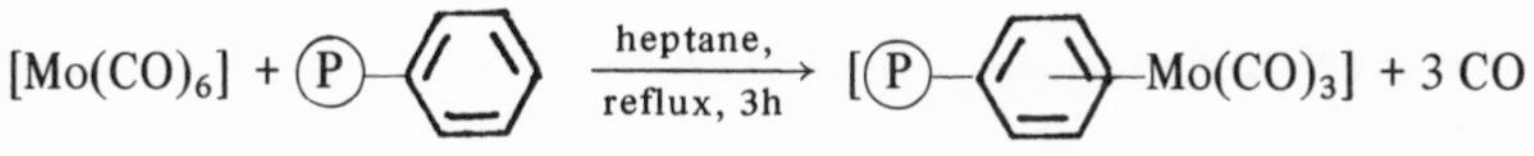

$$[Mo(CO)_6] + \textcircled{P}-\langle \rangle \xrightarrow[\text{reflux, 3h}]{\text{heptane,}} [\textcircled{P}-\langle \rangle -Mo(CO)_3] + 3\, CO \quad (29)$$

8.9. Substitution of Carbonyl Ligands in Metal Carbonyls

Iron pentacarbonyl and substituted iron carbonyls undergo substitution of their carbonyl ligands by isocyanides under relatively mild conditions in the presence of rhodium(I) catalysts (reaction (30)). By using a phosphinated polystyrene supported $[RhCl(PPh_3)_3]$ catalyst which had been thoroughly soxhlet extracted to remove any rhodium(I) that was not firmly bound it was possible to avoid the necessity of the difficult separation of the iron and rhodium complexes at the end of the reaction [141].

$$[Fe(CO)_4(\text{maleic anhydride}] + t\text{-BuNC}$$

$$\xrightarrow[n = 1-3]{Rh(I)cat} [Fe(CO)_{4-n}(t\text{-BuNC})_n(\text{maleic anhydride})] \quad (30)$$

References

1. J. Falbe: *Carbon Monoxide in Organic Synthesis*, Springer-Verlag, Berlin (1970).
2. J. Falbe (Ed.): *New Syntheses with Carbon Monoxide*, Springer-Verlag, Berlin (1980).
2a. D. L. King, J. A. Cusumano and R. L. Garten: *Cat. Rev. Sci. Eng.* **23**, 233 (1981).
2b. J. Evans: *Chem. Soc. Rev.* **10**, 159 (1981).
3. D. T. Thompson in: *Catalysis and Chemical Processes* (Ed. by R. Pearce and W. R. Patterson) Blackie and Son, Glasgow, Chapter 8 (1981).
4. G. W. Parshall: *Homogeneous Catalysis*, John Wiley and Sons, New York, Chapter 5 (1980).
4a. P. J. Davidson, R. R. Hignett and D. T. Thompson in: 'Catalysis', *Chem. Soc. Spec. Per. Rep.* **1**, 369 (1977).
4b. R. A. Sheldon: *Chemicals from Synthesis Gas*, D. Reidel. Dordrecht, Holland (1983).
4c. W. Keim (Ed.): *Catalysis in C_1 Chemistry*, D. Reidel, Dordrecht, Holland (1983).
5. B. Cornils: Chapter 1 of Reference 2.
6. C. Masters: *Homogeneous Transition-Metal Catalysis*, Chapman and Hall, London, (a) pp. 102–135, (b) pp. 227–239 (1981).
7. C. E. Loeffler, L. Strautzenberger and J. D. Unruh in: *Encyclopaedia of Chemical Processing and Design* (Ed. by J. J. McKetta and W. A. Cunningham) Marcel Dekker, Vol. 5, p. 358 (1977).
8. W. O. Haag and D. D. Whitehurst: *Proc. 2nd North American Meeting of the Catalysis Society*, Houston, Texas, pp. 16–18 (1971).
9. W. O. Haag and D. D. Whitehurst: *Proc. 5th Int. Congr. Catal* ('Catalysis, Vol. 1' (Ed. by J. H. Hightower) North-Holland, Amsterdam p. 465 (1973).
10. W. O. Haag and D. D. Whitehurst: *US Patent* 4,098,727 (1978); *Chem. Abs.* **91**, 19901 (1979).
11. L. H. Slaugh and R. D. Mullineaux: *J. Organometal. Chem.* **13**, 469 (1968).
12. A. T. Jurewicz, L. D. Rollmann and D. D. Whitehurst: *Adv. Chem. Ser.* **132**, 240 (1974).
13. B. Heil and L. Markó: *Acta Chim. Acad. Sci. Hung.* **55**, 107 (1968); *Chem. Abs.* **68**, 77430 (1968).
14. L. Markó: *Proc. Chem. Soc.* 67 (1962).
15. A. J. Moffat: *J. Catal.* **18**, 193 (1970).
16. A. J. Moffat: *J. Catal.* **19**, 322 (1970).
17. F. R. Hartley, D. J. A. McCaffrey, S. G. Murray and P. N. Nicholson: *J. Organometal. Chem.* **206**, 347 (1981).
18. A. Gupta, A. Rembaum and H. B. Gray in: *Organometallic Polymers* (Ed. by C. E. Carraher, J. E. Sheats and C. U. Pittman) Academic Press, New York (1978).
19. S. C. Tang, T. E. Paxson and L. Kim: *J. Mol. Catal.* **9**, 313 (1980).
20. G. Gubitosa, M. Boldt and H. H. Brintzinger: *J. Amer. Chem. Soc.* **99**, 5174 (1977).
21. G. Gubitosa and H. H. Brintzinger: *Colloq. Int. CNRS* **281**, 173 (1978); *Chem. Abs.* **91**, 17442 (1979).
21a. L. Verdet and J. K. Stille: *Organometallics* **1**, 380 (1982).
22. W. Cornely and B. Fell: *J. Mol. Catal.* **16**, 89 (1982).
23. L. J. Boucher, A. A. Oswald and L. L. Murrell: *Prepr. Div. Petr. Chem., Amer. Chem. Soc.* **19**, 162 (1974); *Chem. Abs.* **83**, 198255 (1975).
24. V. A. Semikolenov, B. L. Moroz, V. A. Likholobov and Yu. I. Yermakov: *React. Kinet. Catal. Lett.* **18**, 341 (1981).

24a. H. Fu. Y. Luo, Z. Yang, Y. Wang, N. Wu and A. Zhang: *Youji Huaxue* 6, 421 (1981); *Chem. Abs.* 96, 85009 (1982).
25. G. O. Evans, C. U. Pittman, R. McMillan, R. T. Beach and R. Jones: *J. Organometal. Chem.* 67, 295 (1974).
26. M. Ichikawa: *J. Catal.* 56, 127 (1979).
27. M. Ichikawa: *J. Catal.* 59, 67 (1979).
28. G. E. Hartwell and P. E. Garrou: *US Patent* 4,144,191 (1979); *Chem. Abs.* 90, 203478 (1979).
29. Dow Chemical Co., *Japanese Patent* 80109447 (1980); *Chem. Abs.* 94, 53632 (1981).
30. G. E. Hartwell and P. E. Garrou: *Brazilian Patent* 7901172 (1980); *Chem. Abs.* 94, 139208 (1981).
31. P. R. Rony and J. F. Roth: *J. Mol. Catal.* 1, 13 (1975).
31a. B. L. Moroz, V. A. Semikolenov, V. A. Likholobov and Yu. I. Yermakov: *JCS Chem. Commun.* 1286 (1982).
32. K. G. Allum, R. D. Hancock, S. McKenzie and R. C. Pitkethly: *Proc. 5th Int. Cong. Catal. (Catalysis, Vol. 1*: Ed. by J. H. Hightower) North-Holland, Amsterdam, p. 477 (1973).
33. F. R. Hartley, S. G. Murray and P. N. Nicholson: *J. Mol. Catal.* 16, 363 (1982).
34. F. R. Hartley, S. G. Murray and P. N. Nicholson: *J. Polymer Sci., Polymer Chem. Ed.* 20, 2395 (1982).
35. C. U. Pittman and R. Hanes: *Ann. New York Acad. Sci.* 239, 76 (1974).
36. C. U. Pittman, A. Hirao, C. Jones, R. M. Hanes and Q. Ng: *Ann. New York Acad. Sci.* 295, 15 (1977).
37. C. U. Pittman and W. D. Honnick: *J. Org. Chem.* 45, 2132 (1980).
38. C. U. Pittman, W. D. Honnick and J. J. Yang: *J. Org. Chem.* 45, 684 (1980).
39. C. U. Pittman, L. R. Smith and R. M. Hanes: *J. Amer. Chem. Soc.* 97, 1742 (1975).
40. D. Evans, J. A. Osborn and G. Wilkinson: *J. Chem. Soc. (A)*, 3133 (1968).
41. A. R. Sanger: *Preprints of 5th Canadian Symposium on Catalysis*, Calgary, p. 281 (1977); *Chem. Abs.* 89, 179250 (1978).
42. A. R. Sanger and L. R. Schallig: *J. Mol. Catal.* 3, 101 (1977).
43. W. H. Lang, A. T. Jurewicz, W. O. Haag, D. D. Whitehurst and L. D. Rollman: *Organometallic Polymers* (Ed. by C. E. Carraher, J. E. Sheats and C. U. Pittman) Academic Press, New York, p. 145 (1978).
44. W. H. Lang, A. T. Jurewicz, W. O. Haag, D. D. Whitehurst and L. D. Rollman: *J. Organometal. Chem.* 134, 85 (1977).
45. K. G. Allum, R. D. Hancock, I. V. Howell, R. C. Pitkethly and P. J. Robinson: *J. Catal.* 43, 322 (1976).
46. C. U. Pittman, S. E. Jacobson, R. M. Hanes, L. R. Smith and G. Wilemon: *7th Int. Conf. Organometal. Chem.*, Venice, p. 187 (1975).
47. C. U. Pittman and R. M. Hanes: *J. Amer. Chem. Soc.* 98, 5402 (1976).
48. M. Matsumoto and M. Tamura: *J. Mol. Catal.* 16, 195 (1982).
49. M. Matsumoto and M. Tamura: *J. Mol. Catal.* 16, 209 (1982).
50. M. Matsumoto and M. Tamura: *UK Patent Appl.* 2056874A (1980).
51. M. Matsumoto and M. Tamura: *UK Patent Appl.* 2014138A (1980).
52. C. U. Pittman and A. Hirao: *J. Org. Chem.* 43, 640 (1978).
53. C. U. Pittman and C.-C. Lin: *J. Org. Chem.* 43, 4928 (1978).
54. E. Bayer and V. Schurig: *Angew. Chem. Int. Ed.* 14, 493 (1975).
55. V. Schurig and E. Bayer: *ChemTech.* 6, 212 (1976).

56. M. O. Farrell: *Diss. Abs.* **B39**, 1754 (1978).
57. R. D. Hancock, I. V. Howell, R. C. Pitkethly and P. J. Robinson in: *Catalysis, Heterogeneous and Homogeneous* (Ed. by B. Delmon and G. Jannes) Elsevier, p. 361 (1975).
58. L. F. Wieserman: *Diss. Abs.* **B42**, 3253 (1982).
58a. F. Jiao, W. Xie, Z. Ma, Y. Yin and Z. Yang: *Ranliao Huaxue Xuebao* 9, 47 (1981); *Chem. Abs.* 96, 52483 (1982).
58b. A. Luchetti and D. M. Hercules: *J. Mol. Catal.* 16, 95 (1982).
59. P. Panster, W. Bender and P. Kleinschmit: *German Patent* 2834691 (1980); *Chem. Abs.* 93, 32413 (1980).
60. A. T. Jurewicz, L. D. Rollmann and D. D. Whitehurst: *Adv. Chem. Ser.* 132, 240 (1974).
60a. R. T. Smith, R. K. Ungar and M. C. Baird: *Trans. Met. Chem.* 7, 288 (1982).
61. H. Hirai, S. Komatouzaki, S. Hamasaki and N. Toshima: *Nippon Kagaku Kaishi*, 316 (1982); *Chem. Abs.* 96, 199018 (1982).
62. A. Sekiya and J. K. Stille: *J. Amer. Chem. Soc.* 103, 5096 (1981).
63. H. Krausę and H. Mix: *East German Patent* 133199 (1978); *Chem. Abs.* 91, 63336 (1979).
64. A. Bortinger: *Diss. Abs.* 38, 1714 (1977).
65. C. U. Pittman, A. Hirao, J. J. Y. Q. Ng, R. Hanes and C.-C. Lin: *Preprints Div. Petrol. Chem. Amer. Chem. Soc.* 22, 1196 (1977).
66. C. U. Pittman, N. Quoc, A. Hirao, W. Honnick and R. Hanes: *Colloq. Int. CNRS* 281, 49 (1977); *Chem. Abs.* 92, 136004 (1980).
67. S. J. Fritschel, J. J. H. Ackerman, T. Keyser and J. K. Stille: *J. Org. Chem.* 44, 3152 (1979).
68. J. K. Stille: *27th Int. Macromol. Symp.* (Ed. by H. Benoit and P. Rempp) Pergamon, Oxford, p. 99 (1982).
69. C. U. Pittman, Y. Kawabata and L. I. Flowers: *J. C. S. Chem. Commun.* 473 (1982).
70. C.-P. Lau: *Diss. Abs.* **B38**, 4804 (1978).
71. C. U. Pittman and G. Wilemon: *J. Org. Chem.* 46, 1901 (1981).
72. P. F. Jackson, B. F. G. Johnson, J. Lewis, R. Ganzerla, M. Lenarda and·M. Graziani: *J. Organometal. Chem.* 190, C1 (1980).
73. T. Mason, D. Grote and B. Trivedi: *Catalysis in Organic Synthesis* (Ed. by G. V. Smith) Academic Press, New York, p. 165 (1977).
74. R. F. Love, E. R. Kerr and J. F. Knifton: *US Patent* 4,147,730 (1979); *Chem. Abs.* 91, 38926 (1979).
75. P. L. Ragg: *UK Patent* 1,249,033 (1971).
76. D. Forster: *Adv. Organometal. Chem.* 17, 255 (1979).
77. F. E. Paulik and J. F. Roth: *J.C.S. Chem. Commun.* 1578 (1968).
78. J. F. Roth, J. H. Craddock, A. Hershman and F. E. Paulik: *ChemTech.* 600 (1971).
79. M. S. Scurrell: *Platinum Metals Review* 21, 92 (1977).
80. M. S. Jarrell and B. C. Gates: *J. Catal.* 40, 255 (1975).
81. S. Franks, F. R. Hartley and J. R. Chipperfield: *Inorg. Chem.* 20, 3238 (1981).
82. K. M. Webber, B. C. Gates and W. Drenth: *J. Mol. Catal.* 3, 1 (1977).
83. B. C. Gates in: *Chemistry and Chemical Engineering of Catalytic Processes* (Ed. by R. Prins and G. C. A. Schuit) Sijthoff and Noordhoff, The Netherlands, p. 437 (1980).
84. K. M. Webber, B. C. Gates and W. Drenth: *J. Catal.* 47, 269 (1977).
85. R. S. Drago, E. D. Nyberg, A. E. A'mma and A. Zombeck: *Inorg. Chem.* 20, 641 (1981).
86. R. G. Schultz and P. D. Montgomery: *J. Catal.* 13, 105 (1969).

87. R. G. Schultz and P. D. Montgomery: *Amer. Chem. Soc., Div. Petrol. Chem., Preprints* 17, B13 (1972).
88. K. K. Robinson, A. Hershman, J. H. Craddock and J. F. Roth: *J. Catal.* 27, 389 (1972).
89. M. S. Scurrell: *Platinum Metals Review* 21, 92 (1977).
90. A. Krzywicki and G. Pannetier: *Bull. Soc. Chim. France* 1093 (1975).
91. M. S. Scurrell: *J. Res. Inst. Catal. (Japan)* 25, 189 (1977).
92. B. K. Nefedov, N. S. Sergeeva, T. V. Zueva, E. M. Shutkina and Ya. T. Eidus: *Izv. Akad. Nauk SSSR, Ser. Khim.* 582 (1976).
93. M. S. Scurrell and R. F. Howe: *J. Mol. Catal.* 7, 535 (1980).
94. M. S. Scurrell: *J. Mol. Catal.* 10, 57 (1981).
94a. S. L. T. Andersson and M. S. Scurrell: *J. Catal.* 71, 233 (1981).
94b. J. C. Conesa, M. T. Sainz, J. Soria, G. Munuera, V. Rives-Arnau and A. Munoz: *J. Mol. Catal.* 17, 231 (1982).
95. P. Sabatier and J. B. Senderens: *Hebd. Seances Acad. Sci.* 134, 514 (1902).
96. F. Fischer and H. Tropsch: *Brennst. Chem.* 4, 276 (1923).
97. F. Fischer and H. Tropsch: *German Patent* 484,337 (1925).
97a. M. E. Dry and J. C. Hoogendoorn: *Catal. Rev. Sci. Eng.* 23, 265 (1981).
98. G. Henrici-Olivé and S. Olivé: *Angew. Chem. Int. Ed.* 15, 136 (1976).
99. M. A. Vannice: *Catal. Rev. Sci. Eng.* 14, 153 (1976).
100. I. Wender: *Catal. Rev. Sci. Eng.* 14, 97 (1976).
101. C. D. Frohning in: *New Syntheses with Carbon Monoxide* (Ed. by J. Falbe) Springer-Verlag Berlin, p. 309 (1981).
101a. V. Ponec in: *Catalysis, Chem. Soc. Spec. Per. Rep.* 5, 48 (1982).
102. C. Masters: *Adv. Organometal. Chem.* 17, 61 (1979).
103. R. Nakamura, R. G. Bowman and R. L. Burwell: *J. Amer. Chem. Soc.* 103, 673 (1981).
104. R. G. Bowman and R. L. Burwell: *J. Catal.* 63, 463 (1980).
105. D. Ballivet-Tkatchenko, N. D. Chau, H. Mozzanega, M. C. Roux and I. Tkatchenko: *Amer. Chem. Soc., Symp. Ser.* 152, 187 (1981).
106. A. L. Lapidus, A. Yu. Krylova and L. T. Kondrat'ev: *Izv. Akad. Nauk SSSR, Ser. Khim.* 1432 (1980).
107. A. L. Lapidus, A. Yu. Krylova and L. T. Kondrat'ev: *Neftekhimiya* 21, 397 (1981).
107a. K. Lazar, Z. Schay and L. Guczi: *J. Mol. Catal.* 17, 205 (1982).
108. D. Vanhoue, P. Makambo and M. Blanchard: *J.C.S. Chem. Commun.* 605 (1979).
109. M. Deeba, J. P. Scott, R. Barth and B. C. Gates: *J. Catal.* 71, 373 (1981).
110. V. L. Kuznetsov, A. T. Bell and Yu. I. Yermakov: *J. Catal.* 65, 374 (1980).
111. J. Robertson and G. Webb: *Proc. Roy. Soc. (London)* A341, 383 (1974).
112. J. G. Goodwin and C. Nacoache: *J. Mol. Catal.* 14, 259 (1982).
113. A. K. Smith, A. Theolier, J. M. Basset, R. Ugo, D. Commereuc and Y. Chauvin: *J. Amer. Chem. Soc.* 100, 2590 (1978).
114. R. Psaro, R. Ugo, G. M. Zanderighi, B. Besson, A. K. Smith and J. M. Basset: *J. Organometal Chem.* 213, 215 (1981).
115. K. P. C. Vollhardt and P. Perkins, *US Patent Appl.* 39,986 (1979); *Chem. Abs.* 94, 128169 (1981).
116. P. Perkins and K. P. C. Vollhardt: *J. Amer. Chem. Soc.* 101, 3985 (1979).
117. L. S. Benner, P. Perkins and K. P. C. Vollhardt: *Amer. Chem. Soc., Symp. Ser.* 152, 165 (1981).

118. F. Hugues, A. K. Smith, Y. B. Taarit, J. M. Basset, D. Commereuc and Y. Chauvin: *J.C.S. Chem. Commun.* 68 (1980).

118a. R. Hemmerich, W. Keim and M. Roeper: *J.C.S. Chem. Commun.* 428 (1983).

119. F. Hugues, B. Besson and J. M. Basset: *J.C.S. Chem. Commun.* 719 (1980).

120. F. Hugues, P. Bussiere, J. M. Basset, D. Commereuc, Y. Chauvin, L. Bonneviot and D. Oliver: *Stud. Surf. Sci. Catal.* 7A, 418 (1981).

121. F. Hugues, B. Besson, P. Bussiere, J.-A. Dalmon, M. Leconte, J. M. Basset, Y. Chauvin and D. Commereuc: *Amer. Chem. Soc., Symp. Ser.* 192, 255 (1982).

122. M. Ichikawa and K. Shikakura: *Stud. Surf. Sci. Catal.* 7B, 925 (1981).

122a. M. Ichikawa: *ChemTech.* 12, 674 (1982).

123. M. Ichikawa, K. Sekizawa, K. Shikakura and M. Kawai: *J. Mol. Catal.* 11, 167 (1981).

123a. H. A. Dirkse, P. W. Lednor and P. C. Versloot: *J.C.S. Chem. Commun.* 814 (1982).

124. A. K. Smith, F. Hugues, A. Theolier, J. M. Basset, R. Ugo, G. M. Zanderighi, J. L. Bilhou, V. Bilhou-Bougnol and W. F. Graydon: *Inorg. Chem.* 18, 3104 (1979).

125. J. L. Bilhou, V. Bilhou-Bougnol, W. F. Graydon, A. K. Smith, G. M. Zanderighi, J. M. Basset and R. Ugo: *J. Organometal. Chem.* 153, 73 (1978).

126. R. Ugo, R. Psaro, G. M. Zanderighi, J. M. Basset, A. Theolier and A. K. Smith in: *Fundamental Research in Homogeneous Catalysis* (Ed. by M. Tsutsui) Plenum Press, N.Y. 3, 579 (1979).

126a. J. M. Basset, B. Besson, A. Choplin and A. Theolier: *Phil. Trans. Roy. Soc. London* A308, 115 (1982).

127. P. L. Watson and G. L. Schrader: *J. Mol. Catal.* 9, 129 (1980).

128. B. Besson, B. Morawek, A. K. Smith, J. M. Basset, R. Psaro, A. Fusi and R, Ugo, *J.C.S. Chem. Commun.* 569 (1980).

129. A. K. Smith, B. Besson, J. M. Basset, R. Psaro, A. Fusi and R. Ugo: *J. Organometal. Chem.* 192, C31 (1980).

130. D. J. Hunt, S. D. Jackson, R. B. Moyes, P. B. Wells and R. Whyman: *J.C.S. Chem. Commun.* 85 (1982).

131. S. Bhaduri, H. Khwaja and K. R. Sharma: *Indian J. Chem.* 21A, 155 (1982).

131a. R. Ganzerla, F. Pinna, M. Lenarda and M. Graziani: *J. Organometal. Chem.* 244, 183 (1983).

131b. J. J. Verdonck and P. A. Jacobs: *J.C.S. Chem. Commun.* 18 (1979).

131c. J. J. Verdonck, R. A. Schoonheydt and P. A. Jacobs in: *Catalysis* (Ed. by T. Seiyama and K. Tanabe) Elsevier, Amsterdam, part B, p. 911 (1980).

131d. J. J. Verdonck, R. A. Schoonheydt and P. A. Jacobs: *J. Phys. Chem.* 87, 683 (1983).

131e. T. Okano, T. Kobayashi, H. Konishi and J. Kiji: *Bull. Chem. Soc. Japan* 55, 2675 (1982).

132. C. U. Pittman and Q. Y. Ng: *US Patent* 4,258,206 (1981); *Chem. Abs.* 94, 208332 (1981).

133. J. D. McClure and G. W. Conklin: *German Patent* 2,165,355 (1972); *Chem. Abs.* 77, 114022 (1972).

134. R. Braunstein, R. Bender and J. Kervennal: *Organometallics* 1, 1236 (1982).

135. T. Yamahara and M. Usui: *US Patent* 3,925,436 (1975).

135a. B. K. Nefedov, V. I. Manov-Yuvenskii, V. A. Semikolenov, V. A. Likholobov and Yu. I. Yermakov: *Kinet. Katal.* 23, 1001 (1982).

136. K. Schwetlick, K. Unverferth, R. Tietz and J. Pelz: *Proc. Symp. Rhodium in Homog. Catal.*, Veszprem, Hungary, p. 14 (1978); *Chem. Abs.* 90, 61803 (1979).

137. G. Cainelli, F. Manescalchi, A. Umani-Ronchi and M. Panunzio: *J. Org. Chem.* 43, 1598 (1978).

138. C. U. Pittman and R. M. Hanes: *J. Org. Chem.* **42**, 1194 (1977).
139. V. L. Magnotta, B. C. Gates and G. C. A. Schuit: *J.C.S. Chem. Commun* 342 (1976).
140. C. P. Tsonis and M. F. Farona: *J. Organometal. Chem.* **114**, 293 (1976).
141. M. O. Albers, N. J. Coville, C. P. Nicolaides and R. A. Webber: *J. Organometal. Chem.* **217**, 247 (1981).

CHAPTER 9

DIMERISATION, OLIGOMERISATION, POLYMERISATION, DISPROPORTIONATION AND ISOMERISATION OF OLEFINS AND ACETYLENES

The present chapter is concerned with reactions involving olefins and acetylenes without the presence of extra reagents. The early Sections are concerned with dimerisation, oligomerisation and polymerisation, all of which involve the creation of new carbon—carbon bonds and require catalysts. These reactions, particularly polymerisation, are major commercial reactions, which has spurred interest in looking for new and better catalysts. In describing developments in these reactions we have not attempted to be comprehensive, but rather to give an outline of more recent developments together with associated references which the reader can use as an introduction to the more recent literature.

Olefin metathesis represents a remarkable example of carbon—carbon bond breaking and bond making, although supported metal complex catalysts have found less application than the complex supported oxides. The latter part of the chapter is concerned with olefin isomerisation, a reaction that is of widespread importance both when it occurs intentionally and unintentionally during the course of other reactions such as hydroformylation. Recent developments in the valence isomerisation of quadricyclane and norbornadiene, a system which may have applications in solar energy storage, are described. The chapter concludes with two examples of the use of supported catalysts to promote Grignard cross-coupling reactions, which are included here because they are concerned with carbon—carbon bond formation.

9.1. Olefin Dimerisation

Olefin dimerisation is one of the simplest and most widely studied oligomerisation reactions. Although ethylene dimerisation has been widely investigated it is of limited commercial value because the resulting butenes are generally cheaper than ethylene itself [1a]. However dimerisation of propylene is a valuable way of converting an inexpensive material into such useful C_6 compounds as dimethylbutenes. Supported olefin dimerisation catalysts have generally been based on nickel or palladium. Catalysts have involved, for example $[Ni(acac)_2)]$ $-$ Et_2AlCl $-$ PPh_3, or $[Ni(cod)_2]$ $-$ Et_2AlCl or $[Ni(CO)_4]$ $-$ $EtAl_2Cl_3$ $-$ PPh_3 supported on metal oxides such as alumina or silica [2], polymers with 4-vinylpyridine grafted side chains [3–4], aminophosphines such as I where R = Me, Bu, Ph and CH_2Ph [5] and phosphinated polystyrenes [6].

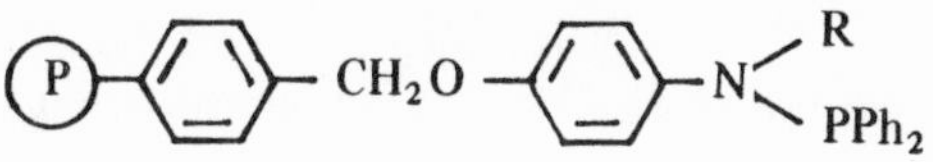

1

A further series of catalysts has been prepared by the oxidative-addition of nickel(0) complexes to halogenated polystyrenes (reaction (1)) [7–9].

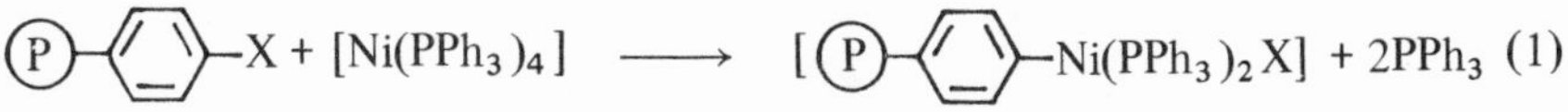

$(X = Br, I)$

These catalysts require activation with $BF_3 \cdot Et_2O$ and water and are then virtually as active as their homogeneous counterparts for ethylene and propylene dimerisation. The solvent fulfils two roles: it swells the polymer and acts as a competing ligand at nickel(II). n-Hexane is the most effective among n-hexane, benzene, dichloromethane and diethylether.

The intimate mechanisms of these nickel catalysed dimerisations are believed to be very similar to those of their homogeneous counterparts. The catalytically active species seems to consist of a square-planar nickel atom with a hydride or alkyl ligand, a phosphine or amine group that provides the link to the support, an electronegative group X and an olefin. The electronegative group X, which may be halide or an aluminium complex such as $R_2Al_2Cl_3$ which is bound to nickel through a chloride bridge, decreases the charge on the nickel. The first stage (reaction (2)) is kinetically controlled and yields initially the Ni-n-propyl

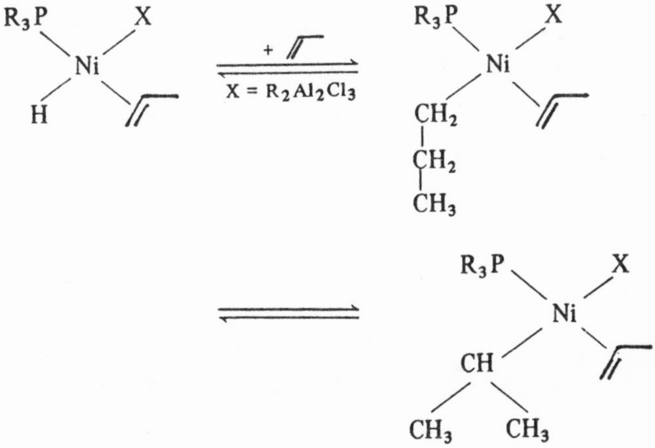

(2)

species which rearranges to the thermodynamically more stable Ni-*i*-propyl species. The same choice exists in the second insertion (Scheme (1)) [10]. The supporting phosphine or amine group exerts a steric influence on the reaction: the first step is largely independent of phosphine, giving a 20 : 80 ration of linear and branched alkyl, but the second step is sensitive to the ligand bulk, more sterically demanding ligands promoting Ni $\rightarrow$ C$_1$ coordination.

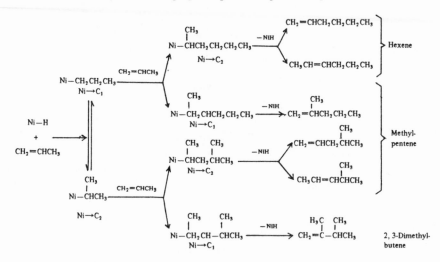

Scheme 1. Stepwise dimerisation of propylene (from [10]).

In addition to olefin dimerisation, supported nickel catalysts have been used for ethylene-propylene codimerisation [11, 12, 12a]. [Pd(dmso)$_2$Cl$_2$] on silica gel promotes ethylene dimerisation [13]. RhCl$_3$ supported on silica gel promotes both ethylene dimerisation, for which it is 10^4 times more active per g atom of rhodium than is its homogeneous counterpart, suggesting the formation of surface bound [{Si}—ORhCl$_2$] groups, and ethylene-propylene codimerisation [14, 15]. The rate-determining step in these reactions is the insertion of the second molecule of olefin.

Reaction of [Ni(η^3-C$_3$H$_5$)$_2$] with silica gel or silica-alumina displaces one η^3-allyl ligand and forms a nickel-oxygen link (reaction (3)).

$$[Ni(\eta^3\text{-}C_3H_5)_2] + M—OH \longrightarrow [M—ONi(\eta^3\text{-}C_3H_5)] + C_3H_6 \qquad (3)$$

(M = silica or alumina)

supported nickel(II) complexes, on activation with aluminium alkyls such as AlR$_n$Cl$_{3-n}$ (R = Me, Et; n = 1, 1.5, 2) catalyse the oligomerisation of propylene primarily to dimers [16–19].

9.2. Olefin Trimerisation

Uncrosslinked phosphinated polystyrene treated with palladium chloride and then a silver salt such as $AgBF_4$, $AgPF_6$ or $AgClO_4$ to remove the chloride ligands gives an active catalyst for the cotrimerisation and codimerisation of ethylene and styrene (reaction (4)) [20].

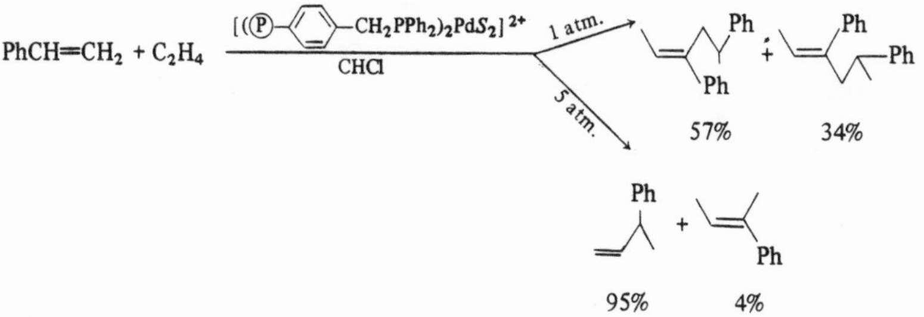

$$PhCH=CH_2 + C_2H_4 \xrightarrow{\quad}$$

(4)

The active catalyst is shown in reaction (4) as having two S ligands; S will be either a solvent molecule or an olefin. Such species have been shown to be remarkably active catalysts because of the ease with which they coordinate to olefins [21–23].

9.3. Oligomerisation and Cyclooligomerisation of Dienes

$[Ni(\eta^3\text{-}C_3H_5)_2]$ supported on alumina or silica catalyses the stereospecific dimerisation of 1,2-butadiene at temperatures below $0°C$ [16]. Phosphinated polystyrene supported nickel carbonyl, $[(P)\text{-}C_6H_4\text{-}PPh_2\text{-}p)_2 Ni(CO)_2]$ swollen in benzene catalysed the cyclooligomerisation of butadiene to give exactly the same product distribution as its homogeneous analogue (reaction (5)) [24].

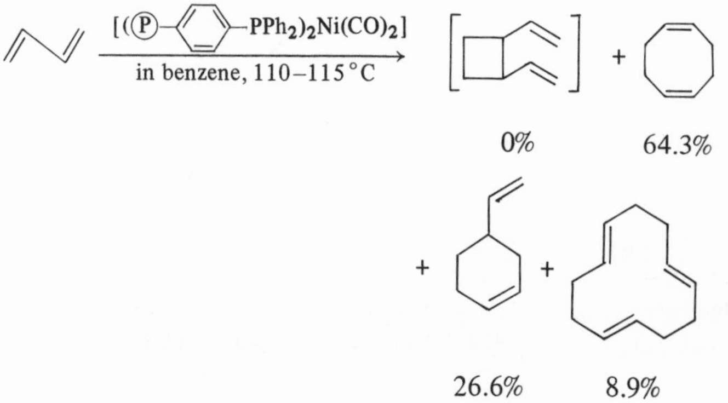

(5)

The molar turnover limits of the supported catalyst, which can be recycled, and the homogeneous catalyst, which cannot, are identical. However the supported catalyst is less active than the homogeneous catalyst, although if the former is used at 115°C it achieves the same rate of reaction as the latter at 90°C.

Palladium acetate anchored to diphenylphosphinated crosslinked polystyrene promotes the oligomerisation of butadiene in thf/HOAc at 100°C (reaction (6)) [25].

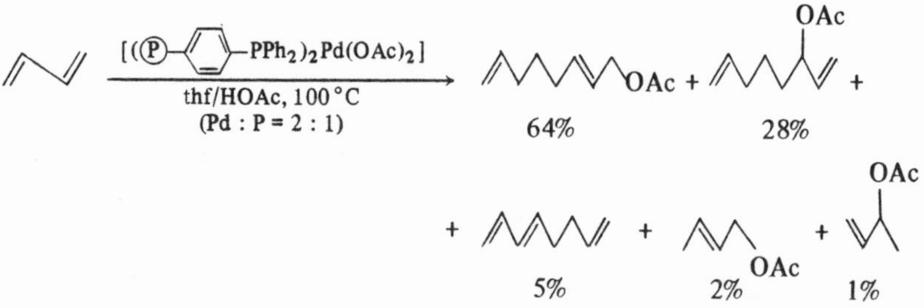

$$\text{\Large(reaction scheme 6)}$$

$$[(\text{P})-\!\!\langle\text{aryl}\rangle\!-PPh_2)_2Pd(OAc)_2]$$
$$\xrightarrow[\substack{thf/HOAc, 100\,°C \\ (Pd : P = 2 : 1)}]{}$$

64% OAc + 28% (with OAc branch) +

+ 5% + 2% (OAc) + 1% (OAc) (6)

The product yield, as with the corresponding homogeneous catalyst varied only modestly with the palladium : phosphorus ratio. The catalysts could be reused and were very resistant to palladium leaching; indeed no mirgration of palladium was observed when unpalladated beads were refluxed with palladated beads [25]. Other palladium species such as $PdCl_2$ and $[Pd(PPh_3)_4]$ on phosphinated polystyrene also promote reaction (6), but palladium leaching is a serious problem in these cases [26]. $[Pd(PPh_3)_4]$ supported on phosphinated polystyrene is more effective than when unsupported in promoting reaction (7).

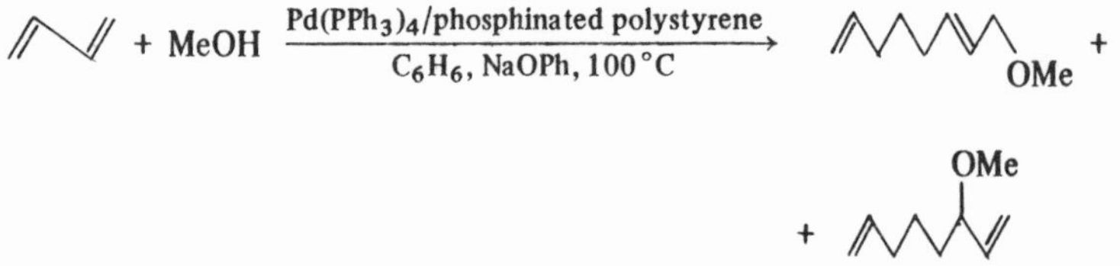

$$\text{(reaction scheme 7)}$$

+ MeOH $\xrightarrow[\substack{C_6H_6, NaOPh, 100\,°C}]{Pd(PPh_3)_4/phosphinated\ polystyrene}$ (product with OMe) +

+ (product with OMe) (7)

This is because supporting the catalyst promotes coordinative unsaturation without the subsequent agglomeration to form inactive species that occurs in homogeneous solution [27–29]. Not only does supporting prevent deactivation, it also allows larger catalyst charges to be used so further accelerating the rate. When the proton donor in the butadiene dimerisations is formic acid, palladium acetate supported on phosphinated polystyrene promotes the formation of 1,7-octadiene in high selectivity (86%–87%) (reaction (8)) [30, 31].

$$/\!\!\!\diagdown\!\!\diagup + HCOOH \xrightarrow[\text{dmf, 90 °C}]{[((P)\!\!-\!\!\langle\diagup\rangle\!\!-\!\!PPh_2)_2Pd(OAc)_2]} /\!\!\!\diagdown\!\!\diagup\!\!\diagdown\!\!\diagup +$$

$$+ /\!\!\!\diagdown\!\!\diagup\!\!\diagdown\!\!\diagup + CO_2 \qquad (8)$$

Phosphinated polystyrene bound palladium(0) catalysts have been used to telomerise butadiene in the presence of amines as well as oxygenated compounds [32]. With secondary amines the 2 : 1 adducts of butadiene and amine are formed selectively in a reaction whose activity decreases with increasing steric bulk about the amine: morpholine > piperidine > Et_2NH > iPr_2NH [32]. Primary amines give 4 : 1 adducts in 60–90% selectivity. Although the products of the direct reaction of nickel, palladium and platinum bis-η^3-allyls with silica and alumina are inactive in promoting the telomerisation of butadiene with diethylamine, active catalysts can be made either by the interation of palladium η^3-allyl and tin acetate with the metal oxide or by introducing a ligand such as $-CS_2^-$, m-pyridyl or p-cyanophenyl onto the surface of the metal oxide and coordinating the palladium to that [33, 34].

9.4. Oligomerisation of Acetylenes

Only a very few reports have described oligomerisations of acetylenes promoted by supported metal complex catalysts. Palladium chloride supported on polymeric $(P)-C_6H_4-CH_2PPh_2$-p catalyses the codimerisation of acetylenes and allylhalides (reaction (9)) without the acetylene polymerisation that accompanies the use of a homogeneous catalyst [35].

$$CH_2=CHCH_2X + PhC\equiv CR \longrightarrow CH_2=CHCH_2CR=CPhX +$$

$$+ CH_2=CHCH_2C(Ph)=CRX \qquad (9)$$

(X = Cl, Br; R = H, Me)

Supported cyclopentadienyl-cobalt- and rhodium-dicarbonyl catalyse the cyclotrimerisation of ethylpropiolate (CH≡CCOOEt) to give a mixture of 1,2,5-tricarbethoxybenzene (17.5%, Rh; 2.4%, Co) and 1,2,4-tricarbethoxy-benzene (52.4%, Rh; 18.1% Co). The catalysts in which the trunk polymer is polystyrene can be reused with only slight loss of activity [36].

Polystyrene supported molybdenum tricarbonyl $((P)-C_6H_4-Mo(CO)_3$-$p)$ promotes the 2 + 2 cycloaddition of phenylacetylene (reaction (10)) and di-phenylacetylene (reaction (11)) in the presence and absence of carbon monoxide [37–39].

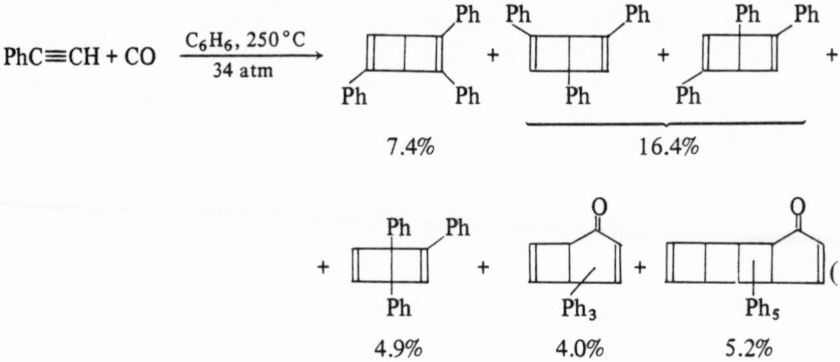
PhC≡CH + CO
C6H6, 250°C
34 atm
Ph
Ph
Ph
Ph
Ph
Ph
Ph
Ph
Ph
7.4%
16.4%
Ph
Ph
Ph
O
Ph3
O
Ph5
4.9%
4.0%
5.2%

(10)

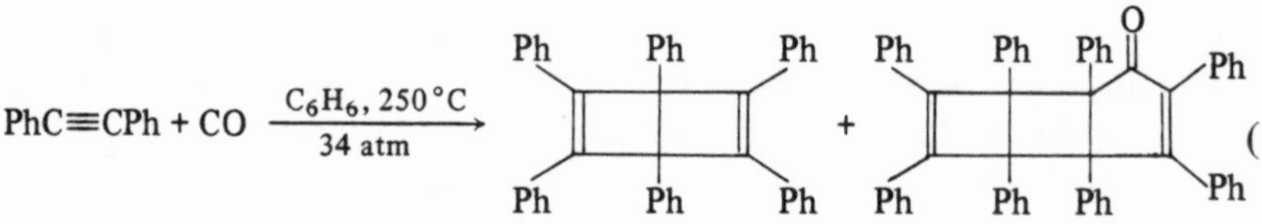
PhC≡CPh + CO
C6H6, 250°C
34 atm
Ph
Ph
Ph
Ph
Ph
Ph
Ph
Ph
Ph
Ph
Ph
Ph
O
Ph
Ph

(11)

9.5. Polymerisation of Olefins

Olefin polymerisation is the largest single part of the chemical industry world-wide with polyethylene the largest volume polymer. Accordingly an enormous amount of research has been undertaken to find catalysts for olefin polymerisation and to optimise those catalysts [16, 40–50]. It would be inappropriate within this book to attempt to do more than summarise very briefly the salient developments concentrating on supported metal complex catalysts.

Catalysts are essential for the polymerisation of olefins to yield useful polymers. Without them the reaction either does not take place at all (polymerisation of ethylene or propylene), or an uncontrollable – sometimes explosive – polymerisation occurs giving a virtually useless product. The catalyst, then has two roles. It must initiate the reaction and it must control it to give the desired end product. This subtle control to give stereochemically reproducible polymers is particularly important in the case of polypropylene, but for all olefins control of polymer chain length is important.

Supported metal complex catalysts are formed by the interaction of transition metal complexes with supports. The latter are usually, but not exclusively, oxide supports such as silica, alumina or silica-alumina. Although all the commercially used supported catalysts are supported on metal oxides and halides, there has been a growing research interest in polymeric supports. We consider the inorganic oxide supported catalysts first.

9.5.1. INORGANIC OXIDE SUPPORTED OLEFIN POLYMERISATION CATALYSTS

The earliest catalyst for ethylene polymerisation — basically MoO_3 supported on alumina — was discovered in the early 1950s by Standard Oil of Indiana [51] and molybdenum catalysts on both metal oxide and polymeric supports are still being investigated today [52, 53]. The first commercial process was achieved by Phillips Petroleum Company at about the same time as the discovery of the Ziegler catalysts. The Phillips catalyst is prepared by impregnating silica or silica-alumina with an aqueous solution of a chromium salt such as H_2CrO_4 followed by calcination in air to form a product in which the chromium is coordinated to oxygen and stabilised in either the +4 or +2 oxidation state [54–58]. A possible mechanism for the polymerisation is shown in Scheme 2.

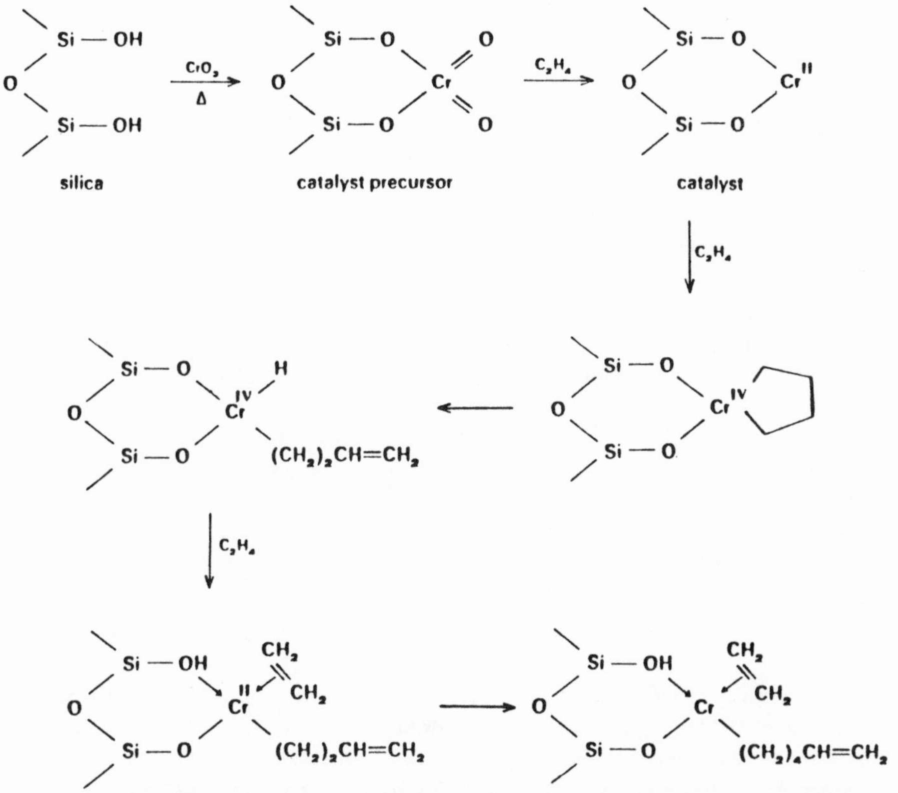

Scheme 2.
Polymerisation of ethylene on a silica supported chromium oxide catalyst (from [49]).

These catalysts are generally non-corrosive and are left in the polymer product; a big advantage, since removal of the catalyst from a viscous polymer solution is very difficult. In addition these catalysts can be used in a 'gas phase' process in which polyethylene is grown directly on the surface of the catalyst in the absence of solvent. Polymerisation pressures are fairly low, in the range 10–30 atmospheres.

A further series of chromium based catalysts was developed by the Union Carbide Corporation [59–63]. These organochromate esters such as $(Ph_3SiO)_2$-CrO_2 when reacted with a silica surface form ethylene polymerisation catalysts after activation with aluminium alkyls. Trisallylchromium and chromocene interact with silica to form active catalysts that require no further activation (reaction (12)).

$$\begin{array}{c}\diagdown\text{Si—OH}\\ \text{O}\diagup\diagdown\\ \diagup\text{Si—OH}\end{array} + \text{Cr}(C_5H_5)_2 \longrightarrow \begin{array}{c}\diagdown\text{Si—O}\diagdown\\ \text{O}\diagup\qquad Cr^{IV}\!-\!H\\ \diagup\text{Si—O}\diagup\end{array} + C_5H_6 \qquad (12)$$

Polymer propagation occurs by insertion of the ethylene into the chromium–hydride bond, not the chromium–cyclopentadienyl bond, which is retained throughout the polymerisation. Many other groups have since studied these catalysts, some recent references being [64–67].

The most important olefin polymerisation catalysts are the Ziegler-Natta catalysts based on titanium trichloride, for which the two scientists were awarded the Nobel Prize in chemistry in 1963. The early catalysts had low surface areas ($\sim 10 m^2/g$) and a great deal of work was done to find ways of increasing the surface area and hence activity. A series of catalysts was made in which inorganic oxides, usually silica and alumina, were treated with titanium tetrachloride and subsequently activated with aluminium alkyls [42]. Some of the more recent references describing the characterisation and use of catalysts based on titanium and vanadium halides and alkoxides are listed in [68–76]. Although supporting the active metal on an oxide support did enhance its activity, it was not until magnesium compounds were used as the support together with aluminium cocatalysts, that activities in excess of a thousand times that of the original catalyst were obtained. Initially magnesium oxide and hydroxide were used, sometimes alone and sometimes as additives to other oxides [77–85], but the really active catalysts, the so-called 'high mileage' Ziegler catalysts, are those in which titanium and vanadium halides and alkoxides are supported on magnesium chloride, which yield highly isotactic polypropylene

[86–108]. It is far from clear how magnesium compounds achieve this dramatic enhancement of activity, particularly as similar calcium derivatives are less effective. Presumably the origin lies in a particularly favourable electron balance in the intermediate Mg—Cl—Ti surface species. Attempts to support vanadium compounds on polymeric supports, such as allyl alcohol grafted on to polyethylene, have been described, but the activities of the resulting catalysts are only modest [109, 110].

By the mid-1960s it was apparent that organotitanium species were the true catalysts formed by reaction of titanium compounds with aluminium alkyls [111–113]. This led to the development of a series of catalysts in which the tetraallyls, tetrabenzyls and tetraalkyls of titanium, zirconium, hafnium, vanadium and chromium as well as cyclopentadienyl compounds such as Cp_2TiCl_2 and $CpTiCl_3$ were supported directly on to high surface area inorganic oxides such as silica, magnesium oxide and $Mg(OH)Cl$ (reactions (13) and (14)) [114–130].

$$[Ti(CH_2Ph)_4] + Mg(OH)Cl \longrightarrow Mg\left[\begin{array}{c} -O \\ -O \end{array}\right]Ti\left\langle\begin{array}{c} CH_2Ph \\ CH_2Ph \end{array}\right. + 2\,PhCH_3 \qquad (13)$$

$$[Cp_2TiCl_2] + \{Al\}-OH \longrightarrow [\{Al\}-OTiClCp_2] + HCl \qquad (14)$$

A great deal of this work has been undertaken in the Soviet Union [40, 41, 48, 131]. With many of these systems the unsupported complexes are inactive. It is often found that one particular combination of organometallic compound and support is more effective than others, e.g. Cr organometallics on SiO_2, Ti and Zr organometallics on Al_2O_3 or Ti organometallics on MgO. The reasons for this are unclear, although they clearly indicate the importance of achieving just the right electronic balance at the active centre. Although all these catalysts are very active for ethylene polymerisation, they are generally unsuitable for propylene because they yield polymers of low stereoregularity.

Uranium and thorium methyl complexes, $[M(\eta^5\text{-}C_5Me_5)_2Me_2]$, react with alumina to liberate methane and yield supported $[M(\eta^5\text{-}C_5Me_5)_2Me_2]$. They also react with alumina to liberate methane and yield supported $[(\{Al\}-O)_2M(\eta^5\text{-}C_5Me_5)_2]$ which catalyses ethylene polymerisation [132].

9.5.2. POLYMER SUPPORTED OLEFIN POLYMERISATION CATALYSTS

A number of polymer supported catalysts have been described in which metal complexes such as $[Cp_2TiCl_2]$, $[Ti(OBu)_4]$, $TiCl_4$, VCl_4 and $[VO(OEt)_3]$ have been supported on polymers such as polyethylene and polypropylene to which have been grafted functionalised olefins such as allyl alcohol, allylamine, vinyl acetate, acrylonitrile and acrylic acid [133–138]. These catalysts all have higher activities than their homogeneous counterparts due to the polymer suppressing bimolecular deactivation. Other polymers used have included ethylene-vinyl alcohol and styrene-vinyl acetate copolymers [139, 140]. Where the nature of the polymer formed from propylene is reported, it is generally found to be atactic differing little from that formed in the presence of the corresponding homogeneous catalyst [139].

Aluminium trichloride reacts with polystyrene to form, after treatment with aqueous ethanol, a selective catalyst for the cationic polymerisation of olefins such as *iso*-butylene [141]. The advantages of supporting on polystyrene are that it enhances the catalysts' resistance to thermal dehydrochlorination, promotes their activity and leads to poly-*iso*-butylene with a narrower molecular weight distribution than the analogous homogeneous catalysts.

9.6. Diene Polymerisation

Many of the desirable properties of rubber arise from the presence of the double bonds along the chains, $(CH_2CH=C(CH_3)CH_2)_n$. Accordingly a great deal of work has been devoted to the polymerisation of dienes, particularly 1,3-butadiene and isoprene. Four major catalyst systems have been developed that give the desired *cis*-1,4-polymerisation of dienes [142].

(i)   Alkyllithiums, usually BuLi.
(ii) Iodide-modified titanium Ziegler catalysts.
(iii) Cobalt based Ziegler catalysts.
(iv) Allylnickel complexes.

Of these, only the cobalt and nickel systems have been studied as supported catalysts. When $CoCl_2$ or $[Co(acac)_2]$ are supported on polyethylene to which 4-vinylpyridine, polyvinylimidazole or polyacrylic acid has been grafted, the resulting catalysts give higher yields of polybutadiene and polyisoprene than their homogeneous counterparts [143]. This is believed to be due to the inhibition of cobalt(II) reduction. Another very active catalyst was formed by supporting $[Co(pyr)_2Cl_2]$ on silica and activating with a very large excess of $AlEt_2Cl$ [144]. Like their homogeneous counterparts both series of catalysts yield >95% of the desired *cis*-polymer.

Nickel based catalysts can be prepared in several ways. When the η^3-allyl complexes are supported on metal oxides (reactions (15) and (16)) the reactivities of 2 decreases with increasing halogen atomic weight (X = Cl > Br > I)

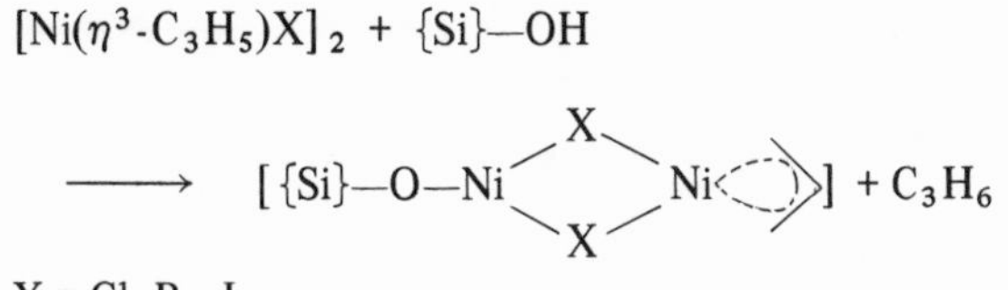
[Ni(η3-C3H5)X]2 + {Si}—OH
[{Si}—O—Ni
X
X
Ni
] + C3H6

(15)

X = Cl, Br, I

2

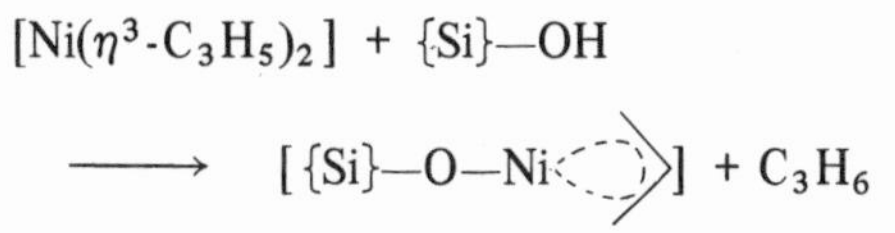
[Ni(η3-C3H5)2] + {Si}—OH
[{Si}—O—Ni
] + C3H6

(16)

3

[145, 146]. The product of reaction (16), **3** is more reactive than **2** formed in reaction (15) [147]. The change of activity across the series **3** > **2** (X = Cl > Br > I) is accompanied by a change in stereospecificity from 1,2-*trans*- formed by **2** to 1,4-*cis*-polydiene formed in the presence of **3** [148]. Similarly the metal oxide used to support $[Ni(\eta^3\text{-}C_3H_5)_2]$ has a pronounced effect on the stereochemistry of the product: SiO_2 gives 96% 1,4-*cis*, but only 2.5% 1,4-*trans* and 1.5% 1,2; Al_2O_3 gives 98% 1,4-*trans* and 2% 1,2; TiO_2 gives 85.5% 1,4-*cis*, 12.5% 1,4-*trans* and 2% 1,2; MgO gives oligomers only [145]. The homogeneous catalyst gives essentially only the 1,4-*trans*-polydiene.

A series of soluble polymers and crosslinked resins containing difluoroacetic acid side groups have been used to support η^3-allylnickel(II) complexes (reaction (17)). The polymers were prepared by copolymerising or terpolymerising ethyl-2,2-difluoro-3-butenoate with styrene and divinylbenzene. The

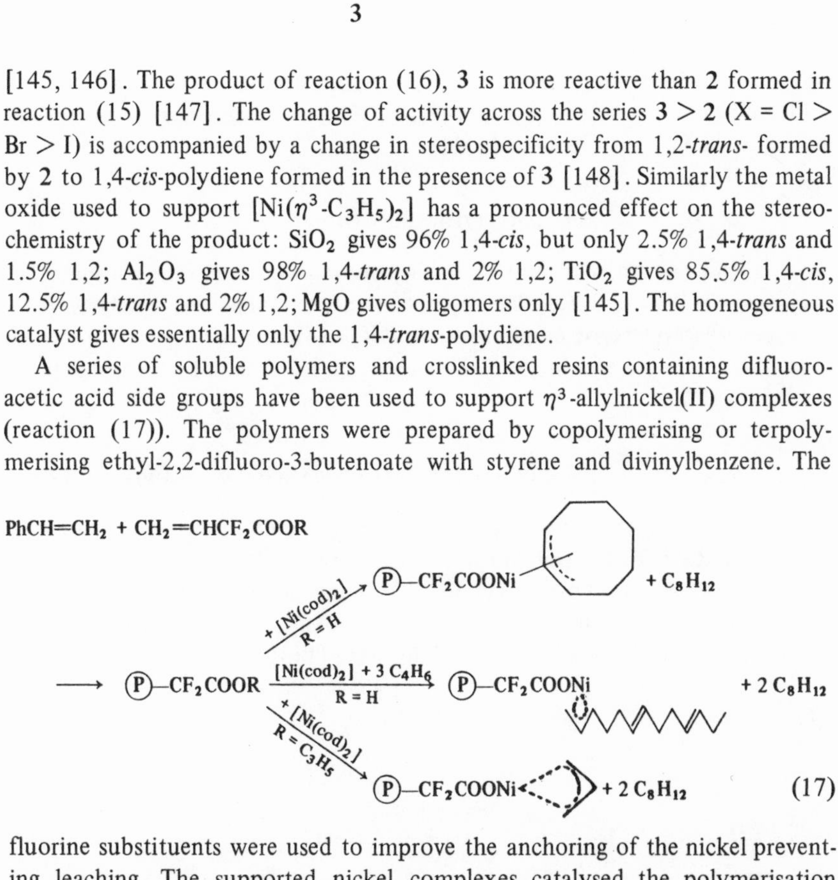
PhCH=CH2 + CH2=CHCF2COOR
P—CF2COONi
+ C8H12
+ [Ni(cod)2]
R = H
P—CF2COOR
[Ni(cod)2] + 3 C4H6
R = H
P—CF2COONi
+ 2 C8H12
+ [Ni(cod)2]
R = C3H5
P—CF2COONi
+ 2 C8H12

(17)

fluorine substituents were used to improve the anchoring of the nickel preventing leaching. The supported nickel complexes catalysed the polymerisation of 1,3-butadiene much more slowly than their homogeneous counterparts to yield about 90% *cis*-1,4-polybutadiene [149]. The molybdenum analogues formed as in reaction (18) (dme = dimethoxyethane) are less active and less specific giving 75% 1,2-polybutadiene and only 25% *cis*-1,4-polybutadiene [149].

$$\text{(P)}-CF_2COOC_3H_5 + [Mo(CO)_6]$$

$$\longrightarrow \quad [(\text{(P)}-CF_2COO)Mo(C_3H_5)(CO)_{6-n}L] + n\,CO \qquad (18)$$

L = thf ($n = 3$)
L = dme ($n = 4$)

Highly dispersed catalysts prepared by supporting $CoCl_2$, $CoBr_2$ or $NiCl_2$ on silica, magnesia, alumina and $MgCl_2$ have high activities for the *cis*-polymerisation of butadiene after their actvation with $AlEt_2Cl$ or $Al_2Et_3Cl_3$ [150]. ESR studies of the nickel catalysts suggest that the catalytically active species are surface {M}—ONiCl groups [151].

9.7. Acetylene Polymerisation

When anhydrous $NiCl_2$ is refluxed in suspension in butanol with phosphinated polystyrene, $\text{(P)}-C_6H_4CH_2PPh_2\text{-}p$, the resulting supported catalysts react with phenylacetylene or ethylpropiolate ($HC{\equiv}COOEt$) in thf in the presence of sodium borohydride to yield a mixture of 1,2,4- and 1,3,5-trisubstituted benzenes [152].

9.8. Copolymerisation of Propylene Oxide and Carbon Dioxide

$ZnEt_2$ reacts with the surface hydroxyl groups of Al_2O_3, SiO_2, MgO, $Mg(OH)_2$, $Mg(OH)Cl$, ZnO, ThO_2, TiO_2, Cr_2O_3, ZrO_2 and $Ca(OH)_2$ to liberate ethane and form a supported zinc ethyl (reaction (19)). These products catalyse the copolymerisation of propylene oxide and carbon dioxide (reaction (20)), the most active being that supported on magnesium oxide [153].

$$\{M\}-OH + ZnEt_2 \quad \longrightarrow \quad \{M\}-OZnEt + C_2H_6 \qquad (19)$$

$$CH_3-\overset{O}{\overset{/\,\backslash}{CH-CH_2}} + CO_2 \xrightarrow{\text{cat.}} \left(CH_2-\overset{CH_3}{\underset{|}{CH}}-O-\overset{}{\underset{\overset{\|}{O}}{C}}-O\right)_{\!n} \qquad (20)$$

9.9. Olefin Metathesis

The olefin metathesis reaction, also known as olefin dismutation or disproportionation, is best represented by reaction (21).

$$\text{(21)}$$

The reaction was first discovered by workers at Phillips Petroleum in the late
1950s [154, 155] and was quickly developed into a commercial process. Only
in the late 1970s did a real understanding of the mechanism begin to emerge
[156–163]. Many applications have been reported and several have been
developed to pilot scale but only two are currently in large-scale commercial
operation [164–166]. These involve the production of neohexene, a monomer
for speciality plastics, from ethylene and di-*iso*-butylene (reaction (22)) and the
'Shell Higher Olefin Process' [164, 165a].

$$Me_2C=CHCMe_3 + CH_2=CH_2 \rightleftharpoons Me_2C=CH_2 + Me_3CCH=CH_2 \quad (22)$$

In the latter ethylene is oligomerised to higher α-olefins, ideally C_8-C_{20} terminal
olefins. However since the distribution of products is largely statistical, substantial amounts of higher and lower molecular weight olefins are also produced.
These are isomerised to their internal isomers and then subjected to metathesis
to produce mixtures of olefins in the $C_{10}-C_{20}$ range.

The applications of olefin metathesis lie in three major areas:

(i) *Increase or decrease in chain length*. This is illustrated by the now defunct
Triolefin process operated by Shawinigan near Montreal to convert a local
surplus of propylene to meet a deficiency of polymerisation grade ethylene
which could be readily separated from the less volatile but-2-ene.

(ii) *Formation of specific diolefins*. α,ω-Dienes are useful comonomers
for cross-linking polymers as well as useful reagents for preparing α,ω-difunc-
tional compounds. They can be prepared by metathetical ring opening of cyclic
olefins with ethylene (reaction (23)).

$$\text{(23)}$$

(iii) *Ring opening polymerisation of cis-cyclic olefins*. A variety of cyclic olefins, but not cyclohexene, undergo ring opening polymerisation to give high molecular weight polymers (reaction (24)). The most significant of these are the polypentenamers which find application as speciality cross-linkable rubbers.

$$(CH_2)_n \quad \overset{CH}{\underset{CH}{\parallel}} \quad \longrightarrow \quad +CH_2)_n CH=CH+_m \tag{24}$$

The catalysts for olefin metathesis fall into four classes, of which we shall only consider further the first two:

(i) *Supported oxide catalysts (heterogeneous catalysts)*. These have been the catalysts of greatest commercial interest. They involve MoO_3, WO_3, Re_2O_7, or mixed oxides such as $CoO \cdot MoO_3$ supported on silica or less frequently alumina or titania [167–172]. They are operated at higher temperature and are sufficiently robust to enable the coke that accumulates on the surface to be burned off at still higher temperatures. They are sensitive to polar species such as nitrogen and oxygen containing compounds, especially water which must be rigorously excluded. The nature of the oxidation states of the active oxides is uncertain although +5 and +6 seem to be favoured for molybdenum and tungsten.

(ii) *Supported metal complexes*. Molybdenum and tungsten carbonyls, allyls and alkyls all form active catalysts when supported on silica or alumina [173–177]. On the whole these catalysts show no real advantages over those of the first type and some have positive disadvantages in that, although activated by stoichiometric quantities of oxygen, they are deactivated in air, thus complicating their preparation and handling.

There has recently been interest in supporting metathesis catalysts on polymers. Many of these are based on the modified Ziegler-Natta catalysts mentioned below. Reaction of phosphinated polystyrene with $[Mo(CO)_6]$ in the presence of UV light gave $[(\text{Ⓟ}-C_6H_4-CH_2PPh_2\text{-}p)Mo(CO)_5]$ which feebly catalysed *cis*-2-pentene metathesis in the presence of $EtAlCl_2$ and molecular oxygen [178]. The tungsten analogue, $[(\text{Ⓟ}-C_6H_4-CH_2PPh_2\text{-}p)W(CO)_5]$, has limited value, but its η^5-cyclopentadienyl derivative, $[(\text{Ⓟ}-C_6H_4-CH_2PPh_2\text{-}p)W(CO)_3(\eta^5\text{-}C_5H_5)]^-$, is of great value because it retains its activity on reuse for longer [179]. $[W(CO)_5]^-$ has been supported on strong trimethylammonium anion exchange resins and found to catalyse 1-octene metathesis after $EtAlCl_2$ activation [180].

Although there is still a very long way to go before polymer supported olefin metathesis catalysts will be of serious value, progress has been made by supporting a molybdenum nitrosyl complex on phosphinated polystyrene (reaction (25)) [181].

$$\text{(P)}\langle\text{\_}\rangle-PPh_2 + [Mo(PPh_3)_2(NO)_2Cl_2]$$

$$\longrightarrow [(\text{(P)}\langle\text{\_}\rangle-PPh_2)_2Mo(NO)_2Cl_2] + 2\,PPh_3 \quad (25)$$

By adding $Me_3Al_2Cl_3$ to the polymer in chlorobenzene a hybrid catalyst active at room temperature for the metathesis of 1,2-octadiene to ethylene and cyclohexene was obtained. When the catalyst had been separated off by filtration it could be reactivated with more aluminium cocatalyst. To obviate the need for reactivation an attempt was made to build the cocatalyst in by simulating the known WCl_6/R_4Sn homogeneous system as in reaction (26).

$$\text{Polystyrene} \xrightarrow[\text{2. Me}_3\text{SnCl}]{\text{1. BuLi, tmeda}} \text{(P)}\langle\text{\_}\rangle-SnMe_3$$

$$\xrightarrow{WCl_6} [\text{(P)}\langle\text{\_}\rangle-SnMe_nWCl_m] \quad (26)$$

Initially this catalyst was very active for the metathesis of 1,7-octadiene but after a few hourse it slowly lost activity [181]. A less sophisticated approach to preparing such a catalyst is claimed to give good activity for 2-pentene metathesis. In this ethylene, isoprene or cyclopentadiene are polymerised in the presence of WCl_6 or $MoCl_5$ and BuLi, BuMgCl or $C_5H_{11}MgBr$. On addition of a Lewis acid such as $AlCl_3$, $SbCl_3$, $SnCl_2$, $TiCl_4$ or $ZnCl_2$ an active metathesis catalyst is obtained [182].

(iii) *Modified Ziegler-Natta Catalysts (homogeneous catalysts)*. A mixture of a transition metal salt, usually the halide of tungsten, molybdenum or rhenium, and a main group metal alkyl or hydride such as $EtAlCl_2$, BuLi, $SnEt_4$, $LiAlH_4$ or $NaBH_4$, gives an active olefin metathesis system. Traces of oxygen are essential for the generation of highly active long-lived catalysts, which are very sensitive to oxygen impurities making assignment of the oxidation state of the active centre virtually impossible.

(iv) *Carbene catalysts (homogeneous catalysts)*. A number of metal-carbene complexes such as $[W{=}CPh_2(CO)_5]$ are active olefin metathesis catalysts. It was, of course, this discovery that did much to open the way for a full understanding of the mechanism of olefin metathesis.

9.10. Olefin Isomerisation

The isomerisation of olefins has been widely studied. It can be a very valuable reaction, whereas it can also be a severe problem when it occurs unintentionally. Many catalysts that promote other reactions of olefins such as hydroformylation, hydrogenation, oligomerisation and polymerisation also promote olefin isomerisation. This is not unexpected since coordination of an olefin to a metal is an essential prerequisite to these other reactions and also facilties double-bond migration. Olefin isomerisation is a kinetic phenomenon. If reactions of olefins are allowed to proceed to completion equilibrium mixtures of olefins result. Thus the ultimate products of 1-butene isomerisation are *trans*-2-butene (69%), *cis*-2-butene (25%) and 1-butene (6%) [165b]. With many catalysts, however, *cis*-2-butene is formed more rapidly than *trans*-2-butene so that it can be isolated in the early stages of the reaction.

Olefin isomerisation is used both industrially and in the laboratory to isomerise olefins that are intermediates. Thus the synthesis of adiponitrile from butadiene involves two olefin isomerisation steps (reaction (27)).

$$CH_2{=}CHCH(Me)CN \;\rightleftharpoons\; CH_3CH{=}CHCH_2CN$$

$$\rightleftharpoons\; CH_2{=}CH(CH_2)_2CN \qquad (27)$$

Double-bond migration is a prerequisite in the hydrogenation of trienes to monoolefins (see Section 6.9.9), whereas double-bond migration is a major problem in the hydroformylation of terminal olefins to terminal aldehydes and alcohols (see Section 8.2).

There are two major mechanisms involved in olefin isomerisation: reaction of a metal-hydride with an olefin to give a metal alkyl (reaction (28)) and a 1,3-hydrogen shift through an η^3-allyl intermediate (reaction (29)).

$$RCH_2CH{=}CHR' + M{-}H \;\rightleftharpoons\; \underset{\overset{|}{M}}{RCH_2CHCH_2R'}$$

$$\rightleftharpoons\; RCH{=}CHCH_2R' + M{-}H \qquad (28)$$

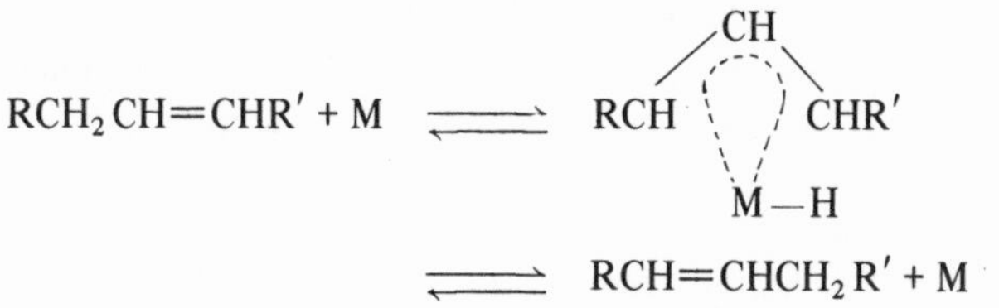

$$RCH_2CH=CHR' + M \rightleftharpoons RCH \overset{\overset{\displaystyle CH}{\diagup \diagdown}}{\underset{\underset{\displaystyle M-H}{\diagdown \diagup}}{}} CHR'$$

$$\rightleftharpoons RCH=CHCH_2R' + M \qquad (29)$$

The former is by far the more common and is typically shown by metal-hydrides, such as $[NiHL_4]^+$, and complexes that readily form hydrides in solution, such as rhodium complexes. The η^3-allyl mechanism is shown by metal complexes such as those of palladium(II) that readily form η^3-allyls. Distinction between these two mechanisms is not easy; deuterium migration studies are typically used. Other mechanisms, particularly those involving radicals or carbene inter-mediates may also be active, but are not well characterised.

Supported olefin isomerisation catalysts may be conveniently considered under six headings:

9.10.1 ZIRCONIUM COMPLEXES

Olefins have been isomerised over silica bound zirconium alkyl and zirconium hydride complexes, $[(\{Si\}-O)_2ZrR_2]$, R = alkyl or H, prepared by reaction of zirconium tetraalkyls with silica [183, 184]. The steric environment of zirconium in $[(\{Si\}-O)_2ZrH_2]$ is much less demanding than in $[ZrCp_2HCl]$ so that double bond migrations past tertiary carbon centres could be accomplished with the supported catalysts.

9.10.2. IRON, RUTHENIUM AND OSMIUM CARBONYL COMPLEXES

Supported iron carbonyl olefin isomerisation catalysts have been prepared by treating phosphinated polystyrene Ⓟ–C_6H_4–PPh_2-p and Ⓟ–C_6H_4–PPh-$(CH_2)_2PPh_2$-p with $[Fe(CO)_5]$ and $[Fe_3(CO)_{12}]$ [185–187]. On irradiation of the supported catalysts $[(Ⓟ–C_6H_4PPh_2\text{-}p)_nFe(CO)_{5-n}]$, n = 1, 2 and $[(Ⓟ–C_6H_4PPh(CH_2)_2PPh_2\text{-}p)_x\{Fe(CO)_n\}_y]$, x = 1, 2; n = 3, 4; y = 1, 2, carbonyl ligands are displaced, in direct contrast to the analogous homogeneous species which lose phosphine rather than carbonyl ligands on irradiation. The photoactivated supported complexes catalyse 1-pentene isomerisation with turnover numbers that sometimes exceed 2×10^4 substrate molecules per iron atom and quantum yields in excess of unity, indicating the generation of a thermal catalyst. However the activity in the dark is only short lived.

The alumina supported anion clusters $\{Al\}^+[H_3RuOs_3(CO)_{12}]^-$ formed by heating $[H_2RuOs_3(CO)_{13}]$ absorbed on γ-alumina to $100-200°C$ under a hydrogen-carbon monoxide atmosphere promotes 1-butene isomerisation at $57°C$ [188]. Refluxing $[Os_3(CO)_{12}]$ and γ-alumina or silica in n-octane under nitrogen yields 4, $\{M\}$ = alumina or silica, which catalyses 1-hexene isomerisation [189]. On heating 4 to $400°C$ a mononuclear osmium(II) species is formed

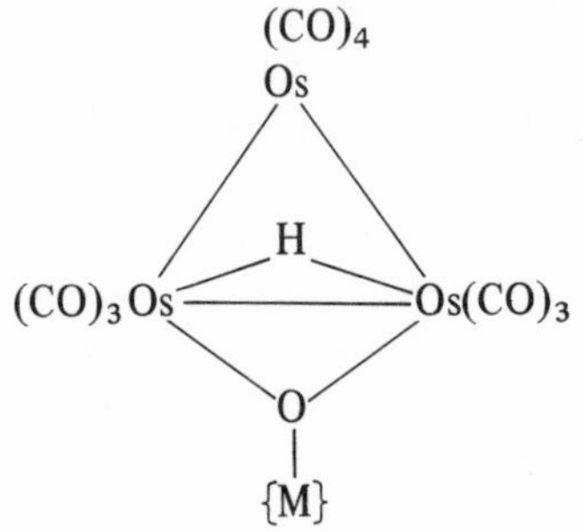

$$
\begin{array}{c}
(CO)_4 \\
Os \\
\end{array}
$$

$(CO)_3Os$ —————— $Os(CO)_3$

with bridging H and O—$\{M\}$

4

which does not isomerise 1-hexene. $[(\textcircled{P}-C_6H_4PPh_2\text{-}p)H_2Os_3(CO)_9]$ in which the support is phosphinated polystyrene, prepared by reaction (30), catalyses 1-hexene isomerisation [190].

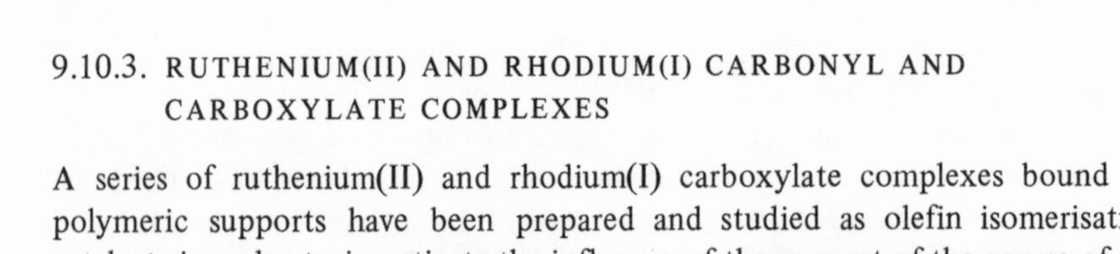

$\textcircled{P}$—⟨ ⟩—PPh_2 + $[H_2Os_3(CO)_{10}]$ ⟶ $[(\textcircled{P}$—⟨ ⟩—$PPh_2)H_2Os_3(CO)_{10}]$

$\downarrow$ mild heating $(-CO)$ (30)

$\textcircled{P}$—⟨ ⟩—PPh_2 + $[H_2Os_3(CO)_9(PPh_3)]$ $\xrightarrow{-PPh_3}$ $[(\textcircled{P}$—⟨ ⟩—$PPh_2)H_2Os_3(CO)_9]$

9.10.3. RUTHENIUM(II) AND RHODIUM(I) CARBONYL AND CARBOXYLATE COMPLEXES

A series of ruthenium(II) and rhodium(I) carboxylate complexes bound on polymeric supports have been prepared and studied as olefin isomerisation catalysts in order to investigate the influence of the support of the course of the reaction [191]. Complexes were prepared by reacting $[RuX(\eta^3\text{-}C_3H_5)(CO)_3]$, X = Cl, Br, with poly-4-vinylpyridine, poly-2-vinylpyridine, poly(p-styryldiphenylphosphine) and styrene/p-styryldiphenylphosphine copolymers to obtain the *cis*-product, 5.

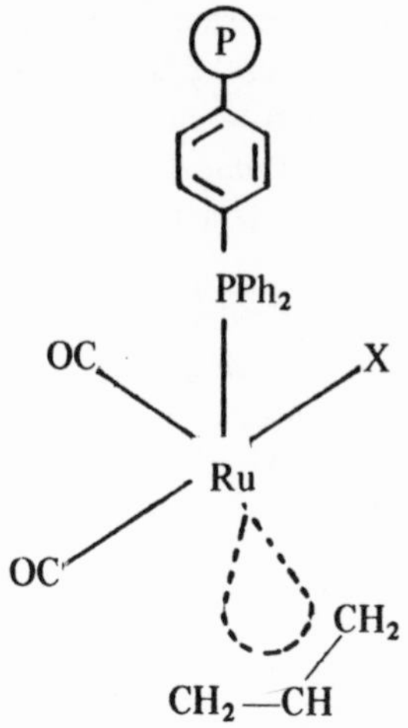

5

Compound **5** catalyses the isomerisation of 1-butene at 100°C in toluene with a lower activity than the corresponding homogeneous analogue. The rate of isomerisation was strongly affected by the nature of the polymeric support, decreasing with increasing molecular weight and tacticity. This arises from the reduction in swelling which reduces the availability of the active sites, as shown by carrying out the same isomerisation on gaseous 1-butene when a large decrease in activity independent of the polymer support was observed. A further demonstration of the role of the primary and secondary structure of the polymeric support was obtained using stereoregular optically active polymers such as copolymers of (R)- or (S)-3,7-dimethyl-1-octene and styrene, which were functionalised as in reaction (31) to yield stereoregular polymers of high conformational homogeneity [194].

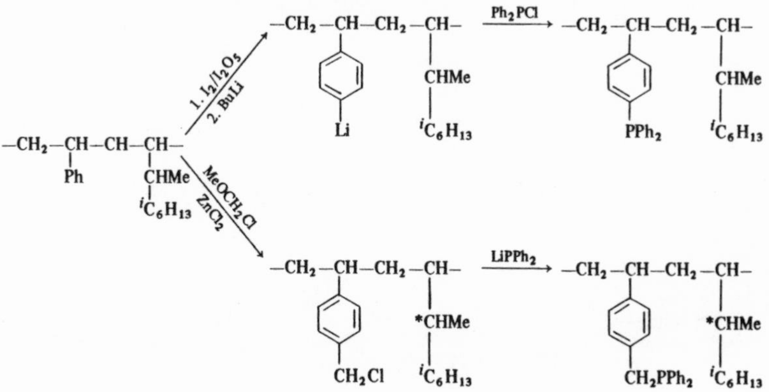

(31)

These were then used to prepare complexes by reaction with $[RhBr(\eta^3\text{-}C_3H_5)\text{-}(CO)_3]$, which were used as catalysts in the isomerisation of racemic 4-methyl-1-hexene to a mixture of *cis*- and *trans*-4-methyl-2-hexene [195]. The very low optical yield of about 0.1% indicated that the asymmetric centre in the polymer in these system is too remote from the active site to induce asymmetry in the catalysis.

A series of ruthenium(II) and rhodium(I) supported catalysts have been prepared using carboxylate supports such as poly(acrylic acid), poly(methacrylic acid), alternating copolymers of maleic acid with vinyl monomers such as ethylene, vinylbenzylether and vinyl alcohol, and polymethacroyl acetone (reactions (32–34)) [196–200].

$$
\xrightarrow[n = 0, 1, 2]{L = CO, \; PPh_3} \qquad + \; (3 - n)PPh_3 + H_2 \qquad (32)
$$

$$
\xrightarrow[n = 1, 2]{} \qquad\qquad\qquad\qquad\qquad\qquad\qquad (33)
$$

$$
\xrightarrow[n = 1, 2, 3]{} \qquad\qquad\qquad + \; (4 - n)PPh_3 + H_2 \qquad (34)
$$

The carboxylate groups bind ruthenium in a bidentate fashion but rhodium only in a monodentate fashion. The supported complexes are active olefin isomerisation catalysts. They are more active in swelling solvents such as toluene than in non-swelling solvents such as heptane. The catalysts derived from $[RuH_2(PPh_3)_4]$ are less active than their homogeneous analogues, whereas the reverse is true for those derived from $[RuH_2(CO)(PPh_3)_3]$. This is because the latter more readily lose triphenylphosphine to yield coordinatively unsaturated species that are stabilised by the presence of the polymeric support. Ruthenium catalysts derived from vinylether-maleic acid copolymers showed a remarkable drop in activity on recycling, which was greater in benzene than in 2-propanol. It was shown to arise from the formation of catalytically inactive bis-carboxylate species 6.

6

9.10.4. SILICA SUPPORTED RHODIUM CATALYSTS

The product obtained by impregnating rhodium trichloride onto silica gel is 100 times more active per rhodium atom than its homogeneous counterpart for 1-butene isomerisation [201]. This suggests replacement of at least one chloride ligand by a surface oxygen. Complexes such as $[Rh(SnCl_3)(PPh_3)_3]$ supported on silica become active 1-hexene isomerisation catalysts on treatment with an alkali metal alkyl or a Grignard reagent [202]. The catalytic activity for 1-pentene isomerisation of a series of rhodium complexes linked to silica by $-OSiMe_2(CH_2)_nPPh_2$ groups increased as n increased in the order $n = 1 < 3 < 7, 9$ [203].

9.10.5. NICKEL CATALYSTS

Although homogeneous nickel olefin isomerisation catalysts have been studied extensively $[\{(\eta^5\text{-}C_5H_5)Ni(CO)\}]_m[\{Ge(OMe)_n\}]$, $m = 1$, $n = 3$ and $m = 2$, $n = 2$, supported on silica are the only 1-hexene isomerisation catalysts we have noted [204].

9.10.6. PALLADIUM CATALYSTS

$[Pd(PhCN)_2 Cl_2]$ supported on silica gel is 100 times more active for 1-heptene isomerisation than in the unsupported form [205]. This implies a specific role for the silica, which is consistent with the observation that the UV spectra of butenes on $PdCl_2$ in diglyme and on silica gel are different [206]. $[Pd(PhCN)_2$-$Cl_2]$ has been impregnated in a triglyme solution on to celite to give an olefin isomerisation catalyst suitable for use in a gas chromatography column [207]. Palladium(II) isomerisation catalysts have been linked to silica gel through a series of amine, thioether, and dithiocarboxylate ligands [208]. The activity of $PdCl_2$ in catalysing both *cis-trans*-isomerisation and double bond migration in heptenes when supported on poly-1,2-butadiene grafted with acrylic acid increased with increasing swelling of the gelatinous polymer [209, 210].

9.11. Quadricyclane-Norbornadiene Isomerisation

The isomerisation of norbornadiene in the presence of light yields quadricyclane, which on reverse yields a considerable amount of energy (reaction (35)). This system is of some interest as a possible solar energy store. Both components

$$\hspace{5cm} (35)$$

are liquids, norbornadiene is readily available and relatively inexpensive. The system can potentially store 1.15×10^6 J l^{-1} [211] and the two isomerisations are very clean, producing only very minor amounts of by-products. Accordingly there has been considerable interest in finding catalysts to promote both isomerisations. Initial searches led to the development of homogeneous catalysts but these are of no value in an actual energy storage device where the catalyst must be immobilised to prevent its dispersal throughout the system.

9.11.1. QUADRICYCLANE TO NORBORNADIENE ISOMERISATION

Whilst several homogeneous catalysts have been developed for quadricyclane to norbornadiene isomerisation only three immobilised systems have so far been reported, one based on polystyrene-anchored cobalt(II) porphyrins and the others based on palladium(II). The preparation of the cobalt(II)-anchored porphyrins is shown schematically in Scheme 3 [212]. Whilst these catalysts are effective in promoting the isomerisation they are very sensitive to deactivation through oxidation of cobalt(II) to cobalt(III). This can be partially reversed by treatment with a strong reducing agent such as titanium(III).

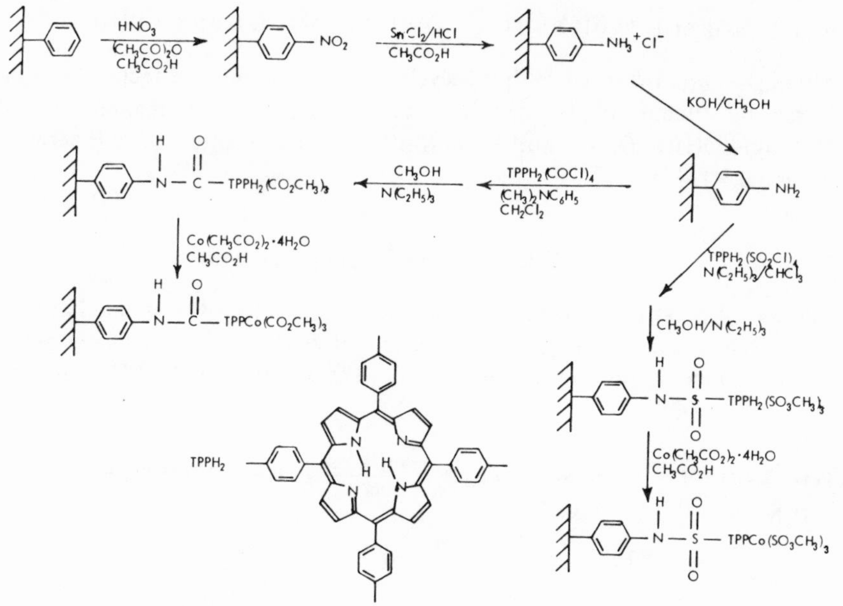

Scheme 3. The synthesis of polystyrene anchored cobalt(II) tetraporphyrins (from [212]).

The first palladium system described, based on a polystyrene supported palladium(0) bipyridyl complex (reaction (36)) also suffers from deactivation on re-use, partly due to oxidation and partly due to loss of palladium by leaching [213]. Initially this catalyst is 30 times more active, on a gram for gram

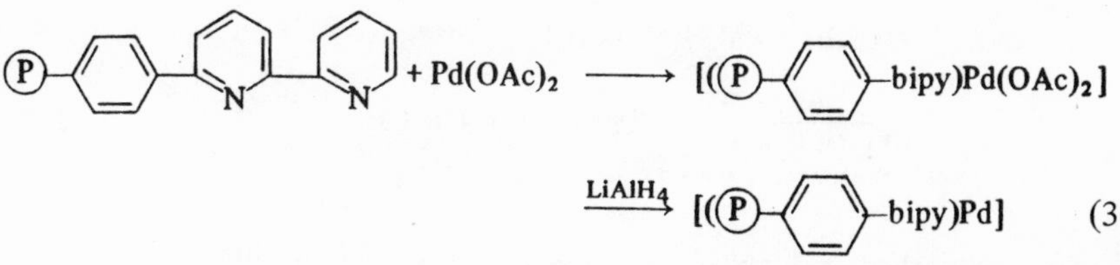

for gram basis, than palladium on charcoal, and even after two cycles it is still as active as palladium on charcoal. An attempt to link palladium to a diphenylphosphinated macroreticular polystyrene using $[Pd(MeCN)_2Cl_2]$ as the palladium(II) salt was less successful. The supported catalyst was between 2 and 85 times less active than soluble $[Pd(PPh_3)_2Cl_2]$ and also lost some of its activity on recycling [214].

9.11.2. NORBORNADIENE TO QUADRICYCLANE ISOMERISATION

Although norbornadiene to quadricyclane isomerisation is effected by light it can be effectively photosensitised by carrying it out in the presence of $[Ir(bipy)_3(OH)](NO_3)_2$ absorbed onto silica gel. The quantum yield is between 0.7 and 0.8 [215].

9.12. Grignard Cross-Coupling Reactions

The use of nickel(II) complexes to promote Grignard cross-coupling has been extensively studied [216]. By supporting nickel(II) on an optically active tertiary amine **7**, prepared as in reaction (37), $PhCH(Me)CH=CH_2$ has been

$$(37)$$

7

obtained in 49% enantiomeric excess by reaction (38) [217]. Similarly β-bromostyrene reacts with methyl magnesium iodide to give $PhCH=CHCH_3$

$$CH_2=CHBr + PhCH(Me)MgCl$$

$$\xrightarrow[\substack{Et_2O,\ -78\,°C \\ \text{then } 0\,°C,\ 2\ \text{days}}]{NiCl_2 + 7} PhCH(Me)CH=CH_2 \qquad (38)$$

in 98% stereoselective yield in the presence of poly(4-styrylPPh$_2$) supported palladium(0) [218]. This same complex also catalyses reaction (39) in better molar turnover than its homogeneous analogue.

$$PhCH=CH_2 + PhI \longrightarrow PhCH=CHPh + HI \qquad (39)$$

References

1. G. W. Parshall: *Homogeneous Catalysis*, Wiley, New York (a) p. 59 (b) Chapter 4 (1980).
2. E. Angelescu, A. Angelescu, S. Nenciulescu and I. V. Nicolescu: *Rev. Chim. (Bucharest)* **32**, 559 (1981); *Chem. Abs.* **95**, 149504 (1981).
3. Kh. I. Areshidze, V. A. Kabanov, V. I. Smetanyuk and R. V. Chediya: *Soobshch. Akad. Nauk Gruz. SSR* **101**, 589 (1981); *Chem. Abs.* **95**, 61348 (1981).
3a. V. A. Kabanov and V. I. Smetanyuk: *Macromol. Chem. Phys. Suppl.* **5**, 121 (1981).
4. D. B. Furman, N. V. Volchkov, L. A. Makhlis, P. E. Matkovskii, G. P. Belov, V. E. Vasserberg and O. V. Bragin: *Izv. Akad. Nauk. SSSR* 573 (1983).
5. X. Cochet, A. Mortreux and F. Petit: *Comp. Rendu* **C288**, 105 (1979).
6. V. S. Aliev, A. A. Khanmetov, R. Kh. Mamedov, R. K. Kerimov and V. M. Akhmetov: *Zh. Org. Khim.* **18**, 265 (1982).
7. N. Kawata, T. Mizoroki, A. Ozaki and M. Ohkawara: *Chem. Lett.* 1165 (1973).
8. T. Mizoroki, N. Kawata, S. Hinata, K. Maruya and A. Ozaki: *Catalysis, Heterogeneous and Homogeneous* (Ed. by B. Delmon and G. Jannes) Elsevier, Amsterdam, p. 319 (1975).
9. N. Kawata, T. Mizoroki and A. Ozaki: *J. Mol. Catal.* **1**, 275 (1976).
10. P. W. Jolly and G. Wilke: *The Organic Chemistry of Nickel*, Academic Press, New York, Vol. 2, p. 1 (1975).
11. V. A. Kabanov and V. I. Smetanyuk: *Sov. Sci. Rev.* **B2**, 83 (1980).
12. A. F. Lunin, Z. S. Vaizin, V. M. Ignatov, V. I. Smetanyuk and A. I. Prudnikov: *Neftekhimiya* **21**, 199 (1981); *Chem. Abs.* **95**, 79995 (1981).
12a. N. V. Volchov, D. B. Furman, L. I. Lafer, S. A. Chernov, A. V. Kudryashev, V. I. Yakerson and O. V. Bragin: *Izv. Akad. Nauk. SSSR.* 507 (1983).
13. Yu. N. Usov, E. D. Chekurovskaya and N. I. Kuvshinova: *Kinet. Katal.* **19**, 1606 (1978).
14. N. Takahashi, I. Okura and T. Keii: *J. Mol. Catal.* **3**, 277 (1978).
15. N. Takahashi, I. Okura and T. Keii: *J. Mol. Catal.* **3**, 271 (1978).
16. S. Malinowski and W. Skupinski: *Rocz. Chem.* **48**, 359 (1974); *Chem. Abs.* **81**, 49189 (1974).
17. W. Skupinski and S. Malinowski: *J. Organometal. Chem.* **99**, 465 (1975).
18. W. Skupinski and S. Malinowski: *J. Organometal. Chem.* **117**, 183 (1976).
19. W. Skupinski and S. Malinowski: *J. Mol. Catal.* **4**, 95 (1978).
20. K. Kaneda. M. Terasawa, T. Imanaka and S. Teranishi: *Tetrahedron Lett.* 2957 (1977).
21. J. A. Davies, F. R. Hartley and S. G. Murray: *J. Chem. Soc. Dalton* 2246 (1980).
22. J. A. Davies, F. R. Hartley and S. G. Murray: *J. Mol. Catal.* **10**, 171 (1981).
23. F. R. Hartley and J. A. Davies: *Rev. Inorg. Chem.* **4**, 27 (1982).
24. C. U. Pittman, L. R. Smith and R. M. Hanes: *J. Amer. Chem. Soc.* **97**, 1742 (1975).
25. C. U. Pittman and S. E. Jacobson: *J. Mol. Catal.* **3**, 293 (1978).
26. C. U. Pittman, S. K. Wuu and S. E. Jacobson: *J. Catal.* **44**, 87 (1976).
27. C. U. Pittman, A. Hirao, C. Jones, R. M. Hanes and Q. Ng: *Ann. N.Y. Acad. Sci.* **295**, 15 (1977).
28. C. U. Pittman and Q. Ng: *J. Organometal. Chem.* **153**, 85 (1978).
29. C. U. Pittman, N. Quoc, A. Hirao, W. Honnick and R. Hanes: *Colloq. Int. CNRS* **281**, 49 (1977); *Chem. Abs.* **92**, 136004 (1980).

30. C. U. Pittman and R. M. Hanes: *US Patent* 4,243,829 (1981); *Chem. Abs.* 95, 6473 (1981).
31. C. U. Pittman, R. M. Hanes and J. J. Yang: *J. Mol. Catal.* 15, 377 (1982).
32. K. Kaneda, H. Kurosaki, M. Terasawa, T. Imanaka and S. Teranishi: *J. Org. Chem.* 46, 2356 (1981).
33. A. M. Lazutkin, V. A. Likholobov, A. I. Lazutkina, V. L. Kuznetsov, V. A. Semikolenov and Yu. I. Yermakov: *Kinet. Katal.* 19, 591 (1978).
34. A. M. Lazutkin, A. I. Lazutkina and Yu. I. Yermakov: *React. Kinet. Catal. Lett.* 8, 353 (1978).
35. K. Kaneda, T. Uchiyama, M. Terasawa, T. Imanaka and S. Teranishi: *Chem. Lett.* 449 (1976).
36. S. H. Chang, R. H. Grubbs and C. H. Brubaker: *J. Organometal. Chem.* 172, 81 (1979).
37. S. Vatanathan and M. F. Farona: *J. Catal.* 61, 540 (1980).
38. S. Vatanathan: *Diss. Abs.* B39, 5915 (1979).
39. S. Vatanathan and M. F. Farona: *J. Mol. Catal.* 7, 403 (1980).
40. Yu. I. Yermakov and V. A. Zakharov: *Adv. Catal.* 24, 173 (1975).
41. Yu. I. Yermakov: *Catal. Rev. Sci. Eng.* 13, 77 (1976).
42. F. J. Karol in: *Encyclopaedia of Polymer Science and Technology*, Wiley, N.Y., Supplement Volume 1, p. 120 (1976).
43. A. D. Caunt, in: 'Catalysis', *Chem. Soc. Spec. Per. Rep.* 1, 234 (1977).
44. F. J. Karol in: *Organometallic Polymers* (Ed. by C. E. Carraher, J. E. Sheats and C. U. Pittman) Academic Press, N.Y., p. 135 (1978).
45. V. A. Zakharov and Yu. I. Yermakov: *Catal. Rev. Sci. Eng.* 19, 67 (1979).
46. B. C. Gates, J. R. Katzer and G. C. A. Schuit: *Chemistry of Catalytic Processes*, McGraw-Hill, N.Y., p. 150 (1979).
47. H. Sinn and W. Kaminsky: *Adv. Organometal. Chem.* 18, 99 (1980).
48. Yu. I. Yermakov, B. N. Kuznetsov and V. A. Zakharov: *Catalysis by Supported Complexes*, Elsevier, Amsterdam, Chapters 2–5 (1981).
49. J. P. Candlin in: *Catalysis and Chemical Processes* (Ed. by R. Pearce and W. R. Patterson) Blackie & Son Ltd., Glasgow, Chapter 10 (1981).
50. J. Candlin, A. D. Caunt and J. Segal in: *The Chemistry of the Metal-Carbon Bond*, volume 3 (Ed. by S. Patai and F. R. Hartley) Wiley Chichester, in press.
51. E. F. Peters and B. L. Evering: *US Patent* 2,820,835 (1958); *Chem. Abs.* 52, 9580f (1958).
52. A. A. Olsthoorn and J. A. Moulijn: *J. Mol. Catal.* 8, 147 (1980).
52a. C. P. Cheng and G. L. Schrader: *Stud. Surf. Sci. Catal.* B7, 1432 (1981).
53. Zh. S. Kiyashkina, A. D. Pomogailo, A. I. Kuzayev, G. V. Lagodzinskaya and F. S. Dyachkovskii: *J. Polymer. Sci., Polymer Symp.* 68, 13 (1980).
54. J. P. Hogan and R. L. Banks: *US Patent* 2,825,721 (1958); *Chem. Abs.* 52, 8621h (1958).
55. J. P. Hogan: *J. Poly. Sci., Polymer Chem.* 8, 2637 (1970).
56. A. Clark: *Catal. Rev.* 3, 145 (1970).
57. K. Schulze: *German Patent*, 2,734,928 (1979); *Chem. Abs.* 90, 169325 (1979).
58. D. D. Beck and J. H. Lunsford: *J. Catal.* 68, 121 (1981).
59. G. L. Karapinka: *US Patent* 3,709,853 (1973); *Chem. Abs.* 78, 85087 (1973).
60. F. J. Karol, G. L. Karapinka, C. Wu, A. W. Dow, R. N. Johnson and W. L. Carrick: *J. Polym. Sci., Polymer Chem.* 10, 2621 (1972).
61. W. L. Carrick, R. J. Turbett, F. J. Karol, G. L. Karapinka, A. S. Fox and R. N. Johnson: *J. Polym. Sci., Polymer Chem.* 10, 2609 (1972).

62. F. J. Karol and R. N. Johnson: *J. Polym. Sci., Polym. Chem. Ed.* 13, 1607 (1975).
63. F. J. Karol, W. L. Munn, G. L. Goeke, B. E. Wagner and N. J. Maraschin: *J. Polym. Sci., Polym. Chem. Ed.* 16, 771 (1978).
64. B. E. Nasser and J. A. Delap: *US Patent* 4,188,471 (1980); *Chem. Abs.* 92, 164523 (1980).
65. D. Stein, R. Bachl and K. Richter: *German Patent* 3,030,916 (1982); *Chem. Abs.* 96, 218423 (1982).
66. S. S. Stanovaya, A. V. Shagilova, V. A. Grigor'ev, N. M. Korobova and N. P. Karandashova: *Plast. Massy.* 9 (1981); *Chem. Abs.* 95, 62737 (1981).
67. B. Rebenstorf, B. Jonson and R. Larsson: *Acta Chem. Scand.* A36, 695 (1982).
68. V. A. Zakharov, V. N. Druzhkov, E. G. Kushnareva and Yu. I. Yermakov: *Kinet. Katal.* 15, 446 (1974).
69. J. C. W. Chien and J. T. T. Hsieh: *Coord. Polym.* 305 (1975); *Chem. Abs.* 85, 47151 (1976).
70. M. J. Todd: *British Patent* 1,484,254 (1977); *Chem; Abs.* 89, 60255 (1978).
71. P. Damyanov, M. Velikova and L. Petkov: *Europ. Polym. J.* 15, 233 (1979).
72. K. Soga, K. Izumi, M. Terano and S. Ikeda: *Makromol. Chem.* 181, 657 (1980).
73. D. R. Fahey and M. B. Welch: *US Patent* 4,199,475 (1980); *Chem. Abs.* 93, 47510 (1980).
74. E. G. Howard and W. Mahler: *US Patent* 4,304,685 (1981); *Chem. Abs.* 96, 52907 (1982).
75. J. G. Speakman and N. P. Wilkinson: *European Patent* 22658 (1981); *Chem. Abs.* 94, 175856 (1981).
76. A. Munoz-Escalona, A. Martin and J. Hidalgo: *Europ. Polym. J.* 17, 367 (1981).
77. A. Delbouille and H. Toussaint: *S. African Patent* 6804338 (1968); *Chem. Abs.* 70, 115706 (1969).
78. K. Soga, M; Akiyoshi and T. Kagiya: *Chem. Lett.* 833 (1973).
79. K. Soga, S. Katano, Y. Akimoto and T. Kagiya: *Polym. J.* 5, 128 (1973).
80. D. D. Eley, D. A. Keir and R. Rudham: *JCS Faraday Trans I* 72, 1685 (1976).
81. D. D. Eley, D. A. Keir and R. Rudham: *JCS Faraday Trans I* 73, 1738 (1977).
82. J. G. Speakman: *British Patent* 1,554,710 (1979); *Chem. Abs.* 92, 111521 (1980).
83. A. G. Rodionov, N. M. Domareva, A. A. Baulin, E. L. Ponomareva and S. S. Ivanchev: *Vysokomol. Soedin* A23, 1560 (1981); *Chem. Abs.* 95, 204510 (1981).
84. A. A. Baulin, I. S. Asinovskaya, A. G. Rodionov, S. S. Ivanchev and A. L. Gol'denberg: *Plast. Massy* 9 (1980); *Chem. Abs.* 94, 31120 (1981).
84a. A. G. Rodionov, A. A. Baulin and S. S. Ivanchev: *Plast. Massy* 8 (1981); *Chem. Abs.* 94, 157384 (1981).
85. A. A. Baulin, I. P. Sidorova and A. L. Gol'denberg: *Plast. Massy* 22 (1982); *Chem. Abs.* 98, 17092 (1983).
86. L. Luciani, N. Kashiwa, P. C. Barbe and A. Toyota: *Ger. Offen* 2,643,143 (1977); *Chem. Abs.* 87, 68893 (1977).
87. S. I. Makhtarulin, E. M. Moroz, E. E. Vernel and V. A. Zakharov: *React. Kinet. Catal. Lett.* 9, 269 (1978).
88. M. J. Todd: *British Patent,* 1,525,693 (1978); *Chem. Abs.* 90, 204863 (1979).
89. J. G. Speakman: *British Patent* 1,539,900 (1979); *Chem. Abs.* 91, 57835 (1979).
90. Y. Yamada and S. Kawai: *Japanese Patent* 79,148,091 (1979); *Chem. Abs.* 92, 129647 (1980).
91. K. Soga, M. Terano and S. Ikeda: *Polym. Bull. (Berlin)* 1, 849 (1979).
92. U. Giannini: *Macromol. Chem. Phys. Suppl.* 5, 216 (1981).

93. A. A. Baulin: *Zh. Prikl. Khim.* **54**, 2257 (1981).
94. Mitsui Petrochemical Ind. Ltd.: *Japanese Patent* 81,67,311 (1981); *Chem. Abs.* **95**, 170082 (1981).
95. S. N. Shepelev, G. D. Bukatov, V. A. Zakharov and Yu. I. Yermakov: *Kinet. Katal.* **22**, 258 (1981).
96. K. Soga and M. Terano: *Makromol. Chem.* **182**, 2439 (1981).
97. B. Keszler and A. Simon: *Polymer* **23**, 916 (1982).
98. S. Cai, H. Lui, H. Wang and S. Xiao: *Cuihao Xuebao* **3**, 7 (1982); *Chem. Abs.* **97**, 128113 (1982).
99. Y. Doi, M. Murata, K. Yano and T. Keii: *Ind. Eng. Chem., Prod. Res. Dev.* **21**, 580 (1982).
100. D. He, C. Chang, Y. Hu and G. Xie: *Gaofenzi Tongxun* 38 (1982); *Chem. Abs.* **97**, 198601 (1982).
101. J. C. W. Chien, J. C. Wu and C. I. Kuo: *J. Polym. Sci., Polym. Chem. Ed.* **20**, 2019 (1982).
102. T. Keii, E.' Suzuki, M. Tamura, M. Murata and Y. Doi: *Makromol. Chem.* **183**, 2285 (1982).
103. J. C. W. Chien and J. C. Wu: *J. Polym. Sci., Polym. Chem. Ed.* **20**, 2461 (1982).
104. J. C. W. Chien and J. C. Wu: *J. Polym. Sci., Polym. Chem. Ed.* **20**, 2445 (1982).
105. W. E. Smith, J. P. Candlin and D. R. Wilson: *British Patent* 2,096,122 (1982); *Chem. Abs.* **98**, 90088 (1983).
106. J. P. Candlin, D. R. Wilson and W. E. Smith: *British Patent* 2,096,123 (1982); *Chem. Abs.* **98**, 90089 (1983).
107. R. Invernizzi, F. Ligorati, M. Fontanesi and R. Catenacci: *European Patent Appl.* 65,700 (1982); *Chem. Abs.* **98**, 90097 (1983).
107a. K. Soga, R. Ohnishi and T. Sano: *Polym. Bull. (Berlin)* **7**, 547 (1982).
107b. M. Fontanesi, R. Invernizzi, F. Ligorati and R. Cantenacci: *British Patent* 2,094,318 (1982); *Chem. Abs.* **98**, 4900 (1983).
107c. Mitsui Toatsu Chemicals Inc.: *Japanese Patent* 57,135,807 (1982); *Chem. Abs.* **98**, 17200 (1983).
107d. A. J. Hartshorn and E. Jones: *British Patent* 2,090,841 (1982); *Chem. Abs.* **98**, 4897 (1983).
107e. C. C. Greco and K. B. Triplett: *US Patent* 4,350,612 (1982); *Chem. Abs.* **98**, 126908 (1983).
107f. Y. Doi, R. Ohnishi and K. Soga: *Makromol. Chem., Rapid Commun.* **4**, 169 (1983).
107g. P. Galli, P. Barbe, G. Guidetti, R. Zannetti, A. Martorana, A. Marigo, M. Bergozza and A. Fichera: *Eur. Polym. J.* **19**, 19 (1983).
107h. J. C. W. Chien, J. C. Wu and C. I. Kuo: *J. Poly. Sci., Poly. Chem. Ed.* **21**, 737 (1983).
107i. J. C. W. Chien, J. C. Wu and C. I. Kuo: *J. Poly. Sci., Poly. Chem. Ed.* **21**, 725 (1983).
108. P. Pino, G. Fochi, A. Oschwald, O. Piccolo, R. Muelhaupt and U. Giannini: *Poly. Sci. Technol.* **19**, 207 (1983).
109. F. S. Dyachkovskii and A. D. Pomogailo: *J. Polym. Sci., Polym. Symp.* **68**, 97 (1981).
110. A. D. Pomogailo, E. Baishiganov and G. M. Khvostik: *Kinet. Katal.* **21**, 1535 (1980).
111. P. Cossee: *J. Catal.* **3**, 80 (1964).
112. P. Cossee: *Trans. Faraday Soc.* **58**, 1226 (1962).
113. P. Cossee in: *The Stereochemistry of Macromolecules* (Ed. by A. D. Ketley) Marcel Dekker, New York, Vol. 1, Chapter 3 (1967).

114. Yu. I. Yermakov, A. M. Lazutkin, E. A. Demin, Yu. P. Grabovskii and V. A. Zakharov: *Kinet. Katal.* 13, 1422 (1972).
115. J. P. Candlin and H. Thomas: *Adv. Chem; Ser.* 132, 212 (1974).
116. V. A. Zakharov, G. D. Bukatov, V. K. Dudchenko, A. I. Minkov and Yu. I. Yermakov: *Makromol. Chem.* 175, 3085 (1974).
117. O. A. Efimov, A. I. Min'kov, V. A. Zakharov and Yu. I. Yermakov: *Kinet. Katal.* 17, 995 (1976).
118. V. A. Zakharov, V. K. Dudchenko, A. I. Min'kov, O. A. Efimov, L. G. Khomyakova, V. P. Babenko and Yu. I. Yermakov: *Kinet. Katal.* (Eng. Trans.) 17, 643 (1976).
119. J. C. W. Chien and J. T. T. Hsieh: *J. Poly. Sci., Poly. Chem. Ed.* 14, 1915 (1976).
120. V. K. Dudchenko, V. A. Zakharov, N. G. Maksimov and Yu. I. Yermakov: *React. Kinet. Catal. Lett.* 7, 419 (1977).
121. N. G. Makismov, G. A. Nesterov, V. A. Zakharov, P. V. Stechastnev, V. F. Anufrienko and Yu. I. Yermakov: *J. Mol. Catal.* 4, 167 (1978).
122. N. G. Maksimov, V. K. Dudchenko, V. F. Anufrienko, V. A. Zakharov and Yu. I. Yermakov: *Teor. Eksp. Khim.* 14, 153 (1978).
123. N. G. Maksimov, G. A. Nesterov, V. A. Zakharov, V. F. Anufrienko and Yu. I. Yermakov: *React. Kinet. Catal. Lett.* 8, 81 (1978).
124. V. K. Dudchenko, V. A. Zakharov, L. G. Echevskaya, G. D. Bukatov and Yu. I. Yermakov: *Kinet. Katal.* 19, 354 (1978).
125. V. K. Dudchenko, V. A. Zakharov, Z. K. Bukatova and Yu. I. Yermakov: *Kinet. Katal.* 19, 584 (1978).
126. W. Skupinski, I. Cieslowska and S. Malinowski: *J. Organometal. Chem.* 182, C33 (1979).
127. D. Slotfeldt-Ellingsen, I. M. Dahl and O. H. Ellestad: *J. Mol. Catal.* 9, 423 (1980).
128. L. T. Finogenova, V. A. Zakharov, A. A. Buniyat-Zade, G. D. Bukatov and T. K. Plaksunov: *Vysokomol. Soedin* A22, 404 (1978).
129. G. A. Nesterov, V. A. Zakharov, Yu. I. Yermakov, K. H. Thiele, M. Schlegel and H. Drevs: *React. Kinet. Catal. Lett.* 13, 401 (1980).
130. ICI Ltd.: *Japanese Patent* 80,147,511 (1980); *Chem. Abs.* 94, 122879 (1981).
131. Yu. I. Yermakov: *Stud. Surf. Sci. Catal.* 7A, 57 (1981).
132. R. G. Bowman, R. Nakamura, P. J. Fagan, R. L. Burwell and T. J. Marks: *JCS Chem. Comm.* 257 (1981).
133. A. D. Pomogailo, D. A. Kritskaya, A. P. Lisitskaya and A. N. Ponomarev: *Dokl. Akad. Nauk SSSR* 232, 391 (1977).
134. A. D. Pomogailo, Yu. N. Kolesnikov, S. S. Shishlov, A. G. Sokolova, V. S. Os'kin, D. A. Kritskaya, A. N. Ponomarev and F. S. D'yachkovskii: *Plast. Massy* 30 (1977); *Chem. Abs.* 88, 23496 (1978).
135. O. S. Roshchupkina, A. P. Lisitskaya, A. D. Pomogailo, F. S. D'yachovskii and Yu. G. Borod'ko: *Dokl. Akad. Nauk SSSR* 243, 1223 (1978).
136. V. A. Kabanov, V. G. Popov, V. I. Smetanyuk and L. P. Kalinina: *Vysokomol. Soedin* B23, 693 (1981).
137. I. N. Kolotsei, V. G. Popov, S. L. Davydova and V. A. Kabanov: *Vysokomol. Soedin* B23, 368 (1981).
138. O. S. Roshchupkina, A. P. Lisitskaya and N. D. Golubeva: *Kinet. Katal.* 23, 1208 (1982).
139. T. Suzuki, S. Izuka, S. Kondo and Y. Takegami: *J. Macromol. Sci. Chem.* A11, 633 (1977).

140. A. D. Pomogailo, A. P. Lisitskaya, N. S. Gor'kova and F. S. D'yachkovskii: *Dokl. Akad. Nauk SSSR* **219**, 1375 (1974).
141. Yu. A. Sangalov, Yu. B. Yasman, I. F. Gladkikh and K. S. Minsker: *Dokl. Akad. Nauk. SSSR* **265**, 671 (1982).
142. W. Copper in: *The Stereo Rubbers* (Ed. by W. M. Saltman) Wiley, New York, p. 21 (1977).
143. N. D. Golubeva, A. D. Pomogailo, A. I. Kuzaev, A. N. Ponomarev and F. S. D'yachkovskii: *J. Polym. Sci., Polym. Symp.* **68**, 33 (1980).
144. K. Soga, K. Yamamoto and S.-I. Chen: *Polym. Bull. (Berlin)* **5**, 1 (1981).
145. Yu. I. Yermakov, Yu. P. Grabovskii, A. M. Lazutkin and V. A. Zakharov: *Kinet. Katal.* **16**, 787 (1975).
146. Yu. I. Yermakov, Yu. P. Grabovskii, A. M. Lazutkin and V. A. Zakharov: *Kinet. Katal.* **16**, 911 (1975).
147. Yu. I. Yermakov, B. N. Kuznetsov, Yu. P. Grabovskii, A. N. Startsev, A. M. Lazutkin, V. A. Zakharov and A. I. Luzutkina: *Proc. Int. Symp. Cat.* (Ed. by B. Delmon and G. Jannes) Elsevier, Amsterdam, p. 145 (1975).
148. Yu. I. Yermakov, B. N. Kuznetsov, Yu. P. Grabovskii, A. N. Startsev, A. M. Lazutkin, V. A. Zakharov and A. I. Lazutkina: *J. Mol. Catal.* **1**, 93 (1976).
149. F. Dawans and D. Morel: *J. Mol. Catal.* **3**, 403 (1978).
150. K. Soga and K. Yamamoto: *Polym. Bull. (Berlin)* **6**, 263 (1982).
151. K. Soga and K. Yamamoto: *Polym. Bull. (Berlin)* **4**, 33 (1981).
152. K. G. Allum and R. D. Hancock: *British Patent* 1,295,674 (1972).
153. K. Soga, K. Myakkoku and S. Ikeda: *J. Poly. Sci., Poly. Chem. Ed.* **17**, 2173 (1979).
154. R. L. Banks and G. C. Bailey: *Ind. Eng. Chem., Prod. Res. Dev.* **3**, 170 (1964).
155. R. L. Banks: *ChemTech.* **9**, 496 (1979).
156. N. Calderon: *Acc. Chem. Res.* **5**, 127 (1972).
157. R. J. Haines and G. J. Leigh: *Chem. Soc. Rev.* **4**, 155 (1975).
158. T. J. Katz: *Adv. Organometal. Chem.* **16**, 283 (1977).
159. J. J. Rooney and A. Stewart: 'Catalysis', *Chem. Soc. Spec. Per. Rep.* **1**, 277 (1977).
160. R. H. Grubbs: *Prog. Inorg. Chem.* **24**, 1 (1974).
161. N. Calderon, J. P. Lawrence and E. A. Ofstead: *Adv. Organometal. Chem.* **17**, 449 (1979).
162. H. Pines: *The Chemistry of Catalytic Hydrocarbon Conversions*, Academic Press, New York, Chapter 7 (1981).
163. R. L. Banks in: 'Catalysis', *Chem. Soc. Spec. Per. Rep.* **4**, 100 (1981).
164. R. Pearce in: *Catalysis and Chemical Processes* (Ed. by R. Pearce and W. R. Patterson) Blackie and Son Ltd., Glasgow, Chapter 9 (1981).
165. G. W. Parshall: *Homogeneous Catalysis*, John Wiley and Sons, New York (a) Chapter 9, (b) Chapter 3 (1980).
166. *J. Mol. Catal.* **15**, Nos. 1 and 2 (1983) contains papers from the Fourth Int. Symp. on Metathesis, held in Belfast in September 1981.
167. K. Tanaka, K. Tanaka and K. Miyahara: *J.C.S. Chem. Commun.* 666 (1980).
168. K. Tanaka, K. Miyahara and K. Tanaka: *Stud. Surf. Sci. Catal.* **B7**, 1318 (1981).
169. K. Tanaka, K. Tanaka and K. Miyahara: *J. Catal.* **72**, 182 (1981).
170. K. Tanaka, K. Tanaka and K. Miyahara: *J.C.S. Chem. Commun.* 314 (1979).
171. K. Tanaka, K. Miyahara and K. Tanaka: *J. Mol. Catal.* **15**, 133 (1982).
172. A. Andreini and J. C. Mole: *J. Colloid Interface Sci.* **84**, 57 (1981).
173. Yu. I. Yermakov, B. N. Kuznetsov and V. A. Zakharov: *Catalysis by Supported Complexes*, Elsevier, Amsterdam, Chapter 6 (1981).

174. L. Bencze and J. Engelhardt: *J. Mol. Catal.* **15**, 123 (1982).
175. Yu. I. Yermakov, B. N. Kuznetsov and A. N. Startsev: *Kinet. Katal.* **15**, 539 (1974).
176. A. N. Startsev, B. N. Kuznetsov and Yu. I. Yermakov: *React. Kinet. Catal. Lett.* **3**, 321 (1975).
177. J. Schwartz and M. D. Ward: *J. Mol. Catal.* **8**, 465 (1980).
178. J. Basset, R. Mutin, G. Descotes and D. Sinon: *Compt. Rend.* **C280**, 1181 (1975).
179. S. Warwel and P. Buschmeyer: *Angew. Chem. Int. Ed.* **17**, 131 (1978).
180. J. A. K. Du Plessis, P. J. Heenop and J. J. Pienaar: *S. Afr. J. Chem.* **35**, 42 (1982).
181. R. H. Grubbs, S. Swetnick and S. C.-H. Su: *J. Mol. Catal.* **3**, 11 (1977).
182. A. D. Shebaldova, V. I. Mar'in, M. L. Khidekel, I. V. Kalechits and S. N. Kurskov: *Izv. Akad. Nauk SSSR* **2509** (1976).
183. M. D. Ward: *Diss. Abs.* **B42**, 2366 (1981).
184. J. Schwartz and M. D. Ward: *US Patent* 4,260,842 (1981); *Chem. Abs.* **95**, 42323 (1981).
185. C. U. Pittman, W. D. Honnick, M. S. Wrighton, R. D. Sanner and R. G. Austin: *Fundamental Research Homogeneous Catalysis* (Ed. by M. Tsutsui) Plenum, New York, **3**, 603 (1979).
186. R. D. Sanner, R. G. Austin, M. S. Wrighton, W. D. Honnick and C. U. Pittman: *Inorg. Chem.* **18**, 928 (1979).
187. R. D. Sanner, R. G. Austin, M. S. Wrighton, W. D. Honnick and C. U. Pittman: *Adv. Chem. Ser.* **184**, 13 (1980).
188. J. R. Budge, J. P. Scott and B. C. Gates: *J.C.S. Chem. Commun.* 342 (1983).
189. M. Deeba, B. J. Streusand, G. L. Schrader and B. C. Gates: *J. Catal.* **69**, 218 (1981).
190. J.-B. N'G. Effa, J. Lieto and J.-P. Aune: *J. Mol. Catal.* **15**, 367 (1982).
191. C. Carlini and G. Sbrana: *Advances in Organometallic and Inorganic Polymer Science* (Ed. by C. E. Carraher, J. E. Sheats and C. U. Pittman) Marcel Dekker, p. 323 (1982).
192. G. Braca, G. Sbrana, C. Carlini and F. Ciardelli: *Catalysis, Heterogeneous and Homogeneous* (Ed. by B. Delmon and G. Jannes) Elsevier, Amsterdam, p. 307 (1975).
193. C. Carlini, G. Braca, F. Ciardelli and G. Sbrana: *J. Mol. Catal.* **2**, 379 (1977).
194. E. Chiellini and C. Carlini: *Makromol. Chem.* **178**, 2545 (1977).
195. F. Ciardelli, E. Chiellini, C. Carlini and R. Nocci: *Amer. Chem. Soc., Polymer Preprints* **17**, 188 (1976).
196. G. Braca, C. Carlini, F. Ciardelli and G. Sbrana: *6th Int. Conf. Catal.*, Imperial College, London, A43 (July, 1976).
197. G. Sbrana, G. Braca, G. Valentini, G. Pazienza and A. Altomare: *J. Mol. Catal.* **3**, 111 (1977).
198. G. Braca, C. Carlini, F. Ciardelli and G. Sbrana: *Chim. Ind.* **59**, 592 (1977).
199. G. Braca, F. Ciardelli, G. Sbrana and G. Valentini: *Chim. Ind.* **59**, 766 (1977).
200. G. Valentini, G. Sbrana and G. Braca: *J. Mol. Catal.* **11**, 383 (1981).
201. N. Takahashi, I. Okura and T. Keii: *J. Mol. Catal.* **4**, 65 (1978).
202. S. A. Panichev, G. V. Kudryavtsev and G. V. Lisichkin: *Vestn. Mosk. Univ. Khim.* **19**, 730 (1978); *Chem. Abs.* **90**, 137313 (1979).
203. V. M. Vdovin, V. E. Fedorov, N. A. Pritula and G. K. Fedorova: *Izv. Akad. Nauk SSSR* 181 (1982).
204. G. V. Lisichkin, A. Ya. Yuffa, A. V. Gur'ev and F. S. Denisov: *Vestn. Mosk. Univ. Khim* **17**, 467 (1976); *Chem. Abs.* **86**, 55029 (1977).
205. Yu. M. Zhorov, A. V. Shelkov and G. M. Panchenkov: *Kinet. Katal.* **15**, 1091 (1974).
206. E. Tijero, F. Castano and E. Hermana in: *Catalysis* (Ed. by J. Hightower) North Holland, Amsterdam, pp. 32–505 (1972).

207. A. I. Zakhariev and V. V. Ivanova: *Dokl. Bolg. Akad. Nauk* 28, 1059 (1975).
208. V. A. Semikolenov, V. A. Likholobov and Yu. I. Yermakov: *Kinet. Katal.* 20, 269 (1979).
209. G. P. Potapov, V. G. Lukina and Z. K. Retunskaya: *Neftekhimiya* 20, 361 (1980).
210. L. B. Sukhobok, G. P. Potapov, V. G. Luksha, V. N. Krutii and B. D. Pokovnikov: *Izv. Akad. Nauk SSSR* 2307 (1982).
211. K. B. Wiberg and H. A. Connon: *J. Amer. Chem. Soc.* 98, 5411 (1976).
212. R. B. King and E. M. Sweet: *J. Org. Chem.* 44, 385 (1979).
213. R. J. Card and D. C. Neckers: *J. Org. Chem.* 43, 2958 (1978).
214. R. B. King and R. M. Hanes: *J. Org. Chem.* 44, 1092 (1979).
215. P. A. Grutsch and C. Kutal: *J.C.S. Chem. Commun.* 893 (1982).
216. M. Kumada: *Pure Appl. Chem.* 52, 669 (1980).
217. T. Hayashi, N. Nagashima and M. Kumada: *Tetrahedron Lett.* 21, 4623 (1980).
218. M. Terasawa, K. Kaneda, T. Imanaka and S. Teranishi: *J. Organometal. Chem.* 162, 403 (1978).

CHAPTER 10

OXIDATION AND HYDROLYSIS

The oxidation of hydrocarbons is an extremely important commercial reaction for functionalising hydrocarbons to yield products that are either important in themselves or are intermediates en route to other chemicals. The oxidative hydrolysis of ethylene to yield acetaldehyde in a single step (the Wacker process) provided the spur for the explosive growth in the study of homogeneous catalysis in the late 1950s. Accordingly, a number of investigations into supporting such catalysts have been undertaken, although with such a volatile product as acetaldehyde separation of the homogeneous catalyst is not really a major problem.

10.1. Hydrocarbon Oxidation

Vanadyl complexes supported on polystyrene are effective for the epoxidation of olefins and cycloolefins [1–3]. The supported catalysts give better conversions than their homogeneous counterparts although the reason for this is not obvious as the mechanism seems to be the same in both cases. Straight chain olefins are less active than branched olefins when supported catalysts are used whereas the reverse is true with homogeneous catalysts [1]. The selectivity depends upon the vanadyl complexes used, $VOSO_4$ being more active and more selective for epoxide formation than [$VO(acac)_2$] [2]. The vanadyl-resin link can be made by having sulphonate, acetylacetonate and ethylenediamine groups on the resin, the latter in particular allowed recycling to be carried out many times before leaching caused a significant loss of activity [3].

A series of manganese, vanadium, iron, cobalt and nickel phthalocyanine complexes supported on polystyrene have been examined as cyclohexene oxidation catalysts. However the oxidations are not very selective, 3-cyclohexene-1-ol (18–30%) and 3-cyclohexene-1-one (*ca*. 30%) being the major products; the catalytic activity was only moderately enhanced over their homogeneous analogues. Supporting was achieved either by sulphonation of the polystyrene with the chlorosulphonated metal phthalocyanine or by the condensation of aminated polymer beads with sulphonated metal phthalocyanines [4]. Cobalt phthalocyanine has also been supported on graphite [4a].

Mixed rhodium(I)-copper(II) complexes catalyse the oxidation of terminal olefins to methylketones by molecular oxygen [5]. However they have relatively short lifetimes. On the assumption that their deactivation was probably multi-ordered in rhodium, supported analogues were prepared as in reaction (1), in the hope that the support would separate the rhodium sites and prevent deactivation [6, 7].

$$SiO_2 + (MeO)_3 Si(CH_2)_3 SH \xrightarrow[\text{reflux}]{\text{xylene}} \{Si\}-(CH_2)_3 SH$$

freshly prepared $[Rh(CO)_2 S'_n] BF_4$, argon
S' = thf or EtOH

$$\begin{bmatrix} OC & & \overset{\displaystyle \{Si\}}{\underset{\displaystyle \{Si\}}{\overset{\displaystyle |}{(CH_2)_3}}} & CO \\ & Rh & Rh & \\ OC & & \underset{\displaystyle (CH_2)_3}{S} & CO \end{bmatrix} + \begin{bmatrix} \{Si\}-(CH_2)_3 S & & S' \\ & Rh & \\ OC & & CO \end{bmatrix} \quad (1)$$

The mononuclear product is the first monomeric rhodium(I)-sulphide complex stable in the presence of only carbonyl ligands and without phosphine or cyclo-pentadienyl stabilising ligands. Its stability is due to site isolation, which prevents dimer formation with associated loss of carbon monoxide. In the presence of copper(II), H^+ and O_2 the monomeric species catalyses the oxidation of 1-hexene to 2-hexanone. Copper(II) supported on poly(2-methyl-5-vinylpyridine) catalyses the oxidation of cumene as well as of the methyl group of the polymer to yield an α-picolinic acid fragment [8].

A supported cobalt(II) catalyst, 1, that is effective for the oxidation of

1

ethylbenzene and which can be reused many times has been prepared by copolymerising the diethylester of vinylphosphenic acid with acrylic acid monomers
followed by treatment with cobalt(II) chloride [9]. Crosslinking is introduced
by ball milling with methylenediacrylamide.

The oxidative-coupling of acetylenes can be promoted by copper complexes
(reaction (2)). 2-Phenylpyridyl substituted atactic polystyrene provides an

$$2\,PhC\equiv CH + \tfrac{1}{2}O_2 \xrightarrow{\text{Cu cat.}} PhC\equiv C-C\equiv CPh + H_2O \qquad (2)$$

effective support for the copper, which is initially in the form of copper(I)
chloride [10]. The reaction is believed to occur as shown in Scheme 1. The
crucial steps in this scheme involve the action of water to form the hydroxy-
bridged copper(II) species. N,N-Dimethylaminomethylated atactic polystyrene

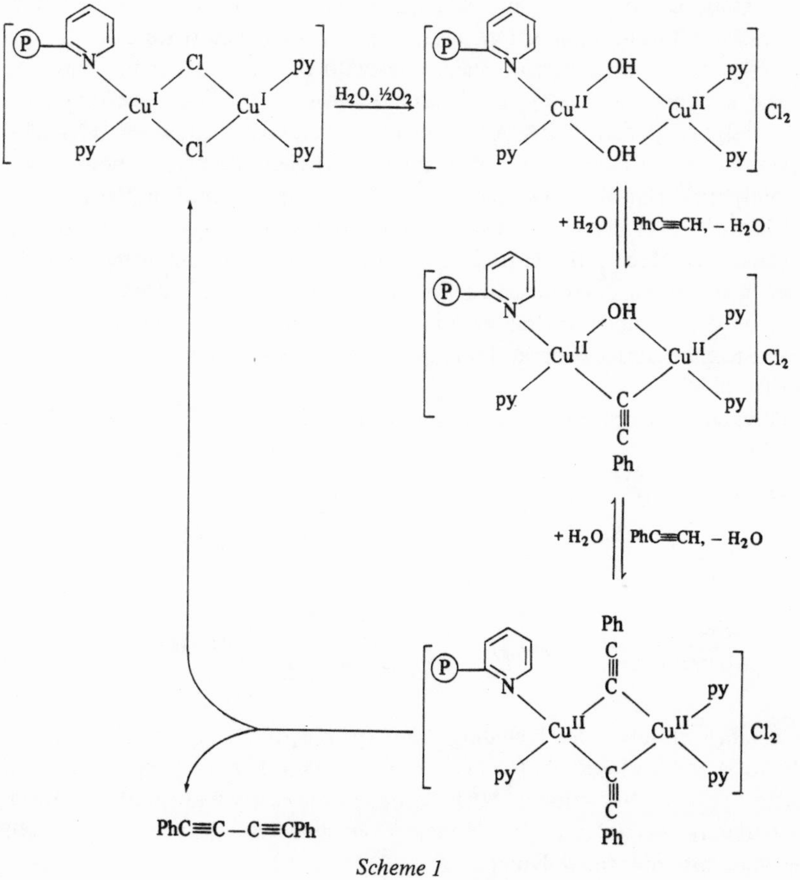

Scheme 1

was found to give considerable protection against water when the copper catalyst was supported on it [11]. As a result higher reaction rates could be achieved. This is probably due to the hydrophobic nature of the polymer reducing the concentration of water, which is formed as a product of the reaction, at the active sites and so reducing the inhibition to phenylacetylene coordination.

K[Pd(dmso)Cl$_3$] and, to a lesser extent, K$_2$PdCl$_4$ and PdCl$_2$ on natural mineral supports are more active than their homogeneous counterparts for the oxidative dehydrogenation of isopentenes to isoprene [12].

10.2. Decomposition of Peroxides

A series of iron(III) complexes cis-[Fe(tetraamine) (OH)$_2$]$^{2+}$, where tetraamine is a tetradentate. amine such as N,N'-bis(2-picolyl)ethylenediamine, bound electrostatically to poly-L-glutamate or dextran sulphate have been studied as catalysts for the decomposition of hydrogen peroxide into water and oxygen [13, 14]. The catalyst performance varies with the mode of binding and nature of the support demonstrating the importance of the local environment of the catalytically active site. Although the apparent activation energies of the supported systems are lower they are less efficient than their homogeneous counterparts. This may also be due to the intimate nature of the active site which in the polymer may not hold the hydrogen peroxide substrate so rigidly in place, so reducing the entropy of activation of the rate limiting step [13]. When a series of metallotetraphenylporphyrins were supported on alumina and NiO, only the cobalt complex had enhanced activity for hydrogen peroxide decomposition relative to its homogeneous analogue; all the others lost their activity [15].

Polystyrene supported vanadyl acetate (reaction (3)) catalyses the oxidation

$$2 \, \text{(P)}-CH_2 \underset{=O}{\overset{=O}{<}} + [VO(acac)_2]$$

$$\xrightarrow[\text{polystyrene}]{\text{(P)} = \text{cross-linked}} \text{(P)}-CH_2-\overset{O}{\underset{O}{<}} \overset{O}{\underset{O}{V}} \overset{O}{\underset{O}{>}}-CH_2-\text{(P)} + 2acacH \quad (3)$$

of a range of substrates including dmso (to Me$_2$SO$_2$), di-n-butylthioether (to ^{n}Bu$_2$SO) and cyclohexene (to cyclohexene epoxide), by t-butylperoxide, which is itself reduced to t-butanol [16]. Although the catalyst loses activity, this loss is markedly less than for its homogeneous analogue. On recycling vanadium becomes lost from the polymer.

A series of trinuclear carboxylate complexes $[M_3O(O_2CR)_6(H_2O)_3]^{n+}$, M = V, Cr, Mo, Mn, Ru, Co, and Rh bound electrostatically to a cation exchange resin catalyse the oxidation of cyclohexene by cumene hydroperoxide (reaction (4)).

$$(\text{by-product}) \qquad (4)$$

The activity depends on the metal in the order Co < V < Rh < Ru < Cr < Mn < Mo [17]. Anchoring (tetraphenylporphinato)manganese(III) acetate to an isocyanide polymeric support considerably enhances its ability to catalyse cyclohexene epoxidation [17a]. The role of the support is to isolate the manganese centres so preventing the formation of relatively inactive (μ-oxo)-manganese(IV) dimeric species. Copper complexes supported on phosphinated pvc $[(\textcircled{P}-PPh_2)_nCu_4Cl_4]$ and $[(\textcircled{P}-PPh_2)_nCu_3Cl_4]$ and phosphinated polymer supported metal dithiolenes catalyse the decomposition of t-butyl-hydroperoxide [18, 19]

10.3. Oxidation of Organic Compounds

The importance of the nature of the support in copper(II) promoted oxidations is well illustrated by the oxidising ability of poly-L-histidine supported copper(II). This support promotes the oxidation of negatively charged and neutral substrates and inhibits the oxidation of positively charged substrates [20, 21]. However an attempt to catalyse stereoselectively the hydrogen peroxide oxidation of L(+)-ascorbic acid using $[Fe(tetrapy)(OH)_2]^{2+}$ supported on partially ordered poly(L-glutamate) or poly(D-glutamate) yields no stereoselectivity in the product [22]. Nevertheless, the second-order rate constant was greater for the complex supported on poly(D-glutamate), suggesting that the chiral polymer residues adjacent to the catalytically active site do play an important role in the catalyst.

Exchanging the chloride ions on Amberlyst A–26 for $HCrO_4^-$ gives a resin that readily oxidises alcohols to aldehydes and ketones [23]. The resin did not noticeably lose activity either on storing in air at room temperature for several weeks or on refluxing for 5 hours in benzene or hexane. It did allow simple isolation of the product.

An effective catalyst for the oxidation of mercaptans should possess both oxidation and basic sites for cooperative interaction because of the oxidation mechanism (reactions (5) and (6)).

$$2\,RSH + OH^- \longrightarrow 2\,RS^- + 2\,H_2O \tag{5}$$

$$2\,RS^- + 2\,H_2O + O_2 \longrightarrow RSSR + H_2O_2 + 2\,OH^- \tag{6}$$

This has been achieved by using poly(vinylamine) or polyacrylamide, which supply the basic sites, as support for cobalt(II)-tetracarboxyphthalocyanine and -tetraaminophthalocyanine [24, 25]. The catalysts do not require additional base and are more active than their homogeneous analogues due to the formation of highly reactive mononuclear cobalt(II)-superoxo species on the support as opposed to the less reactive dimeric cobalt(II)-peroxo species, CoOOCo, in solution [25]. Polymer supported cobalt-tetraphenylporphyrins and phthalocyanines behave similarly [26].

A large enhancement in product selectivity was observed when 2,6-dimethylphenol was oxidised in the presence of support [Co(P—saldpt)] (prepared as in reaction (7)) compared to its homogeneous analogue [27].

$$[Co(P—saldpt)] \tag{7}$$

Thus the ratio of 3,3',5,5'-tetramethyldiphenylquinone (dpq) to 2,6-dimethyl-1,4-benzoquinone (dmbq) is enhanced when the catalyst is supported. This is ascribed to the free radical nature of the reaction. Low concentrations of cobalt in the polymer increase the probability of two radicals combining to form dpq, whereas at higher cobalt concentrations and in solution the probability of organic radical encounters with cobalt(II) is increased and more dmbq is formed [27]. Thus, although free-radical reactions normally have the disadvantage of yielding a variety of products, supporting the catalyst offers considerable potential for altering the selectivity.

The $FeCl_3$ catalysed oxidation of 3,5-di-*t*-butylcatechol to 3,5-di-*t*-butoxybenzoquinone by oxygen in tetrahydrofuran is significantly promoted by the addition of silica, γ-alumina or active carbon [27a]. The $FeCl_3$ is adsorbed on to the support, the adsorption increasing in the order $SiO_2 \ll Al_2O_3 < C$. However since silica is the most active support strength of adsorption cannot be the determining factor; accessibility is certainly important as evidenced by the lower activity of the active carbon where adsorption within the pores renders the catalyst inaccessible.

Supporting copper(II) on polystyrene functionalised with imidazole ligands yielded a catalyst that promoted the oxidation of 2,6-dimethylphenol largely to dpq but at about 1% of the rate of its homogeneous analogue [28]. When the support was altered to a graft copolymer of styrene and 4-vinylpyridine on non-porous silica spheres the activity rose to as much as 70% of the homogeneous catalysts [29]. Copper chloride supported on pyridyl-substituted polymers (reaction (8)) catalyses the oxidation of 1,2-dihydroxybenzene to o-quinone [30].

$$\text{(P)}-py + CuCl \xrightarrow{\text{MeCN}} [(\text{(P)}-py)CuCl] \xrightarrow[\text{CH}_2\text{Cl}_2]{\text{O}_2} [(\text{(P)}-py)CuCl_2] +$$

$$+ [(\text{(P)}-py)Cu\underset{\text{O}}{\overset{\text{O}}{\diagdown\diagup}}Cu(py-\text{(P)})] \tag{8}$$

10.4. Oxidation of Inorganic Compounds

Cross-linked sulphonated polystyrene fibres which have been soaked in aqueous palladium(II) chloride solution catalyse the oxidation of carbon monoxide to carbon dioxide in 98% conversion [31]. Silica-supported copper(II) acetate has an unusually high activity in promoting the oxidation of CO by N_2O at 150° compared to the conventional silica-supported copper(II)-ammine complex [32]. The high activity is due to the dimeric nature of the copper(II) on the support. Cobalt tetraphenylporphyrin supported on TiO_2 is very effective at promoting the oxidation of carbon monoxide at room temperature [32a], whilst rhodium and iridium porphyrin complexes supported on carbon promote the electrochemical oxidation of carbon monoxide [32b].

Copper(II) supported on poly(4-vinylpyridine) and polystyrene bound 2,2'-bipyridine catalyse the oxidation of thiosulphate, trithionate and tetrathionate to sulphate (reaction (9)) [33–37].

$$S_2O_3^{2-} + 2O_2 + H_2O \longrightarrow 2SO_4^{2-} + 2H^+ \tag{9}$$

The effectiveness of the catalyst depends on particle size, increasing with decreasing particle size. Although the initial rates of the polymer supported catalyses are less than of their homogeneous analogues, the rate of the homogeneous catalyses quickly fall, so that the overall rate is greater when the supported catalysts are used [33]. This is due to the reduced stability of the supported complexes, which allows the nitrogen ligands to be displaced by the thiosalt.

This is more readily reversed in the polymer, following thiosalt oxidation, thus preserving the copper(II) salt. Further improvement on activity was obtained by cross-linking poly(4-vinylpyridine) with copper(II) sulphate to obtain a supported system (2) in which copper(II) can readily become coordinatively

$$-(-CH_2-CH-)_n-(-CH_2-CH-)_m-(-CH_2-CH-)_l-$$

2

unsaturated [36]. Poly(4-vinylpyridine) quaternised with 1,2-dibromoethane in the presence of copper(II) ions, which acts as a template, increases the activity of the catalyst for thiosulphate still further [36]. This is probably due to the electrostatic attraction of the quaternary ammonium salts for the thiosalts. However there is a reduction in activity for $S_3O_6^{2-}$ and $S_4O_6^{2-}$, which is probably due to the greater steric hindrance in the cross linked polymer which more than counterbalances the electrostatic attraction of quaternisation in the case of the larger anions.

10.5. Chlorination

The chlorination of alkanes using a mixture of platinum(II) and platinum(IV) complexes as catalyst and oxidant respectively is well-known in homogeneous solution [38]. Recently this system has been shown to be active for the chlorination of methane when supported on silica [39, 39a]. Titanium(IV) supported on silica $[(\{Si\}O)_n TiCl_{4-n}]$, and on cyclopentadiene supported on polystyrene $[(\textcircled{P}-CH_2C_5H_4)TiCl_2Cp]$ and $[(\textcircled{P}-CH_2C_5H_4)_2TiCl_2]$ are effective catalysts for the hydroalumination of olefins by lithium aluminium hydride (reaction (10)) [40].

$$CH_2=CH(CH_2)_nCH=CH_2 \xrightarrow[Ti^{IV}\ cat.]{LiAlH_4\ (al-H)} al(CH_2)_{n+2}CH=CH_2$$

$$\downarrow LiAlH_4,\ Ti^{IV}\ cat. \qquad (10)$$

$$al(CH_2)_{n+4}al$$

Reaction of the product with halogen replaces the aluminium hydride group by halogen so that reaction (10) gives a route to ω-halo-terminal olefins and α,ω-alkanes. Since the progress of reaction (10) depends on the bulkiness of the olefin and the nature of the catalyst support, supporting titanium(IV) can promote the formation of mono-aluminated products and hence formation of ω-halo-terminal olefins [40].

10.6. Ammoxidation

The ammonoxidation of alkanes to nitriles is catalysed by vanadia supported on silica-alumina or γ-alumina [41].

10.7. Hydroxylation of Aromatic Compounds

A silica-supported iron(III)-catechol catalyst is active for the hydroxylation of benzene by hydrogen peroxide to yield to phenol (reaction (11)) [42].

$$PhH + H_2O_2 \xrightarrow{\text{Fe}^{III}\text{ cat.}} PhOH + H_2O \tag{11}$$

Supporting the catalyst obviates the need to use a two phase aqueous-organic solvent mixture. The iron(III) complex of **3** catalyses the hydroxylation of anisole by hydrogen peroxide to give a 95% yield of methoxyphenols [43].

3

10.8. Hydroxylation of Olefins

The hydroxylation, or oxidative hydrolysis, of olefins was one of the earliest homogeneously catalysed reactions to be developed into a commercial process as the Wacker process for the conversion of ethylene to acetaldehyde in a single step [44–48]. In spite of the volatility of the product, and hence its ease of separation, there have been a number of investigations of supported catalysts for this reaction. Such catalysts could be useful if they were very selective with higher olefins.

Polymers with $-CH_2CN$ side chains react with palladium(II) chloride to form $[(\circledP-CH_2CN)_2PdCl_2]$ which in addition to catalysing the oxidation of ethylene to acetaldehyde also promote butanal formation by coupling two ethylene units together [49]. The homogeneous catalysts do not do this. $[PdCl_4]^{2-}$ exchanged on to a quaternary ammonium functionalised polystyrene resin, Amberlyst A–27, has been shown effectively to oxidatively hydrolyse ethylene, in the presence of copper(II) to reoxidise palladium(0) in a full-scale plant [50]. The selectivity was essentially the same as in homogeneous solution and total olefin conversions could be achieved. Palladium(II) sulphate on silica gel showed high catalytic activity for the gas phase oxidation of ethylene to acetaldehyde at 95–155 °C when promoted by nitric acid [51]. Since a number of reagents in addition to copper(II) can be used to reoxidise the palladium(0) formed during the Wacker reaction, an attempt has been made to build the reoxidation agent into the support in the form of p-quinone groups, as in **4** and **5** [52].

Both materials combine with palladium(II) to form catalysts that do not require $CuCl_2$ addition. Reoxidation of the quinone appears to be the rate-limiting step.

10.9. Carboxylation of Olefins and Aromatic Compounds

The carboxylation of olefins (reaction (12)) is formally very similar to the hydroxylation reaction, and is also promoted by palladium(II) catalysts.

$$CH_2{=}CH_2 + RCOOH \xrightarrow{+\frac{1}{2}O_2} CH_2{=}CHOCOR + H_2O \qquad (12)$$

Palladium(II) chloride supported on pvc promotes the hydrocarboxylation (R = H) of 1-nonene to form $n\text{-}C_9H_{19}COOH$ in high yield and selectivity [53]. Phosphinated polystyrene bound palladium(II) complexes catalyse the ethoxycarbonylation of 1-pentene in 1 : 1 ethanol/thf with greater selectivity to ethylhexanoate than the corresponding homogeneous catalyst $[Pd(PPh_3)_2Cl_2]$ [54, 55].

The palladium(II) catalysed acetoxylation of aromatic compounds is of particular interest because it gives an anomalously high *meta*-selectivity. This is retained when palladium(II) acetate is supported on poly(4-vinylpyridine). The polymer supported catalyst can be reused several times with little loss of activity between runs [56].

10.10. Vinyl Ester and Ether Exchange

Palladium(II) complexes readily promote vinyl ester and ether exchange (reactions (13) and (14)) [48]. Palladium(II) complexes supported on activated charcoal [57] or linked to anion-exchange resins catalyse reaction (13) [58] whilst palladium(II) complexes anchored to phosphinated silica readily catalyse reaction (14) [59].

$$CH_2{=}CHCOOR + R'COOH \rightleftharpoons CH_2{=}CHCOOR' + ROH \quad (13)$$

$$CH_2{=}CHOR + R'OH \rightleftharpoons CH_2{=}CHOR' + ROH \quad (14)$$

10.11. Nitrile Hydrolysis

Amides can be obtained by the hydrolysis of nitriles in the presence of palladium(II). Polystyrene anchored bipyridyl palladium(II) complexes such as $[Pd((P){-}bipy)Cl(OH)]_n$ are effective supported catalysts for this reaction [60].

10.12. Nucleophilic Substitution of Acetate Groups

The replacement of acetate groups by other nucleophiles is an important synthetic route in organic chemistry. Palladium(0) complexes may be used as catalysts. When these complexes are supported on phosphinated silica gel or polystyrene not only is their reactivity retained but, because of 'steric steering' they may have greater selectivity than their homogeneous analogues [61]. This is well illustrated by reactions (15) and (16), where the supported catalyst was prepared by treating the phosphinated support with $[Pd(PPh_3)_4]$.

$[Pd(PPh_3)_4]$	69%	31%
Pd⁰ on phosphinated polystyrene	80%	20%

COOMe + Et$_2$NH $\xrightarrow{\text{Pd}^0 \text{ cat. in thf}}$ COOMe + COOMe (16)

OAc → NEt$_2$ ... NEt$_2$

[Pd(PPh$_3$)$_4$]	67%	33%
Pd^0 on phosphinated polystyrene	100%	0%
Pd^0 on phosphinated silica gel	100%	0%

Such catalysts could be stored for up to 2 months in air without loss of activity in contrast to [Pd(PPh$_3$)$_4$] which rapidly decomposes in air.

10.13. Stereoselective Hydrolysis of Esters

If it were possible to hydrolyse enantioselectively one isomer of a D,L-mixture of amino acid esters more rapidly than the other, it would be possible to isolate readily the more useful L-amino acid. The copper and nickel complexes of polystyrene supported $\text{(P)}-(C_6H_4-p)-CH_2\text{-L-Cys}(CH_2CH_2NH_2)OH$ and $\text{(P)}-(C_6H_4-p)-CH_2\text{-L-Cys}(CH_2COOH)OH$ do in fact catalyse the hydrolysis of the methyl esters of L-phenylalanine and L-histidine more rapidly than the D-enantiomers. Thus a facile enantiomeric enrichment can be carried out using the enantioselectivity of the ester hydrolysis reaction [62]. A similar reaction has been effected by preparing chiral complexes by copolymerising $4\text{-CH}_2=\text{CHC}_6H_4CH_2\text{-His-OMe}$ with $CH_2=CH(CH_3)COOCH_2CH_2OH$ and a cross-linking agent $(CH_2=CH(CH_3)COOCH_2)_2$ and complexing the product with nickel(II). The D-isomer of His-OMe was hydrolysed faster than the L-isomer so that a column could be set up in which pure L-isomer could be obtained from a DL-mixture [63, 64].

References

1. G. L. Linden: *Diss. Abs.* **B37**, 2287 (1976).
2. G. L. Linden and M. F. Farona: *Inorg. Chem.* **16**, 3170 (1977).
3. G. L. Linden and M. F. Farona: *J. Catal.* **48**, 284 (1977).
4. M. Gebler: *J. Inorg. Nucl. Chem.* **43**, 2759 (1981).
4a. S. N. Pobedinski, A. A. Trofimenko and L. R. Bychkova: *Izv Vyssh. Uchebn. Zaved.* **26**, 316 (1983); *Chem. Abs.* **98**, 186320 (1981).
5. H. Mimoun, M. M. P. Machirant and I. S. de Roch: *J. Amer. Chem. Soc.* **100**, 5437 (1978).

6. E. D. Nyberg and R. S. Drago: *J. Amer. Chem. Soc.* **103**, 4966 (1981).
7. E. D. Nyberg and R. S. Drago: AD-A 102995 Report (1981); *Chem. Abs.* **96**, 92357 (1982).
8. V. V. Berentsveig, O. E. Dotsenko, A. I. Kokorin, V. D. Kopylova and E. L. Frumkina: *Izv. Akad. Nauk SSSR Ser. Khim.* 2211 (1982).
9. A. A. Efendiev, T. N. Shakhtakhtinsky, L. F. Mustafaeva and H. L. Shick: *Ind. Eng. Chem. Prod. Res. Dev.* **19**, 75 (1980).
10. H. C. Meinders, N. Prak and G. Challa: *Makromol. Chem.* **178**, 1019 (1977).
11. G. Challa and H. C. Meinders: *J. Mol. Catal.* **3**, 185 (1977).
12. V. V. Pogozhil'skii, Yu. N. Usov, T. G. Vaistub and E. V. Skvortsova: *Nauchn. Osn. Pererab. Nefti Gaza Neftekhim.* 142 (1977); *Chem. Abs.* **92**, 59258 (1980).
13. M. Barteri and B. Pispisa: *Gazz. Chim. Ital.* **106**, 499 (1976).
14. M. Barteri, M. Farinella and B. Pispisa: *Biopolymers* **16**, 2569 (1977).
15. I. Mochida, A. Yasutake, H. Fujitsu and K. Takeshita: *J. Phys. Chem.* **86**, 3468 (1982).
16. S. Bhaduri, A. Ghosh, H. Khwaja: *J. Chem. Soc. Dalton* 447 (1981).
17. T. Szymanska-Buzar and J. J. Ziolkowski: *J. Mol. Catal.* **11**, 371 (1981).
17a. A. W. Van der Made, J. W. H. Smeets, R. J. M. Nolte and W. Drenth: *J.C.S. Chem. Commun.* 1204 (1983).
18. K. A. Abdulla, N. P. Allen, A. N. Badran, R. P. Burns, J. Dwyer, C. A. McAuliffe and N. D. A. Toma: *Chem. Ind.* 273 (1976).
19. R. P. Burns, J. Dwyer and C. A. McAuliffe: *17th Int. Coord. Chem. Conf., Hamburg,* 36 (1976).
20. A. Levitzki, I. Pecht and M. Anbar: *Nature* **207**, 1386 (1965).
21. I. Pecht, A. Levitzki and M. Anbar: *J. Amer. Chem. Soc.* **89**, 1587 (1967).
22. M. Barteri, B. Pispisa and M. V. Primiceri: *J. Inorg. Biochem.* **12**, 167 (1980).
23. G. Cainelli, G. Cardillo, M. Orena and S. Sandri: *J. Amer. Chem. Soc.* **98**, 6737 (1976).
24. J. H. Schutten and J. Zwart: *J. Mol. Catal.* **5**, 109 (1979).
25. J. Zwart and J. H. M. C. Van Wolput: *J. Mol. Catal.* **5**, 235 (1979).
26. L. D. Rollman: *7th Int. Conf. Organometal. Chem., Venice* 229 (1975).
27. R. S. Drago, J. Gaul, A. Zombeck and D. K. Straub: *J. Amer. Chem. Soc.* **102**, 1033 (1980).
27a. T. Funabiki, T. Sugimoto and S. Yoshida: *Chem. Lett.* 1097 (1982).
28. F. B. Hulsbergen, J. Manassen, J. Reedijk and J. A. Welleman: *J. Mol. Catal.* **3**, 47 (1977).
29. J. P. J. Verlaan, J. P. C. Bootsma and G. Challa: *J. Mol. Catal.* **14**, 211 (1982).
30. M. Rogic: *12th Sheffield-Leeds Int. Symp. Organometal. Inorg. Cat. Chem.* (1983).
31. K. Sheraine and Y. Tamura: *Japan. Patent* 74 44,992 (1974); *Chem. Abs.* **81**, 126379 (1974).
32. N. Kakuta, A. Kazusaka and K. Miyahara: *Chem. Lett.* 913 (1982).
32a. I. Mochida, K. Suetsugu, H. Fujitsu and K. Takeshita: *Chem. Lett.* 177 (1983).
32b. J. F. Van Baar, J. A. R. Van Veen, J. M. Van der Eijk, T. J. Peters and N. De Wit: *Electrochim. Acta* **27**, 1315 (1982).
33. M. Chanda, K. F. O'Driscoll and G. L. Rempel: *J. Mol. Catal.* **7**, 389 (1980).
34. M. Chanda, K. F. O'Driscoll and G. L. Rempel: *Prepr. Canad. Symp. Catal.* **6**, 44 (1979).
35. M. Chanda, K. F. O'Driscoll and G. L. Rempel: *J. Mol. Catal.* **8**, 339 (1980).
36. M. Chanda, K. F. O'Driscoll and G. L. Rempel: *J. Mol. Catal.* **11**, 9 (1981).
37. M. Chanda, K. F. O'Driscoll and G. L. Rempel: *J. Mol. Catal.* **12**, 197 (1981).
38. D. E. Webster: *Adv. Organometal. Chem.* **15**, 147 (1977).

39. V. P. Tret'yakov, G. P. Zimtseva, E. S. Rudakov and A. N. Osetskii: *React. Kinet. Catal. Lett.* **12**, 543 (1979).

39a. V. P. Tret'yakov and A. N. Osetskii: *Kinet. Katal.* **23**, 1126 (1982).

40. F. Sato, H. Ishikawa, Y. Takahashi, M. Muira and M. Sato: *Tetrahedron Lett.* 3745 (1979).

41. M. Sze: *US Patent* 4,284,781 (1981); *Chem. Abs.* **95**, 157487 (1981).

42. S. Tamagaki, K. Hotta, and W. Tagaki: *Chem. Lett.* 651 (1982).

43. J. Suh and K. Y. Kim: *Bull. Korean Chem. Soc.* **1**, 113 (1980); *Chem. Abs.* **94**, 102388 (1981).

44. J. Smidt: *Chem. Ind. (London)* 54 (1962).

45. J. Smidt, W. Hafner, R. Sieber, J. Sedlmeier and A. Sabel: *Angew. Chem. Int. Ed.* **1**, 80 (1962).

46. A. Aguilo: *Adv. Organometal. Chem.* **5**, 321 (1967).

47. F. R. Hartley: *Chem. Rev.* **69**, 799 (1969).

48. P. M. Henry: *Palladium Catalyzed Oxidation of Hydrocarbons*, D. Reidel, Dordrecht, Holland, Chapter 2 (1980).

49. M. Kraus and D. Tomanova: *J. Poly. Sci., Poly. Chem. Ed.* **12**, 1781 (1974).

50. R. Linarte-Lazcano, J. Valle-Macchorro and D. Cuatecontzi in: *Catalysis, Heterogeneous and Homogeneous* (Ed. by B. Delmon and G. Jannes) Elsevier, Amsterdam, 467 (1975).

51. K. H.-D. Liu, K. Fujimoto and T. Kunugi: *Ind. Eng. Chem. Prod. Res. Dev.* **16**, 223 (1977).

52. H. Arai and M. Yashiro: *J. Mol. Catal.* **3**, 427 (1978).

53. M. R. Popchenko, M. N. Manakov and T. I. Tarasova: *Neftepererab. Neftekhim. (Moscow)*, 37 (1982); *Chem. Abs.* **97**, 181677 (1982).

54. C. U. Pittman, G. M. Wilemon, Q. Y. Ng and L. I. Flowers: *Amer. Chem. Soc. Div. Polym. Chem., Polym. Preprint* **22**, 153 (1981).

55. C. U. Pittman and Q. Y. Ng: *US Patent* 4,258,206 (1981); *Chem. Abs.* **94**, 208332 (1981).

56. L. Eberson and L. Jonsson: *Acta. Chem. Scand.* **B30**, 579 (1976).

57. K. Blum and R. Strasser: *European Pat. Appl.* 54,158 (1982); *Chem. Abs.* **97**, 151500 (1982).

58. N. P. Allen, F. O. Bamiro, J. Dwyer, R. P. Burns and C. A. McAuliffe: *Inorg. Chim. Acta* **28**, 231 (1978).

59. S. I. Chistyakov, V. A. Semikolenov, V. A. Likholobov and Yu. I. Yermakov: *React. Kinet. Catal. Lett.* **13**, 177 (1979).

60. A. Gaset, G. Constant, P. Kalck and G. Villain: *French Patent* 2,419,929 (1979); *Chem. Abs.* **93**, 45991 (1980).

61. B. M. Trost and E. Keinan: *J. Amer. Chem. Soc.* **100**, 7779 (1978).

62. I. A. Yamokov, B. B. Berezin, L. A. Belchich and V. A. Davankov: *Makromol. Chem.* **1980**, 799 (1979).

63. N. Spassky, M. Reix, J. P. Guette, M. Guette, M. O. Sepulchre and J. M. Blanchard: *Comp. Rendu* **287**, 589 (1978).

64. N. Spassky, M. Reix, M. O. Sepulchre and J. P. Guette: *Makromol. Chem.* **184**, 17 (1983).

CHAPTER 11

CONCLUSIONS AND FUTURE POSSIBILITIES

In this chapter we shall draw together some conclusions based on the previous work with supported metal complex catalysts and attempt to indicate some of the future possibilities.

11.1. Sequential Multistep Reactions

Many chemically useful catalytic processes involve the conversion of one compound into another and this into a third. Two quite separate catalysts are used for the two steps. It would be a great advantage if both catalysts could be bound to the same support so that the product of the first reaction could be used as the substrate for the second. This concept has been developed for reactions catalysed by immobilised enzymes [1]. Thus glucose can be converted to glucose 1-phosphate and then glucose 6-phosphate using a polystyrene matrix carrying both hexokinase and glucose 6-phosphate isomerase.

Attempts to link two different metal complex centres to a single support clearly introduce extra complications in comparison with using two separate reaction vessels. These include:

(i) The catalysts may not act individually but may interfere with each other, for example the ligands of one catalyst may affect the functioning of the second.

(ii) It is possible that either catalyst may be destroyed by products formed in the presence of the other.

(iii) Extra side-products may be formed due to one catalyst promoting extra reactions of the products of the other catalyst.

(iv) Only in certain reactions will the conditions for one reaction (temperature, pressure, solvent etc) be identical to those of the other.

A number of sequential multistep reactions have been examined to determine whether or not they are feasible [2–11]. In the sequential cyclooligomerisation – hydroformylation of butadiene (reaction (1)) one of the products of the first reaction, vinylcyclohexene, is then selectively hydroformylated at its terminal double bond to form the linear and branched aldehydes **1** and **2** which may be readily separated from the cyclic polyolefin products of the initial cyclooligomerisation [2–5].

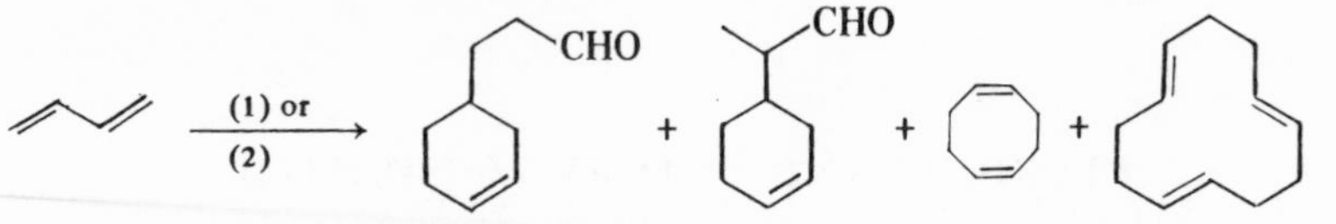

(1)

1 2

(1) $[(\text{Ⓟ} \langle {}^{\text{PPh}_2)_2\text{Ni(CO)}_2}_{\text{PPh}_2)_3\text{RhH(CO)}}]$

(2) $[(\text{Ⓟ}-\text{PPh}_2)_2\text{Ni(CO)}_2]$ and $[(\text{Ⓟ}-\text{PPh}_2)_3\text{RhH(CO)}]$

100 °C 24 h; then 70–80 °C 500 psi 1 : 1 H_2/CO, 18 h

The two catalysts, $[(\text{Ⓟ}-\text{PPh}_2)_2\text{Ni(CO)}_2]$ for the cyclooligomerisation and $[(\text{Ⓟ}-\text{PPh}_2)_3\text{RhH(CO)}]$, can either be bound to the same polystyrene resin or may be bound to two separate resins which are mixed in the reactor [2, 3].

A second sequential multistep reaction that has been demonstrated is the cyclooligomerisation-hydrogenation of butadiene (reaction (2)) [2–5]. The

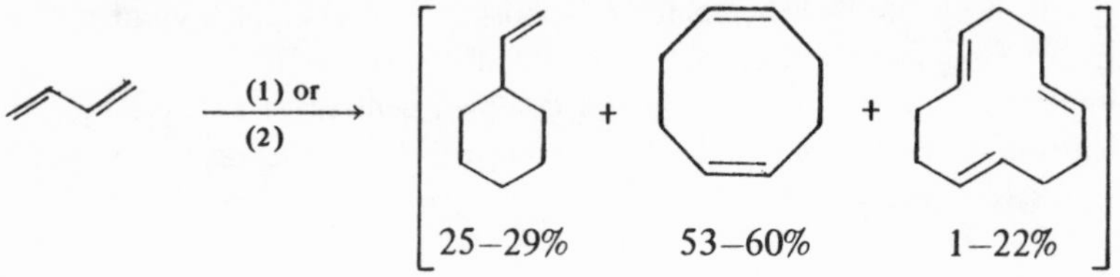

(2)

3 4 5

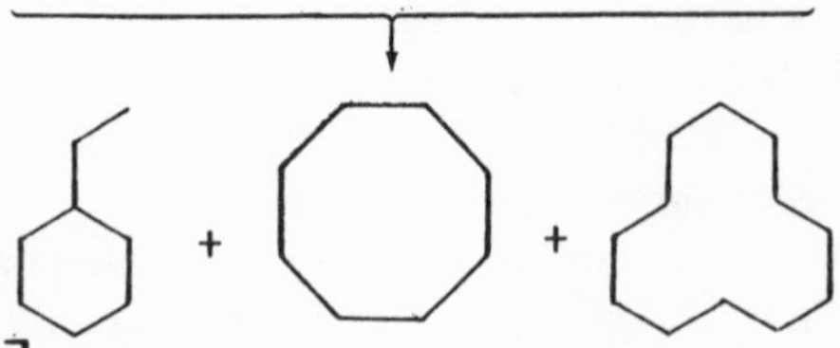

100 °C 24 h; then 50 °C
250 psi H_2, 4h

(1) $[(\text{Ⓟ} \langle {}^{\text{PPh}_2)_2\text{Ni(CO)}_2}_{\text{PPh}_2)_3\text{RhCl}}]$

(2) $[(\text{Ⓟ}-\text{PPh}_2)_2\text{Ni(CO)}_2]$ and $[(\text{Ⓟ}-\text{PPh}_2)_3\text{RhCl}]$

cyclooligomerisation is carried out over $[(\text{P}{-}PPh_2)_2Ni(CO)_2]$ as in reaction (1), and the products then hydrogenated over $[(\text{P}{-}PPh_2)_3RhCl]$. Although the nickel(0) and rhodium(I) did not interfere with each other's action the system lost activity after between 1100 and 1500 turnovers due to deactivation of the nickel(0) catalyst. A further multistep reaction has been the selective hydrogenation of the polyolefins formed initially in reaction (2) to monoolefins using $[(\text{P}{-}PPh_2)_2Ru(CO)_2Cl_2]$ (reaction (3)) [2]. The initial cyclooligomers,

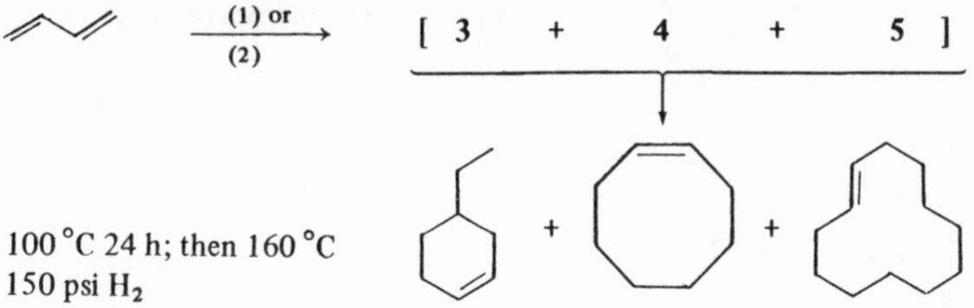

$$\overset{\text{(1) or}}{\underset{(2)}{\longrightarrow}} \quad [\quad 3 \quad + \quad 4 \quad + \quad 5 \quad] \tag{3}$$

100 °C 24 h; then 160 °C
150 psi H_2

(1) $\left[(\text{P})\diagdown{\begin{array}{l} PPh_2)_2Ni(CO)_2 \\ PPh_2)_2Ru(CO)_2Cl_2 \end{array}}\right]$

(2) $[(\text{P}{-}PPh_2)_2Ni(CO)_2]$ and
$[(\text{P}{-}PPh_2)_2Ru(CO)_2Cl_2]$

3, 4 and 5, were selectively hydrogenated to the monoolefins in 95% yield provided the phosphorus : ruthenium ratio was kept high (greater than 8 : 1).

Although both metal complex centres can be attached to the same polymer bead, it is easier to tailor the selectivities in sequential reactions by attaching them to separate beads. This enables both ligand : metal ratios and the nature of the ligands to be varied independently. Thus $[(\text{P}{-}PPh_2)_2Ni(CO)_2]$ is highly selective towards vinylcyclohexene formation in the cyclooligomerisation of butadiene when the carbon monoxide pressure is low. By using $[(\text{P}{-}PPh_2)_3\cdot RhH(CO)]$ with a high phosphorus : rhodium ratio it is then possible to obtain the terminal aldehyde in high selectivity in a sequential cyclooligomerisation-hydroformylation reaction (reaction (4)).

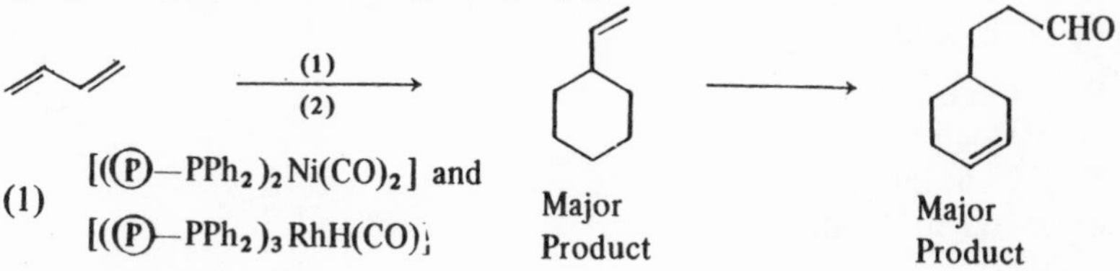

$$\overset{(1)}{\underset{(2)}{\longrightarrow}} \tag{4}$$

Major Product   Major Product

(1) $[(\text{P}{-}PPh_2)_2Ni(CO)_2]$ and
$[(\text{P}{-}PPh_2)_3RhH(CO)];$

115 °C, CO, 24h;
(2) then 250 psi
H_2/CO, 70 °C

Linear butadiene oligomerisation-hydroformylation has been achieved by combining a polymer bound nickel bromide catalyst that had been reduced with sodium borohydride with a rhodium(I) hydroformylation catalyst (reaction 5)

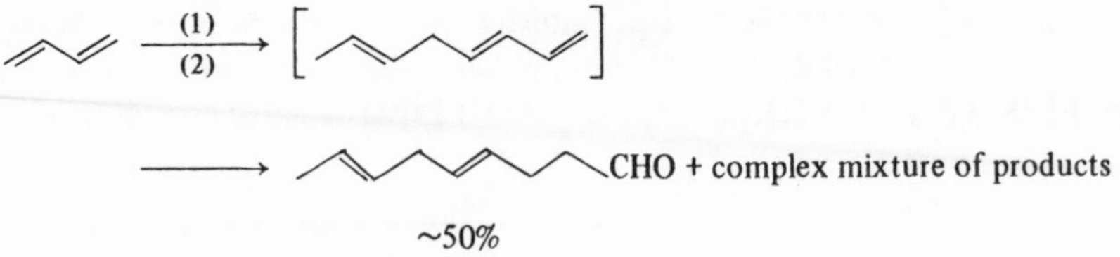

(5)

~50%

(1) $[(\text{P})-PPh_2)_2\,NiBr_2] + NaBH_4$ and
    $[(\text{P})-PPh_2)_3\,RhH(CO)]$

(2) $100\,^\circ C$ 24 h; then $65-70\,^\circ C$
    500 psi H_2/CO

[9]. The sequential linear oligomerisation-acetoxylation-hydrogenation of butadiene has been effected using a supported palladium(II), rhodium(I) catalyst system as in reaction (6) [10].

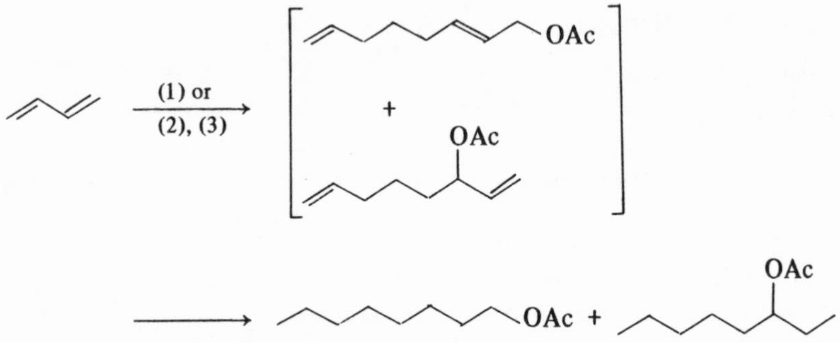

(6)

(1) $\left[(\text{P}) \begin{array}{c} -PPh_2)_2\,PdCl_2 \\ -PPh_2)_3\,RhCl \end{array} \right]$ or

(2) $[(\text{P})-PPh_2)_2\,PdCl_2]$
    $[(\text{P})-PPh_2)_3\,RhCl]$ and
    with HOAc, $Et_3\,N$

(3) $100\,^\circ C$ 24 h; then
    350 psi H_2, $50\,^\circ C$, 4h

The sequential three-step synthesis of propene to 2-ethylhexanal has been accomplished as in reaction (7), which represents all but the last step of the Aldox

$\diagup\!\!\!\diagup\diagdown + H_2 + CO$

Step 1 $\downarrow$

$\diagup\!\!\!\diagdown\!\!\!\diagup CHO \xrightarrow[\text{Step 2}]{\substack{\text{Aldol} \\ \text{Condensation}}} \diagup\!\!\!\diagdown\!\!\!\diagup\!\!\!\diagdown CHO \xrightarrow[\text{Step 3}]{H_2} \diagup\!\!\!\diagdown\!\!\!\diagup\!\!\!\diagdown CHO$ (7)

$+$ (Major)

$\diagup\!\!\!\diagdown CHO \xrightarrow[\text{Step 2}]{\substack{\text{Aldol} \\ \text{Condensation}}} \diagup\!\!\!\diagdown\!\!\!\diagup CHO \xrightarrow[\text{Step 3}]{H_2} \diagup\!\!\!\diagdown\!\!\!\diagup CHO$

process for the conversion of propene nitro-2-ethylhexanol [11]. The catalyst used was the multifunctional rhodium(I)-secondary amine catalyst prepared as in reaction (8).

$(P)\!\!-\!\!\langle\bigcirc\rangle\!\!-\!\!CH_2Cl \xrightarrow[\text{2. [Rh(PPh}_3)_2Cl(CO)]}{\text{1. LiPPh}_2} (P)$ ⟨with⟩ $CH_2PPh_2)_2\,Rh(CO)Cl$ and CH_2Cl

$\xrightarrow{EtNH_2} (P)$ ⟨with⟩ $CH_2PPh_2)_2\,Rh(CO)Cl$ and CH_2NHEt (8)

The initial hydroformylation is catalysed by the rhodium complex. The resulting aldehyde then undergoes the aldol condensation under the influence of the secondary amine groups. The aldol products are then hydrogenated by the rhodium(I). The absence of alcohol products indicates that the amine ligands do not compete with the phosphine ligands for the catalytically active rhodium(I) sites [12, 13]. However the amine groups do not promote loss of rhodium(I) from the resin. An attempt to combat this was made by using two sets of polymer beads, one binding the rhodium complex and the other binding the amine. When the reaction was carried out using the separately functionalised beads the rate was reduced 5-fold for the first step, 15-fold for the second step and 30-fold for the hydrogenation step relative to the doubly functionalised beads. The rate enhancement of the first step is due to the amine groups increasing the polarity inside the doubly functionalised beads. The rate enhancements of the second and third steps, however, are due to the extra diffusion gradients which occur when two sets of resins are used. The slowest step of the process is the aldol condensation.

11.2. Selectivity Enhancement

Although the original motivation for supporting metal complex catalysts was to enable a facile separation of the catalyst from the products at the end of the reaction, it has become apparent that in a number of cases the selectivities of supported catalysts are significantly greater than those of their homogeneous analogues. We have noted many examples of this in Chapters 6–10. There are several factors that contribute to this selectivity enhancement:

1. When the reactant is not a single compound but a mixture of compounds diffusion within the support can control the reactivity of the substrates in such a way that some may not react at all. Thus the size selectivity of supported $[Rh(\text{P}-CH_2PPh_2)_3Cl]$ as compared to $[Rh(PPh_3)_3Cl]$ in the hydrogenation of olefin shows that the polystyrene supported material suppresses the hydrogenation of bulky olefins (Table I) [14–16].

TABLE I
Size Selectivity in the hydrogenation of olefins

Substrate	Rate (relative to cyclohexene) in the presence of	
	$[Rh(\text{P}-CH_2PPh_2)_3Cl]$	$[Rh(PPh_3)_3Cl]$
Hex-1-ene	2.5	1.4
Cyclohexene	1	1
Cyclododecene	0.22(5)	0.67
Δ^2-cholestene	0.03	0.71

2. Similarly when mixtures of reactants are present it is sometimes possible to promote the activity of one relative to the others by using supports of different polarity; a change of polarity can induce a change in the relative reactivities of

the substrates. This can be exemplified by the multistep Fischer-Tropsch reaction of hydrogen and carbon monoxide over $[Fe_3(CO)_{12}]$. When $[Fe_3(CO)_{12}]$ is supported on NaY faujasite, hydrocarbons up to at least C_{11} are formed, whereas on silica gel or alumina only short chain hydrocarbons up to C_4 are formed [17].

3. Polarity changes can be effected by changing the solvent. By combining changes of solvent with changes of substrate the polarity gradient through which the substrate must diffuse on going from the bulk solvent to the active site can be altered so altering the relative reactivities of substrates. This has been exemplified using microporous polystyrene supports which swell less in polar solvents such as ethanol in comparison with benzene. Thus, ethanol decreases the pore size and increases diffusional restrictions. Consequently the rate of cyclohexene reduction in the presence of supported $[(\textcircled{P}-CH_2PPh_2)_3RhCl]$ increased by a factor of 2.4 (after correction of rates for the corresponding homogeneous rate differences) on changing the solvent from benzene to $1:1$ benzene : ethanol. Thus despite a decrease in swelling, a polar gradient is set up which favours a higher concentration of olefin within the resin when ethanol is present [16]. As expected just the reverse trend was observed for the polar olefin allyl alcohol which was reduced 4.3 times more slowly in the presence of $[(\textcircled{P}-CH_2PPh_2)_3RhCl]$ in $1:1$ ethanol : benzene than in pure benzene. Here the polar gradient results in a lower concentration of allyl alcohol in the polymer than when benzene alone is present [16].

4. The steric demands of a catalytic centre bound to a support can be very different to those of its homogeneous analogue. Thus coordinatively unsaturated species, which can make relatively low steric demands can be stabilised by supporting the metal complex. Conversely the three-dimensional structure within a support can create extra steric demands and so enhance selectivity. This is well illustrated by a comparison of the normal : branched selectivity in the hydroformylation of 1-hexene in the presence of $[Rh(\textcircled{P}-PPh_2)(CO)(acac)]$ as compared to its homogeneous counterpart (Table III, Section 8.2.2) which is six times greater in the presence of the polymer supported catalyst at the same phosphorus : rhodium ratio [18]. A similar 'steric steering' is found in the reactions of carbanions with diolefin acetates (reaction (15) Section 10.12) where the supported palladium(0) catalysts give $4:1$ ratios of linear to branched products whereas their homogeneous analogues give a ratio of only $2.2:1$ [19].

5. The support can itself influence the selectivity of reaction at a given site. Thus optical yields of 28% are obtained when an achiral rhodium(I) complex on a chiral support is used to catalyse the hydrogenation of the α-phthalimido-acrylic acid derivative of alanine at 50 °C and 1 atmosphere of hydrogen (reaction (9)) [20].

$$CH_2=C-CO_2Me$$

[Reaction scheme structures showing the starting phthalimide-type structure, the rhodium catalyst intermediate with OPPh$_2$—RhCl(PPh$_3$)$_2$, and the product]

$$\xrightarrow[H_2,\ 50\,°C]{}$$

$$Me-\overset{*}{C}H-CO_2Me$$

(9)

6. When a metal complex is supported on what is effectively a polydentate ligand the position of the equilibrium between the metal ion and its surrounding ligands is significantly different to that of a metal ion in homogeneous solution. This undoubtedly occurs when rhodium(I) hydroformylation catalysts are supported on phosphinated or aminated polymers and accounts, in part, for the fact that much lower phosphorus or nitrogen : rhodium ratios give greater enhancements of selectivities than in their homogeneous analogues (see Table III Section 8.2.2) [18, 21].

11.3. Activity

Many of the factors that lead to selectivity enhancement are negative factors in that they inhibit the reactions of substrates. Selectivity is achieved through greater suppression of the activity of the less desirable reactions. Although it is generally found that reactions at supported metal complex catalysts are slower than in the presence of their homogeneous analogues, we have seen some important exceptions to this in Chapters 6–10. The principle factors that give rise to these exceptions are:

1. Coordinatively unsaturated species are sometimes stabilised by supporting them. This is particularly so where coordination gives rise to site separation and so inhibits deactivation through dimerisation. This is well illustrated by the fact that whereas $[Fe(\eta^5\text{-}C_5H_5)(CO)_2H]$ rapidly loses hydrogen in solution forming dinuclear $[(\eta^5\text{-}C_5H_5)(CO)_2Fe-Fe(CO)_2(\eta^5\text{-}C_5H_5)]$, $[(\text{P})-C_6H_4CH_2C_5H_4)\text{-}Fe(CO)_2H]$ bound to polystyrene is stable for months at room temperature [22]. The ability of a polymeric support to promote coordinative unsaturation is widely exemplified in Chapters 6–10. A particularly good example is the use of a polymer matrix to generate 5-coordinate cobalt(II) porphyrins for use as reversible oxygen carriers [23]. Thus when cobalt(II) tetraphenylporphyrin is exposed to 1-methylimidazole a six-coordinate complex is formed in which the solvent occupies the sixth position (reaction (10)); such complexes do not bond

$$\tag{10}$$

to oxygen whereas the corresponding 5-coordinate complexes, which can be formed using polystyrene-bonded imidazole readily form oxygen complexes on exposure to air (reaction (11)).

$$\tag{11}$$

The classical example of the use of polymeric supports to stabilise highly reactive, coordinatively unsaturated metal sites is the use of supported titanocene as a hydrogenation catalyst (Section 6.9.1) since in homogeneous solution the actual monomer 6 rapidly forms the inactive dimer 7, a reaction that is prevented by the polymer due to the isolation of the titanium sites [24, 25].

6   7

It must always be remembered that site isolation depends on the rigidity of the support. With polymeric supports this in turn depends on the solvent used. Thus although 20% cross-linked polystyrene is normally considered to be a fairly rigid polymer that supports site isolation, this is not so in strongly swelling solvents [26].

2. Species that are unstable in solution, though not necessarily coordinatively saturated, may be stabilised on a support. Thus supported rhodium(I) hydrogenation and hydroformylation catalysts are deactivated more slowly than their homogeneous analogues. In part this is because coordination to a matrix reduces rhodium(I) mobility and prevents dimerisation, which is an important deactivation mechanism [27, 28].

3. Supporting a metal complex can significantly reduce its sensitivity to poisons such as air, water and sulphur-containing compounds. Thus many supported rhodium(I) complexes are less sensitive to air than their homogeneous analogues. Although rhodium(I) hydrogenation catalysts are sensitive to poisoning by thiols, n-butyl thiol reacts with silica supported Wilkinson's catalysts to reduce their activity but enhance their thermal stability [29, 30].

4. When long alkyl chains are used to link the active centre to the support the active centre behaves essentially as if it was freely soluble in solution. As a result supported catalysts of virtually identical activity to their homogeneous counterparts can be obtained [31].

11.4. Organic versus Inorganic Supports

Many of the opportunities for enhanced selectivity described in Section 11.2 are only available on organic as opposed to inorganic supports and accordingly where high selectivity is essential organic supports will be the materials of choice. However high activities will rarely be achieved at catalytically active sites that are buried deep within the matrix. If the catalytic sites are to be on or close to the surface of the support, then inorganic supports would generally be preferred to organic supports. There are several reasons for this:

1. Inorganic supports are generally more robust mechanically than organic supports so avoiding the formation of 'fines' which can clog pipes and pumps. Macroporous styrene-divinylbenzene polymers are particularly notorious for this, especially in swelling solvents. Accordingly there has been interest in developing mechanically stronger organic polymers for use as supports such as poly(phenylene oxides) [32].

2. Inorganic supports are thermally more stable than their organic counterparts. If the metal complexes can withstand them, this enables higher temperatures to be used. Thus polystyrene cannot be used above about 150 °C.

3. Under pumped flow conditions polystyrene gels, which are very popular supports, as well as other organic polymers, can pack down very tightly into a bed so generating very high pressure drops [33]. Inorganic supports are much less susceptible to this problem. There are two ways of overcoming this problem with organic supports: the first is to polymerise the support with an inorganic matrix [33] and the second is to mix the polymer intimately with an inorganic matrix of similar bead dimensions, using the mechanical rigidity of the inorganic polymer to avoid clogging the flow channels [34].

4. Industrial chemical engineers have developed a great deal of experience in the design of equipment for using heterogeneous catalysts based on inorganic supports [35]. This can be transferred directly to the supported metal complex catalysts provided inorganic supports are used.

5. Although the swellability of organic polymers is a very valuable parameter for varying the activity and selectivity of polymer supported catalysts in the laboratory it is a less desirable parameter as far as a chemical engineer is concerned. The dimensional changes consequent to swelling cause severe problems in plant work-up and operation.

11.5. Future Developments

Clearly the major future development will be the application of supported metal complex catalysts in large scale industrial chemical reactions. They have already been used in small scale fine chemical applications. There are potentially two major reasons for using a supported catalyst: Either (i) because it has very high activity and long life which implies no significant deactivation by either poisons or as a result of leaching; or (ii) because it gives rise to very high selectivity.

The first property will be promoted by having a thermally stable, mechanically durable support so that we may expect to see a lot more work on inorganic supports where the inorganic support provides the mechanical strength and the organic anchor the flexible environment around the catalytic site. Inorganic supports that clearly merit further study include zeolites, clays and glasses [36]. Their regular topological structure could be used for the entrapment of the catalyst and thus for the modification of its selectivity.

The importance of high selectivity will lead to far more studies of the detailed three dimensional nature of the active sites. Thus the support will not be used merely as an insoluble support but rather as a material that contributes to the total environment of the active site. A great deal of synthetic work backed up by physical characterisation will be required to reach this goal; the development of solid state NMR spectroscopy (Section 4.3.5) will undoubtedly provide a most important tool for understanding the structures of active sites.

References

1. K. Mosbach: *Scientific American* **224** (March), 26 (1971).
2. C. U. Pittman and L. R. Smith: *J. Amer. Chem. Soc.* **97**, 1749 (1975).
3. C. U. Pittman, L. R. Smith and R. M. Hanes, *J. Amer. Chem. Soc.* **97**, 1742 (1975).
4. C. U. Pittman, L. R. Smith and S. E. Jacobson in: *Catalysis: Heterogeneous and Homogeneous* (Ed. by B. Delmon and G. Jannes) Elsevier, Amsterdam, p. 393 (1975).
5. C. U. Pittman and L. R. Smith in: *Organotransition-Metal Chemistry* (Ed. by Y. Ishii and M. Tsutsui) Plenum Press, New York, p. 143 (1975).
6. W. O. Haag and D. D. Whitehurst: *German Patent* 1,800,379 (1969); *Chem. Abs.* **72**, 31192 (1970).
7. S. E. Jacobson and C. U. Pittman: *J. Chem. Soc., Chem. Commun.* 187 (1975).
8. C. U. Pittman in: *Comprehensive Organometallic Chemistry* (Ed. by G. Wilkinson, F. G. A. Stone and E. W. Abel) Pergamon, Oxford, vol. 8, Chapter 55 (1982).
9. C. U. Pittman and L. R. Smith: *J. Amer. Chem. Soc.* **97**, 341 (1975).
10. C. U. Pittman, S. K. Wuu and S. E. Jacobson: *J. Catal.* **44**, 87 (1976).
11. R. F. Batchelder, B. C. Gates and F. P. J. Kuijpers: *Sixth International Congress on Catalysis*, London, preprint A40 (1976).
12. W. O. Haag and D. D. Whitehurst in: *Catalysis* (Ed. by J. W. Hightower) North-Holland, Amsterdam, Vol. 1, p. 465 (1973).
13. C. U. Pittman and G. M. Wilemon: *J. Org. Chem.* **46**, 1901 (1981).
14. R. H. Grubbs and L. C. Kroll: *J. Amer. Chem. Soc.* **93**, 3062 (1971).
15. R. H. Grubbs, L. C. Kroll and E. M. Sweet: *Amer. Chem. Soc., Polymer Preprints* **13**, 828 (1972).
16. R. H. Grubbs, L. C. Kroll and E. M. Sweet: *J. Macromol. Sci. Chem.* **7**, 1047 (1973).
17. D. Ballivet-Tkatchenko, N. D. Chau, H. Mozzanega, M. C. Roux and I. Tkatchenko: *Amer. Chem. Soc. Symp. Ser.* **152**, 187 (1981).
18. F. R. Hartley, S. G. Murray and P. N. Nicholson: *J. Mol. Catal.* **16**, 363 (1982).
19. B. M. Trost and E. Keinan: *J. Amer. Chem. Soc.* **100**, 7779 (1978).
20. H. Pracejus and M. Bursian: *East German Patent* 92031 (1972); *Chem. Abs.* **78**, 72591 (1973).
21. A. T. Jurewicz, L. D. Rollmann and D. D. Whitehurst: *Adv. Chem. Ser.* **132**, 240 (1974).
22. G. Gubitosa, M. Boldt and H. H. Brintzinger: *J. Amer. Chem. Soc.* **99**, 5174 (1977).
23. J. P. Collman, R. R. Gagne, J. Kouba and H. Ljusberg-Wohren: *J. Amer. Chem. Soc.* **96**, 6800 (1974).
24. R. H. Grubbs, C. Gibbons, L. C. Kroll, W. D. Bonds and C. H. Brubaker: *J. Amer. Chem. Soc.* **95**, 2473 (1973).
25. W. D. Bonds, C. H. Brubaker, E. S. Chandrasekaran, C. Gibbons, R. H. Grubbs and L. C. Kroll: *J. Amer. Chem. Soc.* **97**, 2128 (1975).
26. S. L. Regen and D. P. Lee: *Macromolecules* **10**, 1418 (1977).
27. M. Yagupsky, C. K. Brown, G. Yagupsky and G. Wilkinson: *J. Chem. Soc.* (*A*), 937 (1970).
28. J. A. Belmont: *Diss. Abs.* **B43**, 2543 (1983).
29. K. G. Allum, R. D. Hancock, I. V. Howell, T. E. Lester, S. McKenzie, R. C. Pitkethly and P. J. Robinson: *J. Organometal. Chem.* **107**, 393 (1976).
30. K. G. Allum, R. D. Hancock, I. V. Howell, T. E. Lester, S. McKenzie, R. C. Pitkethly and P. J. Robinson: *J. Catal.* **43**, 331 (1976).
31. J. M. Brown and H. Molinari: *Tetrahedron Lett.* 2933 (1979).

32. L. Verdet and J. K. Stille: *Organometallics* 1, 380 (1982).
33. E. Atherton, E. Brown, R. C. Sheppard and A. Rosevear: *J. Chem. Soc., Chem. Commun.* 1151 (1981).
34. J. H. Barnes, C. Bates and F. R. Hartley: *Hydrometallurgy* 10, 205 (1983).
35. A. Guyot and M. Bartholin: *Prog. Polym. Sci.* 8, 277 (1980).
36. J. M. Basset and J. Norton in: *Fundamental Research in Homogeneous Catalysis* (Ed. By M. Tsutsui and R. Ugo) Plenum Press, New York, p. 215 (1977).

INDEX

acetaldehyde 2, 4, 231, 294
α-acetamidoacrylic acid 183
acetic acid 4, 217, 229–235
acetylacetone 45, 82, 157, 162, 173, 195–196, 223, 226, 228, 234, 262
acetylenes 2, 169, 192–193, 204, 206, 231, 257–258, 264, 287
actinides 132, 194
activation analysis 119
activity 4, 152–153, 164–167, 229, 232–235, 256, 305–309
α-N-acylaminoacrylic acid 180
adiponitrile 268
alcohol formation 217–219, 221
aldehydes 4, 6
    formation 217–229, 244–245, 301, 303
aldol condensation 303
Aldox process 302–303
alkane oxidation 292–293
alkoxycarbonylation of olefins 243
alumina 14, 71–72, 92–93, 95, 97–107, 123–124, 134–135, 162, 173, 189, 193, 229, 234, 236–237, 240–243, 252, 255, 258, 261, 263–264, 266, 270, 288, 290, 305, 309
aluminium chloride 8, 217, 245, 262
ammoxidation 299
anthranilic acid 81, 149–150, 176, 188, 193
asymmetric hydroformylation 228
asymmetric hydrogenation 54, 150–151, 178–185
asymmetric hydrosilylation 211–212
atomic absorption spectroscopy 119

barium 107, 132

benzene, hydrogenation 6, 150, 174–175, 188
bipyridine 46, 85, 173, 192, 275–276, 291, 295
butadiene 205–208, 213, 255–257, 262–263
butene 238–239, 268, 270–271

carbene complexes 238, 267–269
carbon 229, 234, 290
carbon monoxide, oxidation 107, 291
    reactions 216–245
carbonylation, of azides 243–244
    of methanol 4, 217, 229–235
    of nitrocompounds 243–244
carbonyl complexes 90, 95–103, 120, 124, 162, 173–175, 186–187, 222–223, 234, 237, 240–243, 257–258, 266, 269, 289, 305–306
    ligand substitution in 245
carboxylation 294–295
catalysis, definition 1
catalysts, characterisation 9, 34, 118–136
    controllability 4
    efficiency 4
    heterogeneous 2
    homogeneous 2
    phase transfer 16–19
    reproducibility 4
    separation 3
    specificity 4
    supported gas phase 19–23
    supported liquid phase 19–23, 55
    supported metal complex 3
    triphase 16–19
cellulose 56
characterisation of catalysts 9, 34, 118–136

charcoal 295
chiral phosphines 54, 72, 150–151, 178–185, 211–212, 228
chloromethylation 40–41, 70, 89, 135, 155
cholestene 152, 167, 304
chromatography,
    gel 119–120
    temperature programmed decomposition 97, 99, 120–121
chromium 83, 87–94, 97–98, 104, 156–157, 240, 259–261
chromocene 94
clay 14, 309
coal 162, 217, 235
cobalt 7, 52, 55, 81–83, 87, 90–91, 98, 100, 106, 130, 134, 162, 192, 194, 219–223, 231, 236, 238, 257, 262, 264, 266, 274–275, 285–291, 306–307
[60]Co-radiation 57
controllability 4
copper 82, 130, 217, 231, 240, 286–287, 289–292, 294, 296
corrosion 5
cross-linked polymers 36–38, 151, 153–154, 308
cross-polarisation (NMR) 126–129
crown ether 17
cyclododecatriene 160, 162, 255
cyclododecene 151, 160, 167, 304
cycloheptene 169
cyclohexadiene 255
cyclohexene 151–152, 160, 162, 164, 167–169, 267, 289, 304–305
cyclooctadiene 160, 161
cyclooctene 151, 160, 169
cyclooligomerisation of dienes 255–257, 300–301
cyclopentadiene 160, 169, 191–193
cyclopentadienyl 44, 47, 50, 87, 222–223, 228, 237, 257, 292, 306–308
cyclopentene 160

decene 171–172, 206
decyne 171–172
dehydrocyclisation 103
dehydrogenation 95, 107, 177, 189, 288
deuterium/hydrogen exchange 170

diene, cyclooligomerisation 255–257, 300–301
    hydrogenation 171, 188
    oligomerisation 255–257, 302
dimerisation,
    of complexes 6, 153–154, 164, 186, 306–308
    of olefins 213, 252–254
diop 54, 72, 151, 179–183, 211, 228
diphenylacetylene 171
divinylbenzene 37, 50, 80, 126, 130, 135, 155, 160, 167, 175, 177, 205, 213, 263, 308

efficiency 4
electrodes, chemically modified 16
electron microscopy 135–136
electron probe microanalysis 135, 155
enzyme 1, 8, 16, 150, 183, 299
epoxidation 80, 285
ESCA 81, 122, 133–134, 176, 193
ESR 122, 128, 130–131, 153, 155, 171
ester hydrolysis 296
ethylene 4, 5, 107, 158, 169, 188, 231, 238–239, 252, 254, 258, 267, 294
EXAFS 122, 134–135, 165

ferredoxin 157
Fischer-Tropsch 100, 102–103, 217, 235–240, 305
fluidised bed 142, 146
Friedel-Crafts reaction 245
functionalised supports 81–87

gamma-radiation grafting 55–67, 126, 128, 132–133, 220–224
gas phase catalysis, supported 19–23
gellular polymers 12
gels 12, 37–38
glass 14, 71–72, 103, 188, 244, 309
gold 85, 132
graphite 72, 181, 285
Grignard cross-coupling 276

hafnium 92, 132, 154–156
heat transfer 11
hectonite 15, 151, 171
heptene 206, 208–209, 274

hexene 4, 6, 151–152, 158, 167, 206–210, 221, 223, 270, 273, 304–305, 308
hexyne 171
hydrodealkylation 188
hydroformylation 4, 6, 14, 19–22, 24–25, 52, 81, 85, 103, 161, 217–229, 268, 299, 302–303, 305, 307–308
    asymmetric 228
hydrogenation 4, 6, 9, 12, 25, 52, 54, 99–100, 103, 149–196, 300–302, 304–305, 307–308
    asymmetric 54, 150–151, 178–185
    transfer 186
hydrogen/deuterium exchange 170
hydrogenolysis 95, 99, 102
hydrolysis of esters 297
    of nitriles 295
hydrosilylation 204–214
    asymmetric 211–212
hydroxylation of olefins 293–294

imidazole 194, 262, 291, 306
inelastic electron tunnelling spectroscopy 122, 125
infrared spectroscopy 121–124, 229
inorganic supports 14–16, 259–261, 308–309
ion-exchange resins 80–81, 189–190, 229, 244, 289, 291, 295
iridium 83–85, 102, 106–107, 132, 135–136, 153, 185–187, 214, 240, 243, 276
iron 7, 25, 80, 83–84, 87–90, 99, 132, 157, 192, 194, 213, 235–238, 244–245, 269, 285, 288–290, 293, 305–306
isocyanate formation 243–244
isomerisation of olefins 173, 176, 193, 218–219, 268–276
isoprene 262, 288

ketone
    formation 217, 244–245, 286
    hydrogenation 158, 183, 185
    hydrosilylation 204, 208

lanthanum 107

lanthanum oxide 240
lattice metal complexes 23
leaching 84, 166, 226, 230, 234, 256, 309
Linde process 217
Lindlar catalyst 193
linear polymers 34
liquid phase catalysis, supported 19–23, 55, 227
lithiation 39–40

macroporous polymers 12, 81, 308
macroreticular polymers 12, 38
magic angle spinning (NMR) 126–127
magnesium 260
magnesium chloride 264
magnesium oxide 99, 103–104, 106, 123, 229, 234, 236–237, 240–243, 261, 263–264, 305
maleic acid 80
manganese 89–90, 231, 285, 289
mass spectrometry 122, 132–133
mass transfer 141
mechanical strength of supports 9, 11–12, 14, 146, 308–309
melts 23
mercury 2, 82, 132, 231
Merrifield resin 2, 16, 41–43, 170, 179
metal carbonyls 90, 95–103, 120, 124, 162, 173–175, 186–187, 222–223, 234, 237, 240–243, 245, 257–258, 266, 269, 306
    ligand substitution 245
metal complexes, introduction onto supports 80–107
metal vapour chemistry 91
metathesis 107, 134, 265–268
methane synthesis 216–217, 236–239, 261
methanol carbonylation 4, 217, 229–235
    synthesis 217
Michael addition 195–196
microanalysis 118–119, 135
microporous polymers 12, 37–38, 305
moisture sensitivity 5, 308
molecular sieves 14–15, 71–72, 98–100, 103, 106, 134, 162, 217, 228–229, 236, 243, 305, 309
molybdenum 25, 83, 88–89, 91, 95, 97–98, 105, 107, 130, 134, 156–

157, 194, 236, 243–245, 257–
    259, 264, 266–267, 289
montmorillonite 15, 18, 151
Mössbauer spectroscopy 122, 131–132
multistep reactions 299–304

neutron activation analysis 119
nickel 82–83, 88, 92, 95, 102–103, 132,
    136, 150, 187–189, 235, 252–
    255, 262–264, 269, 273, 276,
    285, 296, 300–302
niobium 92, 107
nitric oxide 194
nitrile hydrolysis 295
nitrobenzene, hydrogenation 150, 159, 188,
    193
nitrogen fixation 194
nitrous oxide, decomposium 107
NMR 11, 81, 122, 125–131, 309
norbornadiene, isomerisation 274–276

octadiene 267
octene 173, 207
olefin
    alkoxycarbonylation 243
    carboxylation 294–295
    dimerisation 213, 252–254
    epoxidation 285
    hydroformylation 4, 6, 14, 19–22, 24–
        25, 52, 81, 85, 103, 161, 217–
        229, 268, 299, 302–303, 305,
        307–308
        asymmetric 228
    hydrogenation 4, 6, 9, 12, 25, 52, 54,
        99–100, 103, 149–196, 300–
        302, 304–305, 307–308
        asymmetric 54, 150–151, 178–185
    hydrosilylation 204–214
        asymmetric 211–212
    hydroxylation 293–294
    isomerisation 173, 176, 193, 218–219,
        268–276
    metathesis 107, 134, 265–268
    oxidation 286
    polymerisation 91, 94, 258–264
    trimerisation 255
oligomerisation of acetylenes 257–258
    of dienes 255–257, 302
optimisation 142–145

osmium 85, 99–100, 107, 124, 132, 157–
    158, 236–237, 241, 243, 269–
    270
oxidation 55, 95, 107, 231, 285–293
oxide supports 14–16, 259–261, 308–309
oxychlorination 107
oxygen sensitivity 5, 308

palladium 5, 80, 82–83, 88–89, 106–107,
    119, 134, 149–150, 189–194,
    205–207, 212–213, 231, 243–
    244, 252, 254–257, 274, 276,
    288, 291, 294–296, 302
pentene 159, 209, 225, 269, 294
peroxide decomposition 288–289
phase transfer catalysis 16–19
phenylacetylene 162, 257–258, 264, 287
phosphination, polymers 41–44, 49, 135
    silica 67–71
photoactivation 8, 84
photodehydrogenation 157
phthalocyanine 106, 285, 290
pinene 169, 190
Plackett-Burman matrix 142–145
plasmas 55–56
platinum 2, 6, 80–81, 83, 85, 88–89, 103,
    106, 126–129, 132, 150, 189–
    194, 204–207, 210, 229, 240,
    292
poisoning 5, 8, 167, 308–309
polarity of solvent 14, 151–152, 165–166,
    305
polyacetylene 189
polyacrylamide 290
polyacrylic acid 11, 159, 170, 173, 177,
    262, 272
polyallyl alcohol 52, 130, 173
polyamide 35, 52, 134, 163
polybutadiene 88
polydiacetylene 163
polyester 35, 52
polyethylene 56, 130, 136, 258, 262
polyethylene glycol 52
polyiminoethylene 164
polymaleic acid 170
polymerisation of olefins and acetylenes
    91, 94, 258–264
    of dienes 262–264
polymers 11–14

cross-linked 36−38
linear 34
macroporous 12, 38, 308
macroreticular 12
microporous 12, 37−38, 305
popcorn 13, 38
proliferous 13, 38
supported metal complexes 25−26
swellability 15, 37, 146, 165, 171, 253, 309
poly(phenylene oxide) 11, 51−52, 222−223, 228, 308
polypropylene 11, 56, 63−64, 124, 126, 132−133, 136, 220−224
polystyrene 11, 22, 36, 38−51, 56, 80−81, 83−85, 88−89, 91, 133, 135, 155−157, 164, 174−175, 181, 192, 205, 213, 222−223, 228−229, 234, 237, 262−263, 270, 274−275, 287, 290−292, 295−296, 305−306, 308−309
chloromethylated 40−41, 89, 135, 154, 176, 193, 303
cyclopentadienylated 44, 47, 50, 87, 222−223, 228, 237, 257, 292, 306−308
functionalisation 38−51
lithiation 39−40
phosphinated 41−44, 49, 83−84, 126, 133−135, 150−152, 154, 160, 163, 165, 170, 186−187, 206−209, 226−229, 253, 255, 257−258, 264, 266−267, 269−290, 276, 294, 296, 300−305
polytetrafluoroethylene 61
polyurethane 36, 243
polyvinyl acetate 262
polyvinyl alcohol 52, 262
polyvinyl chloride 11, 52, 56, 61, 150, 163, 294
poly(4-vinylpyridine) 55, 85, 130, 194, 234, 270, 291−292, 295
popcorn polymers 13, 38
pore size 37
porphyrin 46, 82, 106, 194, 274−275, 288−290, 306−307
proliferous polymers 13, 38
propene 238−239, 252, 254, 258
propylene oxide 264

propyne 169

quadricyclane 274−276
quinone 294

radiation grafting 55−67, 126, 128, 132−133, 220−224
Raman spectroscopy 122, 124−125
reactors 141−142
regioselectivity 13, 170
reproducibility 4
rhenium 105−107, 132, 266
rhodium 4, 8−9, 14, 18−22, 24−25, 52, 54, 80−87, 92, 95−96, 100−102, 106−107, 119, 124, 126, 129−130, 133−135, 149−154, 161, 163−185, 204, 206−213, 219−220, 223−235, 240−241, 243−245, 254, 257, 270, 272−273, 286, 289, 300−306, 308
ruthenium 82, 85, 88−89, 106−107, 132, 134, 157−162, 177, 192, 228, 236, 243, 269−273, 289, 301

SASOL process 235
Schiff bases 85, 130
selectivity 9, 13, 153−154, 167−170, 210, 219−229, 234, 243, 301, 304−306
separation of catalyst 3
silica 14, 17, 21, 67−71, 83, 85, 91−94, 96, 98−107, 123, 129, 134−135, 151, 156, 162, 167, 169, 174−175, 177, 188, 193, 229, 234, 236−237, 240−243, 252, 254−255, 258−259, 261, 263−264, 266, 269−270, 273−274, 276, 290−291, 293, 305, 309
functionalisation 67−71
phosphinated 67−71, 85, 135−136, 150, 162−165, 167, 171, 174, 191−192, 206−208, 210, 212, 214, 243, 273, 295, 308
silver 8, 163, 255
silylation 204−214
asymmetric 211−212
site isolation 6−7, 154−155, 306−308
site-site interaction 7, 154−155, 306−308
smectite 15

solvent 5, 151–152, 305
    polarity 14, 151–152, 165–166, 305
soybean esters, hydrogenation 193
specificity 4, 211
spectroscopy
    atomic absorption 119
    ESCA 81, 122, 133–134, 176
    ESR 122, 128, 130–131, 153, 155, 171
    EXAFS 122, 134–135, 165
    inelastic electron tunnelling 122, 125
    infrared 121–124, 229
    Mössbauer 122, 131–132
    NMR 11, 81, 122, 125–131, 309
    Raman 122, 124–125
    UV-visible 122, 125, 274
    XPS 81, 122, 133–134, 176, 193
    X-ray fluorescence 119, 135, 155
stability, thermal 5, 14, 146, 154, 167,
        308–309
stereospecificity 258
steric steering 295–296, 305
strontium 107
p-styryldiphenylphosphine 49, 64, 126,
        128, 167, 171, 223–224, 270, 276
supported gas phase catalysis 19–23
supported liquid phase catalysis 19–23
supported metal complex catalysts 3, 25–26
    characterisation 9, 34, 118–136
    preparation 80–107
    use 141–309
supports
    functionalised 81–87
    introduction of metals onto 80–107
suspension polymerisation 35–38
swelling 15, 37, 146, 165, 171, 253, 309
synthesis gas 216–217, 235–243

tantalum 107, 132
tellurium 107, 132
temperature programmed decomposition
        chromatography 97, 99, 120–121
tetrathionate 291–292
thermal stability 5, 14, 146, 154, 167, 308,
        309
thiosulphate oxidation 291–292
thorium 132, 194, 261
titanium 52, 87, 91–93, 103, 130, 153–
        156, 228, 260–261, 292–293,
        307

titanium dioxide 102, 106, 134, 194, 234,
        263–264, 291
titanocene 153, 307
transfer hydrogenation 186
trimerisation of olefins 255
triphase catalysis 16–19
trithionate oxidation 291–292
tungsten 25, 80, 83, 88–91, 98, 107, 132,
        134, 156–157, 266–267
turnover number 9

uranium 132, 194, 261
UV-radiation grafting 55–67
UV-visible spectroscopy 122, 125, 274

vanadium 91–92, 104, 107, 130, 260–261,
        285, 288–289
Vaska's complex 186
vinyl acetate 5
4-vinylcyclohexene 153, 160, 255
vinyl ester exchange 295
vinyl ether exchange 295
4-vinylpyridine 48, 55–56, 62–63, 130,
        132–133, 136, 220–222, 253,
        262, 291

Wacker process 4, 231, 293–294
water gas shift reaction 25, 240–243
water soluble complexes 24–25
Wilkinson's catalyst 149, 151–152, 154,
        163–170, 204, 245, 304, 308
Wool 56

X-ray fluorescence spectroscopy 119, 135,
        155
X-ray photoelectron spectroscopy 81, 122,
        133–134, 176, 193

zeolite 14–15, 71–72, 98–100, 103, 106,
        134, 162, 217, 228–229, 234,
        236, 243, 305, 309
Ziegler catalysts 259, 262
Ziegler-Natta catalysts 35, 92, 103–106,
        145, 260–261, 266–267
zinc 82, 132, 240
zinc oxide 99, 103, 106, 264
zirconium 10, 92–94, 102, 154–156, 261,
        269
zirconium dioxide 102, 240, 264